UNDERSTANDING CULTURAL GEOGRAPHY

Understanding Cultural Geography offers a comprehensive introduction to perhaps the most exciting and challenging area of human geography. By focusing on the notion of 'place' as a key means through which culture and identity is grounded, the book showcases the broad range of theories, methods, and practices used within the discipline. This book not only introduces the reader to the rich and complex history of cultural geography, but also the key terms on which the discipline is built. From these insights, the book approaches place as an 'ongoing composition of traces', highlighting the dynamic and ever-changing nature of the world around us.

The second edition has been fully revised and updated to incorporate recent literature and up-to-date case studies. It also adopts a new seven-part structure, and benefits from the addition of two new chapters: Place and Mobility, and Place and Language. Through its broad coverage of issues such as age, race, scale, nature, capitalism, and the body, the book provides valuable perspectives into the cultural relationships between people and place. Anderson gives critical insights into these important issues, helping us to understand and engage with the various places that make up our lives.

Understanding Cultural Geography is an ideal text for students being introduced to the discipline through either undergraduate or postgraduate degree courses. The book outlines how the theoretical ideas, empirical foci, and methodological techniques of cultural geography illuminate and make sense of the places we inhabit and contribute to. This is a timely update on a highly successful text that incorporates a vast foundation of knowledge; an invaluable book for lecturers and students.

Jon Anderson is a Reader in Human Geography in the School of Planning and Geography at Cardiff University, UK. His research interests focus on the relations between identity, culture, and place, in particular the actions, practices, and politics that such relations produce.

UNDERSTANDING CULTURAL GEOGRAPHY

Places and traces

Second edition

Jon Anderson

 Routledge
Taylor & Francis Group

LONDON AND NEW YORK

Second edition published 2015
by Routledge
2 Park Square, Milton Park, Abingdon, Oxon OX14 4RN

and by Routledge
711 Third Avenue, New York, NY 10017

Routledge is an imprint of the Taylor & Francis Group, an informa business

First edition published by Routledge 2010

British Library Cataloguing in Publication Data
A catalogue record for this book is available from the British Library

Library of Congress Cataloging in Publication Data
Anderson, Jon, 1973–
Understanding cultural geography: places and traces / Jon Anderson.
— Second edition.
Includes bibliographical references and index.
1. Cultural geography. 2. Human geography. I. Title.
GF41.A475 2015
304.2—dc23   2014043220

ISBN: 978-0-415-73449-3 (hbk)
ISBN: 978-0-415-73450-9 (pbk)
ISBN: 978-1-315-81994-5 (ebk)

Typeset in Galliard
by Keystroke, Station Road, Codsall, Wolverhampton

Printed and bound by CPI Group (UK) Ltd, Croydon, CR0 4YY

Contents

Figures

The Publishers have made every effort to contact authors and copyright holders of works reprinted in *Understanding Cultural Geography*. We would welcome correspondence from individuals or companies we have been unable to trace.

Tables

Boxes

Preface to the second edition

I am very pleased that *Understanding Cultural Geography: Places and Traces* has reached its second edition. As you will already know, human geography is a vital discipline, providing us with important insights into the world around us and helping us to understand how our identities, practices, and actions are thoroughly influenced by the places in which we live our lives. Cultural geography as a sub-discipline focuses specifically on these identities, practices, and actions, and is perhaps the most lively and vibrant sub-discipline within human geography today. Through its broad coverage of issues such as age, race, scale, nature, capitalism, and the body, cultural geography offers invaluable perspectives into the relations between people and place. This book gives critical insights into these important issues, alongside the explanation of key theoretical concepts in human geography. This book not only introduces the reader to the rich and complex history of cultural geography, but also the key terms on which the discipline is built. From these insights, the book approaches place as an 'ongoing composition', highlighting the dynamic and ever-changing nature of the world around us.

The second edition adopts a new structure to present its insight into the cultural world. Within seven parts, fifteen chapters offer comprehensive examination of the ideas and practices of cultural geography. This edition has been completely updated, including up-to-date concepts and empirical examples, whilst two new chapters have also been included – Place and Mobility (Chapter 8), and Place and Language (Chapter 12). By introducing the notion of 'traces' – or the marks, residues, or remnants left in place by cultural life – the book helps us to understand and engage with the various places that make up our lives. I hope you enjoy this book and find it useful not only in 'Understanding Cultural Geography', but also in identifying and re-visioning your own connections and values to the world of which we are a part.

With best wishes

Jon Anderson
Cardiff, UK. July, 2014.

PART

1 INTRODUCING CULTURAL GEOGRAPHY

Introduction

Key

We live in a world of cultural places. You and I, and that person across from you in the library/café/bedroom (delete as appropriate) are in the cultural world. We live in it, we survive it (hopefully), and we contribute to it, every day and night. The purpose of this book is to help us get a handle on why our world is like it is, what our role in it might be, and how we can sustain or change it. To this end, this book outlines how cultural geography explores, investigates, and contributes to these places. It exists to give us insights, feelings, and understandings of our place in the world.

'. . . does not take place in a vacuum'

Working out how cultural geography approaches the world involves thinking about the terms 'culture' and 'geography'. Let's start with 'geography' first. Geography can often seem an abstract term, an abstract discipline even. Useful perhaps for knowing the capital of Iceland for holiday games of Trivial Pursuit, but not much else. This book argues that geography is anything but abstract. One way we can evidence this argument is through using a popular cliché borrowed from high school science classes, 'X [in this case, "cultural life"] does not take place in a vacuum'. Things, ideas, practices, and emotions all occur in a context, in a broader world

that influences, values, celebrates, regulates, criminalises, sneers, or tuts at particular activities and objects. Interest in this context, and how it influences, values, celebrates etc., is one thing that geography and geographers do. As Cook (in Clifford and Valentine, 2003: 127) identifies, 'so much depends on the context'. Context can influence what actions we choose to make and how we choose to make them, it can influence how these actions are judged by ourselves and others, and thus how successful and significant they turn out to be. Context is therefore vital to take notice of and understand, yet in everyday life it is something we often ignore – we are so used to it that it becomes ordinary, obvious, and even natural. Cresswell (2000: 263) describes this through using the South East Asian phrase 'the fish don't talk about the water'; in normal life we are often like fish in that we don't talk about our geographical context. Geographers, however, are weird fish, we seek to sensitise ourselves to the 'water'. Geographers swim in and investigate context.

So, what are the appropriate contexts for geographical study? Context can be thought about in a variety of ways. Geographical context is often thought about in terms of national or political territories, physical landscapes, or exotic places. These contexts are often clear, identifiable spaces which may be hotter or drier, colder or wetter, and defined by particular languages, laws, and customs that may be different or similar to our own. Each context will have an influence – and often be influenced by – the activities occurring in that place. For example: in many countries religious, cultural, or political laws influence what clothes are appropriate for different genders (see Box 1.1 for an example); in others seasonal conditions influence working and sleep patterns; whilst in yet other contexts ideologies of freedom and technologies of movement coincide to influence the degree of mobility within and between territories.

Political territories or physical landscapes are, however, not the only contexts that geographers can study. Any place or area, at any scale, or in any circumstance, could be thought about as a geographical context. For example, on a macro scale, we could think of Planet Earth as a context and how it influences and is influenced by the activities going on within it. At a micro scale, we could think of the room in which I am typing

BOX 1.1 CULTURE IN CONTEXT: THE CASE OF LAOS

In Laos, South East Asia, the Provincial Tourist Department of Luang Prabang offers advice to visitors to sensitise them to the geographical and cultural context of the country. As excerpts from their advice leaflet illustrate, the cultural context of Laos outlines appropriate ways to dress in particular places, and how the dominant religion influences how our bodies should stand in this place.

Luang Prabang Provincial Tourism Department *and* **Big Brother Mouse**

Sabai dee!

Welcome to one of the friendliest countries on the planet. These ten suggestions will help you enjoy your visit, while helping us preserve our culture and traditions.

1 TEMPLES ARE BEAUTIFUL, and interesting, but most of all, they are holy places. Please visit, but dress appropriately, covering shoulders, knees, and everything in between. (Yes, that stomach, too!) Whether in a temple or on the street, women who wear a traditional sihn (long skirt) will find the gesture much appreciated.

IN BUDDHIST CULTURES, the head is high and the feet are low. Use your feet only for walking. (Okay, we'll make an exception if you're a kick-boxer!) Pointing your feet at someone – for example, by putting your feet up on a stool – is rude. So is stepping over someone seated on the floor. 2

Laos

Source: Luan Prabang Provincial Tourist Department

BOX 1.2 NO BIKES ON THE SIDEWALK: ALL GEOGRAPHIES ARE REGULATED

The following images outline the regulations imposed on a range of geographical contexts. In the municipality of Santa Cruz, California, all visitors are reminded that it is considered normal not to start fires or feed wildlife; in Thailand's parks it is accepted that people won't smoke or drink; and in Bangkok's Royal Palace you shouldn't sit on the walls.

Santa Cruz
Source: Author

Thai Parks
Source: Claire Cutforth

Thai Palace
Source: Claire Cutforth

these words as a geographical context, or even the messy desk on which my computer rests – and explore how it influences and is influenced by the activities going on here. Alternatively and imaginatively, we could think of other contexts for geographical exploration. A public square – and how it seems to encourage some uses and users rather than others – a field (ditto, but maybe for non-humans too?), a home, a wall, a coastline? What about the contexts of a classroom, a street, a bar, or a sports field? What about a theatre, a mine, a museum, a library . . .? What activities are accepted as normal in these places? What behaviours are frowned upon? How are they regulated? Do people conform to these regulations, and what happens if they don't? These are all crucial, critical questions that geographers employ to analyse the contexts they study.

Alongside political spaces, physical landscapes, and socially engineered places – at a variety of scales – are other forms of geographical context that we could study. Perhaps we could think about contexts of communication as relevant 'media spaces' for investigation. We could for example study the mainstream media or virtual space, spoken languages, written codes, or even non-verbal communications and how these influence (and are influenced by) particular groups. Other geographical contexts could include places of the body, for example the head, the heart, the skin even, in both material and metaphorical senses. These are perhaps unorthodox but nevertheless fascinating geographical contexts, and maybe you can think of others. In sum, geographical contexts can exist wherever there are human (and non-human) activities; the trick is to acknowledge them, work out what produces them, and what effect they have.

X marks . . .

We have established then that geographers have an interest in context. But if we turn back to our initial cliché, 'X [in this case, "cultural life"] does not take place in a vacuum', we should identify that this statement has other key components to it. We have focused on the vacuum (or context) component, but there is, of course, the initial activity – the *X* – that occurs too. Geography is equally concerned with this *X*, as it marks the spot where this activity interconnects with context. As this book deals with cultural geography, the question arises: what are appropriate cultural activities for geographers to study? The book argues that culture includes the material things, the social ideas, the performative practices, and the emotional responses that we participate in, produce, resist, celebrate, deny, or ignore. Culture is therefore the constituted amalgam of human activity – *culture is what humans do*. In this light culture can include a range, perhaps even an endless range, of things. It includes aspects of society, politics, and the economy, and can be identified or categorised by a range of different (and sometimes overlapping) groups. It is commonplace, for example, to hear the word culture being prefixed by terms such as 'pop', 'global', or 'high'; by 'capitalist', 'democratic', or 'class'; by 'Western', 'dance', or 'nomadic'; by 'Islamic', or 'sporting', or 'youth'. Cultures can be 'mainstream', 'sub-', or 'counter-'. Culture then can be seen as encompassing a wide spectrum of human life; it is not a separate entity from society, politics, or the economy, but influences (and is influenced by) them all.

What cultural geography seeks to do, therefore, is explore the intersections of context and culture. It asks why cultural activities happen in particular ways in particular contexts. It is interested in exploring how cultural activities and contexts interact, influence, and perhaps even become synonymous with one another. It operationalises this interest through identifying that the product of the intersection between context and culture is *place*. As our introductory cliché tells us: 'X [in this case, "cultural life"] does not take place in a vacuum.' In this (final) instance, the key component of this cliché is 'take place'. This component confers two important meanings. The first is that the incident in question took place, in the sense that it occurred, it happened. Secondly, however, there is a clearly geographical element to this phrase: the cultural activity literally *took place*. It aggressively, passively, intentionally, or otherwise, took a place and contributed to its meaning and identity. From the perspective of cultural geography, therefore, places come by their meanings and identities as a result of the complex intersections of culture and context that occur within that specific location. Cultural geography explores place – these confluences of culture and context – to help us to know and act better in the world around us.

Making traces, making places

From the perspective of cultural geography, places are taken and made by intersections of culture and context. In more detail, places are constituted by imbroglios of *traces*. Traces are marks, residues, or remnants left in place by cultural life. Traces are most commonly considered as material in nature (material traces may include 'things' such as buildings, signs, statues, graffiti, i.e. physical additions to our surroundings), but they can also be non-material (non-material traces might include, for example, activities, events, performances, or emotions). We can therefore see visible traces in places, but we can also sense them in other ways (we can hear them, smell them, even taste them, or feel them), as well as being able to think on them, reflect on them, and perhaps – in our more sentimental moments – reminisce about them. Traces can therefore be durable in places both in a material sense (they have longevity due to their solidity and substance as things), but may also last due to their non-material substance (they may leave indelible marks on our memory or mind). As traces are constantly produced they continually influence the meanings and identities of places. In both material and non-material form they function as connections, tying the meaning of places to the identity of the cultural groups that make them. Traces therefore tie cultures and geographies together, influencing the identity of both. As a consequence of the constant production of traces, places become dynamic entities; they are in fluid states of transition as new traces react with existing or older ones to change the meaning and identity of the location. It is argued here, therefore, that places should be understood as *ongoing compositions of traces*. Cultural geography interrogates these traces, their interactions, and repercussions. It critically appraises the cultural ideas and preferences motivating them, and the reasons for their significance, popularity, and effect.

Let's take an example to illustrate the cultural geography approach to place advocated here.

Figure 1.1 is a photograph of a place. More specifically it's a photograph of a place constituted by numerous traces: a statue, on a column, in a public square, with two fountains. Do you recognise where it is? What do you think these traces stand for? This is one of the first questions cultural geographers ask themselves: what do traces stand for? Taken literally, the statue, for example, is clearly standing on a column, which itself is standing in this square. (It is a material piece of cultural life standing in a geographical context.) But taken metaphorically,

cultural geographers are asking what do traces stand for in the sense of what do they represent? Cultural geographers believe that traces can embody or represent 'ideas', and in this case this public statue was commissioned, developed, sited, and maintained to foster particular value systems of particular groups. In other words, cultural geographers argue that traces are not neutral, they stand for particular cultural preferences or ideas about what the world should be like. Through doing this, these traces try to persuade us to agree with and support these cultural values too.

So let's look at this photograph again. It's a statue of Lord Admiral Horatio Nelson, raised to 61.5 metres altitude by the column that bears his name, in Trafalgar Square, London. The naming of the square is by no means incidental. It commemorates Nelson's greatest naval victory over combined French and Spanish fleets in 1805. Why would the city planners of the 1800s want to create this statue and plaza? It has been argued that this place can be understood as a space of empire (Gilbert and Driver, 2000), built to commemorate British leadership and V/victory (Nelson's flagship was called *HMS Victory*). It represents the military might and power of the British as a people and as a state. It seeks to inspire pride and patriotism in the country and demonstrate the values and urban design expertise of a civilised, industrial nation. So when cultural geographers look at this photograph, they do not simply see a statue, on a column, in a public square with two fountains, they see a material cultural trace, tightly bound up with a range of cultural ideas. These traces come together to constitute the place of Trafalgar Square.

In fixing our attention on the traces which constitute a place, geographers are in effect interpreting and translating places. They do so by interpreting and translating the traces (be they material or non-material) into the meanings intended by their 'trace-makers'. However, as we know in our own lives, there are many

Figure 1.1 The place of Trafalgar Square

Source: Author

different opinions, thoughts, and judgements about the meanings associated with any given trace. Translations may not only change with particular cultural groups, they may also change over time. We could ask ourselves, for example, whether the ideas of the nineteenth-century trace-makers in Trafalgar Square still have currency and influence over us today. Do their ideas remain as strong and sturdy as Nelson's Column itself, or do they lurk in the shadows of the Square which has been taken over by new trace-makers?

In 2005, a statue called 'Alison Lapper, Pregnant' by artist Marc Quinn was unveiled on a plinth in Trafalgar Square (see Figure 1.2).

This material trace is a sculpture of a pregnant nude woman who has *phocomelia*, a condition which resulted in her being born with no arms and shortened legs. The conjunction of Alison Lapper and Horatio Nelson can be read as highlighting differences but also continuities between the nineteenth- and twenty-first century traces in Trafalgar Square. The statue of Alison Lapper clearly does not celebrate war and victory (as Lapper herself stated, 'At least I didn't get there [Trafalgar Square] by slaying people' (Ebony, 2005: 37)), it may, therefore, celebrate a victory for a 'post-imperial' Britain (as Pile puts it, 2003: 30), where femininity, maternity, and differences in ability are accepted and publicly endorsed. (It

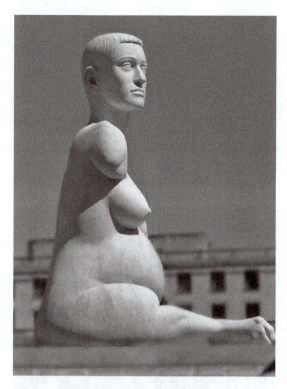

Figure 1.2 The traces of Alison Lapper

Source: David Nurse (Menscrifa) www.flickr.com/photos/81824281@N00/2902047797/

Figure 1.3 Blue cockerel in Trafalgar Square

Source: Picqero

also perhaps draws our attention to the disabling injuries incurred by Nelson that are not conspicuous in his statued form.)

In 2013, a new statue was commissioned for the fourth plinth in Trafalgar Square. German artist Katharina Fritsch designed a large, blue, farmyard cockerel for the site (Figure 1.3). Fritsch wanted this material trace to be ambiguous in its representation. For her it was a 'visual sign' (cited in Barnett, 2013), one she hoped could have 'many meanings; you can play around with it. It's humorous and also serious' (cited in Higgins, 2013a). What do you think this material trace could stand for? Perhaps this trace playfully mocks the vainglorious statues of men like Nelson in the Square? Perhaps the Germanic origins of its designer represent a further move towards London becoming a post-imperial and perhaps even post-national place (see also Chapter 6)? To many the cockerel is a bird long-associated with France; with this in mind does it represent an 'irreverent Gallic response' to Napoleon's defeat by Nelson (see Searle, 2013)? Whatever its meaning, the Thorney Island Society (a society which defends the interests of residents in this area of London) lodged a planning objection to Fritsch's cockerel. They found it 'totally inappropriate' for this place (Higgins, 2013b). What meanings of this trace do you think the Thorney Island Society was objecting to?

Thus in fixing our attention on socially engineered places such as Trafalgar Square and the social ideas motivating them, we begin investigating the value systems and cultural preferences of these places' 'trace-makers'. We come to recognise that these places are made by many agents, not just the original planner, designer, or architect. We see that anyone who uses these places has the capacity to edit and re-edit them, adding their own cultural ideas through their specific cultural actions. In other words, with all the actions we participate in we inevitably take and make place (we, whether we like it or not, inevitably leave traces). In fixing our

Figure 1.4 Criminal Justice Bill demonstrations, Trafalgar Square, 1994

Source: Peter Forsland

Figure 1.5 Iraq War protests, Trafalgar Square, 2004

Source: Austin Taylor Labourn

attention on the ways in which ideas, objects, and actions take and make place we come to realise that places – and what they represent – are not fixed. As actions and ideas change over time, places become dynamic states of transition. In some places the ideas, things, and practices may change very slowly, almost imperceptibly, yet in others changes may occur by the season, the day of the week, or by the hour of the day. One thing is certain, change is inevitable.

Be they relatively dynamic or relatively stable, the meanings and identities of places are often taken-for-granted, something we no longer consider and take notice of. Perhaps out of habit, custom, or apathy, we often act in line with the dominant cultural ideas of a place, thus practising our agreement with or ambivalence towards them. In some cases, however, the cultural ideas motivating places are not taken for granted. In these cases the ideas may be highly contentious, generating disagreement and conflict among different cultural groups. In these cases, particular groups may no longer go along with the existing uses of these places; they may wish to challenge these uses and the ideas they stand for. Here the users of a place overtly change from being passive recipients of dominant place-makers' intent, into actors who purposely seek to edit places in new, contrasting ways. Through leaving their own traces in places, different groups can criticise one set of cultural ideas, and perhaps offer alternatives to them. Figures 1.4–1.7 show some examples from Trafalgar Square.

In each figure, particular traces have made Trafalgar Square into a different place. In each case, members of the public rally against the British state, rather than supporting and celebrating it. Through their actions they take and make a Trafalgar Square that celebrates dissent, popular protest, and even anarchy (see Mills et al., 1990). Some may argue that free speech and assembly are bastions of British democracy, and thus such demonstrations emphasise the strengths of the British political system. However, Figure 1.4 illustrates a rally in 1994 against the Criminal Justice Act, a piece of British legislation which criminalised this precise form of free speech and assembly. In further defiance of this legal statute, Figure 1.5 depicts the Anti-Iraq War demonstration of 2004 where British and American foreign policy was the focus of dissent. In Figure 1.6, 'anonymous' protests in 2013 challenge specific welfare cuts and the broader political system that condones tax avoidance by corporations and elite individuals. (If you're interested, the full slogan on the placard is a quotation from American futurist Jacques Fresco which states: 'Democracy is a con game. It's a word invented to placate people to make them accept a given institution. All institutions sing, "We are free". The minute you hear "freedom" and "democracy", watch out . . . because in a truly free nation, no one has to tell you you're free' (see Venus Project, 2012)). Each of these incidents creates a different challenge to the dominant interpretation of Trafalgar Square as a place of national pride and unity. Taken together they symbolise a place that has come to represent a site of recurrent division and disenfranchisement between the public and politicians, it stands for the types of ideas the former hold and the cultural activities they will employ in order to confront the ideas and activities of the latter. This alternative translation of Trafalgar Square is recognised by the stencil from the political artist Banksy on the plinth of Nelson's Column (Figure 1.7).

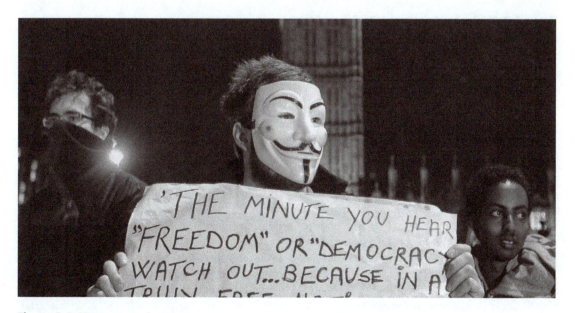

Figure 1.6 'Anonymous' protests

Source: Anonymous

Figure 1.7 'Designated riot area', Trafalgar Square
Source: Banksy/Lazarides

In citing Trafalgar Square as a 'designated' riot area, Banksy intentionally plays with notions of power and authority, highlighting how the public is now authoring the cultural geography of the Square in a way not sanctioned by the State. In sum, therefore, these examples show that the identity and meaning of Trafalgar Square is not straightforward. It is not simply a place of patriotism, but also a place of civil disobedience against the national state. It is, like all other places, a composite of cultural ideas, activities, histories, presents, and possible futures. Due to the multiple traces that come together to make it, Trafalgar Square is, as Gilbert and Driver (2000: 29) put it, a 'contradictory' place.

Traces beget traces . . .

So cultural geographers translate the coming together of multiple traces in places. They interpret the material things, the cultural activities, the social ideas, and the broader contexts that constitute places in order to make sense of the world around us. Although material things (such as Nelson's Column) may leave more durable traces in places, cultural activities such as celebrations or protests, even more mundane everyday activities like commuting or shopping, despite being more transient in nature, are equally important traces to investigate. These activities may themselves leave some material traces (such as property damage after riots, pollution after commuting, litter and waste after shopping), but they also leave non-material traces in people's hearts and minds. (For example, perhaps Fritsch's blue cockerel may prick the egos of patriots and zenophobes of all nationalities.) Such emotional and psychological traces – be they individual or collective – can be emphasised, concealed, or spun by other actions in different, but related, contexts (for example, through reportage and comment on events in media spaces). These 'trace-chains' of things, activities, emotions, and contexts have significance not simply for that generation of place and its identity. As we live in the cultural world, our attachment to places can be strengthened or eroded through these traces. These trace-chains are also significant as they may motivate similar actions in other places. For example, they may prompt law changes which affect the cultural context in future, or may change cultural attitudes among protesters, politicians, and the public which produce new spaces of understanding, repression, or dialogue. Thus when cultural geographers study place they study material objects, cultural activities, social ideas, and geographical contexts. They investigate how these material and non-material traces, these emotions and ideologies come together with spatial contexts to constitute particular places in time, and how these places impact on other sites that may be geographically or temporally disparate.

Non-human traces

However it is not simply humans that contribute traces to places. Non-human actions and interactions also shape the cultural geographies of places. This non-human activity may come in the form of 'natural' disasters, weather events, or animal activity. If we continue our example of Trafalgar Square, pigeons and their interactions with context and culture have played a key role in shaping the cultural geography of place. Pigeons leave their own inimitable traces (guano) on the statues and plinths, and these traces react with

rainwater to produce an acidic fungus which dissolves the masonry (McClure, 2005). How these traces and their perpetrators are dealt with by the authorities highlight key aspects of the cultural relations between humans and non-humans (in this case, living birds and non-living statues). Are these birds tolerated or celebrated, are they protected or killed? In practice, a range of tactics has been used to keep Trafalgar Square a pigeon-free zone. Falcons have been used to discourage pigeons, bird-seed sale and pigeon-feeding have been outlawed, whilst netting and mesh is used to discourage roosting. As a consequence of the removal of the majority of pigeons from Trafalgar Square, not only are the monuments preserved, but the place is easier to market for other uses (such as a setting for feature films).

Summary: ongoing compositions of traces

Cultural geographers therefore analyse and interrogate all the agents, activities, ideas, and contexts that combine together to leave traces in places. In the example above we have focused on a material context in the case of Trafalgar Square, but, as stated, cultural geographers focus on a wide range of contexts, not simply material but also the non-material, for instance, emotional spaces, language spaces, or virtual spaces. The traces left in this range of contexts may be material or non-material, durable or temporary, left by humans or non-humans. They may produce the intentions desired by the trace-makers, or have unintentional effects on the agents and their audiences. These traces may overlap one another, synergise together, or come into conflict with one another. They may change over time, thus making the identities and meanings of places dynamic, but nevertheless these traces often remain in-place as shadows and echoes of places-past. The traces may generate other traces in other places, linking places and traces together in chains that may become entangled and (con) fused. By understanding place as an *ongoing composition of traces* it facilitates the interrogation of these traces and how they come to confer cultural meanings to geographical sites. We can come to understand and critically act in place through interrogating the ideas motivating traces, the frequency and manner in which theses traces are repeated and reinforced, their popularity and persuasiveness, the influence of their composers, and how they react with other competing traces-in-place. Through doing so, this book maps how cultural geography approaches place.

BOX 1.3 PLACE: AN ONGOING COMPOSITION OF TRACES

You are here
Source: Author

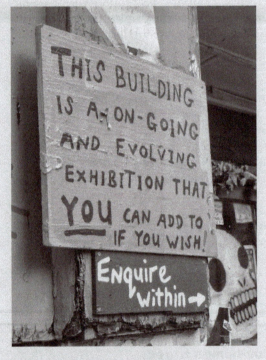

Places as ongoing composition
Source: Author

This book suggests that place can be considered as an 'ongoing composition of traces'. This way of thinking about place resonates with ideas put forward by Bruno Latour in his 'Compositionist Manifesto' (2010). As Latour states, although the word 'compositionist' may be 'a bit too long and windy' it does have its uses. For this book, the idea of 'composition' is useful as it suggests five key qualities to places and our relations with them. Firstly, and most obviously, it suggests that places are 'composed': they are not neutral entities which are 'just there', but rather they are constructed by cultures and peoples, including us. Secondly, considering places as compositions means that they are not wholly dissimilar to other cultural products or processes, such as art, dance, music, or writing. Places are written, practised, and performed, and in each action they are (re)interpreted and undergo change in ways that are both intentional and unexpected. Thirdly, if places can be 'composed', they can also 'decompose'. Places can therefore change, their traces can be removed, eroded, or erased, and their non-material traces mis-remembered or forgotten. Fourthly, and relatedly, places are diverse and fragile. In Latour's words, places are 'immanent' (2010: 475), they are always 'now' and of the moment – in short, they are always in a state of *ongoing composition*. Lastly, this 'compositionist' way of considering the world positions us as both responsive to the influence of place, and able to change it. Framing the world as an ongoing composition suggests a positive approach to geography and culture, allowing us to examine, critique, but also change and transform. This framing tells us to 'be here now', and like the sign on the building above, reminds us we can take responsibility for contributing meaningfully to our existence in the cultural world.

Approaching the cultural geography of place

This approach to the cultural geography of place thus encourages us to take a critical understanding of the places around us, as well as our position within them. It critically evaluates places as ongoing compositions of traces. In summary, this approach asks the following questions:

- What cultural traces dominate a particular place? Who and what do these traces stand for? In other words, whose place is this anyway?
- Are the traces in this place challenged and resisted? If so how?
- What do these alternative traces stand for? Whose places do they seek to make, and what would these places be like?
- What are the consequences of this ongoing composition? What trace-chains are set in motion, and what cultural orders and geographical borders are being established, new or otherwise?

As cultural geography also positions us within our world, two further questions are raised:

- Do these changes have any effect on how we should think about place?

- Do these changes have any effect on how we should act in place?

As you may have identified, the questions raised by cultural geography's approach to place can be remembered through the shorthand of the emboldened acronym found in these queries (TRACE: Traces? Resisted? Alternatives? Consequences? Effects?). In the chapters that follow, this book will further interrogate these traces and places through this cultural geography approach. In each chapter we will look at different cultures and geographical contexts and interrogate the ways in which traces come together to define places and peoples. In each chapter we will answer the questions posed above, and introduce both empirical examples and theoretical ideas which facilitate understanding about how the world around us is generated, and our role within it.

Before we employ our cultural geography approach we begin Part 2 of this book by outlining two chapters on 'where cultural geography has been'. These chapters give us a clear overview of the history of the discipline, covering the development of the definitions of, and approaches to, the terms 'culture' and 'geography'. Chapter 2 outlines the main changes encountered by the discipline up to the late twentieth century. It begins with an investigation of the connections between culture and geography in this history of exploration, and journeys through to the arte-factual approach of Carl Sauer. Chapter 3 charts the emergence of a 'new' cultural geography in the latter decades of the twentieth century. It outlines two contemporary approaches that are often seen in opposition: the representational and the non-representational. The chapter outlines the key differences – but also the impor-tant similarities – between these theories and emphasises how this book's approach to place as an ongoing composition of traces effectively combines them.

In Part 3 we investigate place, power, and scale. In Chapter 4 our focus turns directly to place. It outlines differing definitions and under-standings of the term, and how places are central to human tracings of belonging and community. It will introduce notions of *orders* and *borders* to help explain how places come to unite human identities and cultural geographies, giving us our 'sense of place'. Chapter 5 explores the notion of power. This chapter interrogates conventional understandings of power as 'domination' and 'resistance', and how these are inevitably tied to culture and geography. It will illustrate how, in practice, these conventional terms are often entangled together, producing contestations and multiple traces in places. Chapter 6 explores how senses of place can be created around geo-graphical scales, but also around cultural ideas. It investigates how traces such as patriotism and religion can combine to strengthen but also complicate our own sense of place.

In Part 4 we investigate cultures of capital-ism, mobility, and nature. Chapter 7 focuses on perhaps the dominant culture in the world today: the culture of capitalism. It investigates the contested traces of capitalism through mobi-lising the theoretical instruments of power and place, the geographical connections between places across the globe, and their impact on notions of community and identity. Chapter 8 explore cultures of mobility, investigating how

dependent our modern lives are on how things, ideas, and people move. It outlines how we can consider various 'constellations' of mobil-ity, and explores what mobility means for those in the fast and slow lanes of movement across the world. Chapter 9 explores the relation of cultures to nature. It investigates how different cultures naturalise 'nature', and examines how these relations influence how these cultures act in nature as a consequence.

In Part 5 we explore race and ethnicity, youth cultures, language, and cultures of the body. Chapter 10 investigates how race and ethnicity impact on how places are taken and made. It explores a range of bordering tactics that seek to purify some but partner other ethnic traces to produce the ideal composition of place for particular cultural groups. Chapter 11 centres on the traces motivated and engendered by ideas of age. This chapter examines how adult and youth cultures interact to take and make the places around us, focusing on examples of free running and graffiti. In Chapter 12 we look at cultures of language, exploring how dominant language cultures are able to (b)order not only our words, but also our understandings and relations to the cultural world. It will investigate how language is multisensory, affecting our sound world as well as our material places, as well as arguing that we all speak in multiple tongues. In Chapter 13 we explore the place of the body. Specifically we investigate the manner in which traces can be written on the body through our gender and sexuality. We see how these bodily categories can be culturally and geographically controlled through invented roles and relations, before exploring how these controls are subverted or resisted through processes of individualisation.

From these chapters we see that we are inex-tricably part of the cultural world. We are part of this world as citizens, but also as cultural geog-raphers. As it is important to consider how we relate to and act in this world, Part 6 outlines a range of common methods for *doing* cultural geography. It begins by outlining key issues on ethics and access, before giving a brief introduction to methodologies that seek to explore the language of the word (namely interviews and the analysis of cultural texts), and the language of the world (through embodied practices and ethnography).

Finally, Part 7 (Chapter 15) brings together the threads outlined in *Understanding Cultural Geography: Places & Traces*. It encourages readers to think critically about their own experiences and perceptions of living in cultural places and contributing to cultural geography. Let's begin this book, therefore, by finding out where cultural geography has been.

CULTURAL GEOGRAPHY THEN AND NOW: THE HISTORY OF THE DISCIPLINE

2

The history of cultural geography

Culture has come down to us as a very complex word indeed, highly suggestive in its meanings, and rich in its possibilities. By contrast, its history in cultural geography – at least until quite recently – has been rather more restricted.

(Mitchell, 2000: 16)

This book maps how cultural geography approaches place. As the introduction has outlined, this book's interpretation of cultural geography focuses on the creation and combination of traces in particular geographical sites; it looks at how places are generated through complex amalgams of culture and context. Yet this version of cultural geography isn't how the discipline has always been. Ideas of culture and geography are not something that have been invented by this book, or discovered fully formed by a single individual for us to put into operation; rather cultural geography has a history. Versions of the discipline have developed by various scholars, in various parts of the world, right up to the present day. The purpose of Part 2 is to outline this history; it tells the story of the ideas, practices, and scholars that have come to define past and present versions of cultural geography.

The history of cultural geography and the 'god trick'

On the face of it, charting the history of cultural geography seems to be a relatively straightforward task. However in truth it is a difficult and complex undertaking. Such a task is risky for both the writer and the reader because it invites both to fall for the 'god trick'. As Buttimer (1998: 90) identifies:

> An invitation to speak on the state-of-the-art in the history of geographical thought is surely a temptation to play what Donna Haraway (1988) has called the 'god trick' – to assume an Archimedean or sovereign gaze from nowhere over an the entire array of stories which geographers tell about their practices.

As Buttimer points out, the 'god trick' is an idea conjured up by Donna Haraway and it suggests that we can be tempted to act as if we know everything, that we can see everything, that we can re-write history and even contemporary society because we are god-like in omnipotence and omnipresence. This is a trick because we are, of course, not gods. It is only through reading, experience, intuition, and reason that we build our arguments and evidence our versions of history. Through sleight of hand we create our stories by cutting and splicing together a complex and entangled history into a version of events with clear threads and a persuasive weave.

This is therefore what this chapter does; it takes the history of cultural geography and – with luck – translates its knotty complexities into something we can get to grips with. So let's begin learning the ropes.

A genealogy of cultural geography

There are many ways to splice together the historical threads of cultural geography. Duncan uses the idea of 'heterotopias' (1994: 402), whilst Oakes and Price call the ghostbusters to investigate the 'hauntings' of the discipline (2008: 118). This chapter however adopts the idea of *genealogy* to knot together a *family history* of cultural geography. As with our own families, it is possible to look through history and make out the connections, legacies, and influences that our forebears have had on the way we are today. We can discern who has been important and what significance they have had over our family. We can make out certain patriarchs or mad uncles. We can discover how their ideas influenced, and were influenced by, the ideas and necessities of the time, and we can explore what relevance they have, if any, today. So, as with your own families, even if you don't like your auntie, or agree with her views, you can still be interested in why she thinks like she does, how she was perhaps a child of her time, and what differences there are between your ideas and hers. So it is possible to illuminate aspects of our own ideas and identity through shining a light on our ancestors, and this is the case for the history of cultural geography, as Anderson and Jacobs (1997: 12) outline:

> Exploring the genealogy of one's professional endeavours [in this case the history of cultural geography] offers much more than a stimulating exercise. In looking back, in taking up a reflexive position in relation to one's disciplinary identity, it is possible to excavate the modalities of knowledge which have come to structure perspectives in the present.

In knotting together a family history of cultural geography, we become aware that ideas about what 'culture' and 'geography' could be are undoubtedly influenced by the time and place of their creation. There are, in other words, *historical and geographical signatures* on different branches of the family tree. As Castree identifies, these signatures are carved by a nexus of influence that includes both broader society as well as discussion within the discipline of geography itself: 'Geography has always been a contested enterprise: it has no essence. The discipline, like all others, has constantly changed its spots in response to *external pressures* and *internal debates*' (Castree, in Castree and Braun, 2001: 4, emphasis added).

These external pressures – from broader societal changes and events, and internal debates – arising from intellectual and philosophical disagreement from within the discipline itself, have influenced the way cultural geography has been practised throughout history, dating back to the early threads of the discipline in Western Europe and North America of the eighteenth and nineteenth centuries. It is at this time and in these places that we begin tying together our history of cultural geography.

Geographers, explorations and empiricism

In the eighteenth and nineteenth centuries, geography became instituted in an era of exploration. The novelist Joseph Conrad clearly identified the connection between geography and European exploration by marking three phases in the geographical exploration of the world (1924). As Godlewska and Smith (1994) point out, the first phase – 'Geography Fabulous' – occurred in the sixteenth and seventeenth centuries, where early science combined with mythology and magic to marvel at the new world of discoveries. The second phase, the phase that concerns us here, Conrad named 'Geography Militant'. This involved more systematic and practical explora-tion, and, as we shall see, the political conquest of new territories (after Godlewska and Smith, 1994: 1). The success of this phase led to the third – 'Geography Triumphant'. Sponsored by the growing industrial powers of Western Europe, 'militant' geographers went out in the field and sought to systematically explore 'new' areas of the physical and human world. In this period, physical and human geography were closely aligned, and together they literally cre-ated the maps and encyclopaedia we use today. They navigated the seas and documented the lands and people that had been little known in their home countries.

Thus, in the era of 'Geography Militant', much cultural geography was concerned with filling in the detail on empty maps, indeed, cartographi-cally creating the maps themselves. According to Driver (2000), a key figure and inspiration to the geographers of the time was the explorer and naturalist Alexander von Humboldt. As Pyne (1998: 29) outlines, von Humboldt was akin to geography's very own Indiana Jones:

> What Beethoven was to the music of the Romantic period, what Napoleon was to its politics, Humboldt was to its science . . . Born 1769, Alexander von Humboldt explored South America from 1799–1804. He had paddled up tropical rivers like the Orinoco, scaled peaks like Chimborazo, sketched the ruins of lost civilisations, experimented with electric eels, watched fruit eating bats and carnivorous alligators, measured latitudes and mountain slopes, and obsessively col-lected more than sixty thousand specimens in all . . . Here was the naturalist as Promethean hero.

If von Humboldt was Indiana Jones-like in terms of his style, we can also categorise his endeav-ours in terms of their substance. In mapping the genealogy of the discipline it is important to not only determine *what* geographers studied, but also *how* they studied it. Von Humboldt and his fellow explorers can be categorised as 'empiri-cist' in terms of how they studied the world. But what does 'empiricist' mean? Empiricist geog-raphers observed and documented the world around them. They approached their work from the premise that through conscientious observation and faithful recording the truth of relationships between species and environ-ments could be captured. In short, empiricists felt that the truth of the world was out there, as long as they looked in the right ways and in the right places. Empiricism therefore placed weight on the accuracy, rationality, and neutral-ity of scientists and their fieldwork observations. The popularity of empiricism spread, universi-ties and scientific institutions (such as the Royal Geographical Society inaugurated in 1830 in London) were founded in order to collate, study, and disseminate the knowledge gathered from these explorations.

From empiricism to positivism and Darwin's legacy

From the empiricist data collated from differ-ent places, geographers hypothesised about the nature of the relationships between species and the environments they inhabited. As patterns between observations were identified, theories to explain these relationships were suggested. This move from simple *collation of observations* to *the-orisation on these observations* can be categorised as the shift from *empiricism* to *positivism*. This move was seen as strengthening the geographi-cal enterprise in two interrelated ways. Firstly it made geography appear more 'expert' as more explanations were being suggested and subse-quently tested. Secondly, it created the need for more geographic exploration – if the best theories were to be produced, further fieldwork was essential to make sure they were rigorously developed and accurate. Due to the close align-ment between human and physical geography at this time, the growing geographic enterprise looked to natural science ideas for theoretical inspiration. Of key influence in this area were ideas generated in the fields of ecology, botany, and biology, particularly those of natural scientist Charles Darwin.

Darwin's thinking can be considered a small but important step in instigating the phase of geographic endeavour Conrad identified as 'Geography Militant'. Before Darwin, many thinkers explained environmental conditions through the role of God in nature (Lindberg and Numbers, 1986). Darwin's thinking displaced these 'holistic' and 'divine' ways of understanding and replaced them with 'modern' scientific explanations. This marginalisation of the magical, what Adorno (1991) and Weber (1978) called the 'disenchantment of the world', favoured explanations found through geographical enquiry and scientific logic. Darwin's own explanations came from lengthy explorations in the 'New World' and were published in *The Origin of Species* (1859). At this time Darwin developed laws to explain variation in species through a combination of adaptive mutation, genetic inheritance, and environmental influence. In Darwin's own words (1996: 2):

> we see beautiful adaptions everywhere and in every part of the organic world . . . all th[is] results . . . from the struggle for life. Owing to this struggle, variations, however slight and from whatever cause proceeding, if they be in any degree profitable to the individuals of a species, in their infinitely complex relations to other organic beings and to their physical conditions of life, will tend to the preservation of such individuals, and will generally be inherited by the offspring . . . I have called this principle, by which each slight variation, if useful, is preserved, by the term Natural Selection . . . But the expression often used by Mr Herbert Spencer of the Survival of the Fittest is more accurate and sometimes equally convenient.

In other words, Darwin's hypothesis, which has come to form the basis for the almost universally accepted theory of evolution, emphasised the relations between species and their environment. According to this idea, different species adapt through habit and genetic mutation to best survive the climatic and environmental conditions they exist in. The best adapted species survive and flourish whilst less adapted species struggle, fall victim to other species, and ultimately die out. Darwin's notion drew the attention of geographers interested in human cultures. Perhaps this theory could help explain the variation of human cultures in different places? The geographical application of Darwin's ideas came to be known as environmental determinism.

Human cultures and environmental determinism

> Environmental determinism . . . represented the first attempt . . . at generalisation by geographers during the modern period. Instead of merely presenting information in an organised manner, either topically or by area, geographers sought *explanations* for the patterns of human occupation of the earth's surface. Their major initial source for explanations was the physical environment, and a position was established around the belief that the nature of human activity was controlled by the parameters of the physical world within which it was set.
>
> (Johnston and Sidaway, 2004: 46, my emphasis)

Darwin thus theorised that the environment played a determining role in non-human species' adaptation and survival. As Johnston and Sidaway suggest, this notion of environmental determinism was adopted by geographers (such as Huntingdon (1924) and Griffith Taylor (1947, 1961)) to explain variations in human cultures in different places. For these geographers, particular cultures were produced due to the overwhelming influence of the natural conditions they developed in; in short, cultures themselves were environmentally determined.

Environmental determinists not only accounted for cultural difference on the basis of environmental difference, they also offered judgements on the relative advancement of the cultures discovered through exploration. For them, the strange cultures encountered in the new world were often deemed not simply different from those of the West, but barbaric and uncivilised in comparison to them. Their hypothesis followed that a direct relationship could be identified between the latitude of an environment (a proxy for temperateness of climate) and the civilised nature of the human culture found therein. They reasoned the higher the latitude (or more temperate the environment, such as those found in Western Europe, the home of many of the geographers undertaking exploration), the more evolved, advanced, and civilised the culture produced. The lower the latitude (or more tropical the environment), the more savage, barbaric, and uncivilised the culture would inevitably be. As Livingstone outlines, this crude causal relationship between latitude and cultural advancement was widespread:

> Consider, for instance, the analysis of Edinburgh geologist, geographer, and explorer Joseph Thomson, who, during a short life, travelled extensively in East Africa. In 1886, during a period of convalescence between trips, he spoke to the Birmingham meeting of the British Association about his experience in Niger and Central Sudan, an address that appeared later that year in the Scottish Geographical Magazine. Here he paused to reflect on the moral-evolutionary impact of climatic conditions:
> 'It is fact worthy of our attention that, as the traveller passes up the river [Niger] and finds a continually improving climate . . . he coincidentally observes a higher type of humanity – better ordered communities, more comfort,

with more industry. That these pleasanter conditions are due to the improved environment cannot be doubted. To the student with Darwinian instincts most instructive lessons might be derived from a study of the relations between man and nature in these regions.'
> (Darwin, 1992: 223)

In short, therefore, the environmental determinist thesis came to be interpreted in ways that played to Western perceptions of cultural superiority and advancement. Indeed, it could be argued that its popularity was based in no small part on the way it pandered to Western cultural prejudice. The environmental determinist thesis was also popular because such prejudice also had a political dimension. Due to the relative 'superiority' of Western states and their cultures, a moral justification was installed to 'help' the 'less civilised' cultures 'improve' themselves. Thus through the process of geographical theorisation, a broader social and political project gathered momentum: the project of colonisation of these newly discovered places through the spread of Western culture.

Geography and empire

The opening up of the world through cartographic and geographic exploration thus commenced a process of European colonisation. As geographers identified new resources, labour, and trade routes, states 'imposed political and military control, battled or bargained with local ruling elites, [and] confronted or diverted local opposition' (Godlewska and Smith, 1994: 56). All aspects of Western culture were spread into these new areas. States subjugated and took control of foreign tribes, their land, and resources. Religious, political, and economic influences took and made places in line with Western cultural values. This spread of European culture was sometimes peaceful, but often bloody, and either through the Bible or the bayonet Western powers exercised their control over the new world (see Driver, 2000). Geographers therefore not simply studied and theorised cultures but had a hand in spreading their own – as Godlewska and Smith argue, at this time geographers and the project of empire were tied by an 'umbilical connection' (1994: 2).

Such empire building, underwritten by theories of environmental determinism, thus instigated major economic, social, and political differences to the cultural fabric of the world. This phenomenon itself came to be the focus of geographical investigation as scholars sought to explain, and perhaps even justify, the new process of colonisation that had been set in motion. They did so by (re)turning to Darwinian ideas of survival of the fittest. German theorist Friedrich Ratzel applied notions of environmental determinism in an attempt to not only explain differences between cultural groups, but also their associated political organisations and national states (1896–1898). If, theorised Ratzel, strong organisms needed enlarged ecological spaces to develop within and grow, then why not human cultures and, more specifically, political states too? For Ratzel, some states would be culturally strong and better disposed to thrive. These states would have a growing population and subsequent need for more resources. It made sense, therefore, that strong political states required more territory – or living space (*lebensraum* as Ratzel called it) – in order that they flourish. It was only 'natural' that this living space should be obtained through annexation and subjugation of weaker states. Ratzel thus reasoned that colonisation was a necessary process for all parties as it allowed the strong to thrive, but also 'raised' the status of the weak through inter-breeding (Ratzel, 1896–1898). From this hypothesis, Ratzel not only *explained* the growth of empires, but also *legitimated* them. One can imagine how such ideas were seized upon by those states who saw themselves as strong and worthy of expansion – it was not simply greed or security that justified their growth – from this perspective it was only

'natural' and 'right' that such expansion took place. Indeed, Ratzel's ideas were closely connected to the eugenics and nationalist ideologies of Nazi Germany in the early decades of the twentieth century (and following the Second World War fell into disrepute due to this association (Dodds and Atkinson, 2000)).

However, the notion of environmental determinism did not pass without criticism. Many scholars expressed disquiet over the use of geographical ideas to justify and legitimate empire-building tendencies. For example, the Russian theorist Peter Kropotkin argued that nature was far from being 'red in tooth and claw' (Tennyson, 1850); he postulated that successful species worked synergistically together through mutual aid and co-operation, rather than aggressive competition (Kropotkin, 1915). Other geographers refuted the idea that humans were passive agents solely responding to environmental conditions. These scholars responded to environmental determinism with their own ideas, captured under the heading of environmental possibilism.

Human cultures and environmental possibilism

Environmental possibilists, including geographers such as the Frenchman Paul Vidal de la Blache, Scot Patrick Geddes, and Englishman Richard Hartshorne, explained variations in cultures in different ways to environmental determinists. They argued that cultures and cultural variations were not formed by environmental factors alone. For them, the environment was simply one factor that influenced, rather than determined, the production of cultures. From their observations, environments offered *possibilities* from which human groups could generate their own activities and processes, they did not render inevitable certain destinies and outcomes. From this perspective, therefore, humans were not considered to be passive automatons responding to the natural world, rather they were active agents creating their own cultures and to some extent their environments. These geographers also stopped short of placing negative valuations on the differences identified between these new cultures and their own. They did not resort to crude prejudices of superiority and inferiority; cultures in other places were not better nor worse than those in the West, simply different. For environmental possibilists the nature of the geographic project was also fundamentally changed. For them the most important geographical scale was not the nation state and its expanding colonial potential, rather the sub-national region and its constituent localities. Politically these 'regional geographers' were closer to the mutual aid and civic participation ideas of Kropotkin, rather than the lebensraum of Ratzel. Vidal de la Blache, for example, catalogued the different cultures of the regions of France in order to show the diverse yet common culture of the French people (Vidal de la Blache, 1941). He argued that this culture tied these localities together in the face of their aggressive European neighbours. Geddes, for his part, wanted to encourage civic rehabilitation, devolving power back to local communities so they could have greater participation in the democratic process and political decision-making (Geddes, 1912).

This political project had implications for the way environmental possibilists conducted their geography. For them geography involved a return to the empiricist cataloguing of the era of exploration. The objective was to identify variations in lifestyles in different regions, and document the specific material artefacts produced by each culture in each place. This documentation perhaps included cataloguing agricultural equipments, housing styles, or clothing types – the particular material *things* that were hewn from the physical environment and came to be associated with, perhaps even define, the culture of a region. Environmental possibilism was therefore an essentially descriptive cultural geography which sought to document and catalogue in order to establish the extent, nature, and difference (or indeed similarity) between particular cultures in different places.

Summary: geography, environments, and cultures

It is clear, therefore, that environmental determinists and environmental possibilists had fundamentally different ideas about the relations between cultures and the environment. They disagreed about what produced variations in cultures, what should be investigated to explain these differences, and what political actions these differences legitimated. Table 2.1 generalises these key differences between environmental possibilism and determinism.

Table 2.1 Environments and cultures: determinism and possibilism

	Environmental determinism	Environmental possibilism
Source of key ideas	Natural science	Natural science
Methodology	Positivist law making	Empiricist fieldwork
Cultural agent	Environment	Environment and people
Geographical scale	National/global	Regional/local
Social values	Commerce	People
Political ends	State/military	Mutuality/civic participation

Environmental possibilism thus stood in opposition to the geographical project of positivism, state expansion, and environmental determinism. These possibilist ideas were not limited to Europe, but also took root across the Atlantic in the United States. Here geographers in the Berkeley School at the University of California developed these notions into a particularly American branch of the family tree. It is this branch that we now look to in order to continue our genealogical tracing of cultural geography.

The American branch of the cultural geography family tree

Akin to the regional geographers of Europe, the main branch of cultural geography in America was rooted in a strong antipathy to environmental determinism. Environmental determinism had taken hold in the United States through the work of Ellen Semple (1942) and Ellsworth Huntingdon (1913), but the Berkeley School, led by Carl Sauer, turned to the work of the regionalist tradition in Europe to reassert the empiricist documentation of cultures that they favoured. European *chorological* cataloguing – i.e. the documentation of cultural difference and distribution – led this branch of cultural geography away from positivist theorising and back to fieldwork study, recording specific objects and associating them with particular regions and cultural activity. As Solot explains, Sauer 'believed that geography should be devoted to the collection of areal facts and not to the search for general principles' (Solot, 1986: 509), or in Sauer's own words, 'Underlying what I am trying to say is the conviction that geography is first of all knowledge gained by observation . . .' (Sauer, 1963: 400).

Sauer thus looked to re-establish the empiricist tradition in geography in America, and legitimated this decision by grafting his fieldwork valorisation to a new philosophical lineage. Rather than taking intellectual insight from the

Figure 2.1 Inspirations for the American branch of the cultural geography family tree

'Sauer' source: http://people.cas.sc.edu/ajames/Bing_Symp/1956%20Photos/Sauer.jpg

'Lowie' source: www.americanethnography.com/img/robert-lowie.jpg

'Kroeber' source: www.lib.berkeley.edu/ANTH/emeritus/kroeber/kroeberinfield.jpg

'Boas' source: www.spartacus.schoolnet.co.uk/USAboas.jpg

natural sciences (as environmental determinists and possibilists had done), Sauer looked to anthropology for philosophical impetus. Here he turned to the anthropological work of Robert Lowie (1917) and Alfred Kroeber (1923), who themselves were directly influenced by Frank Boas (1911, 1912) (see Figure 2.1). This anthropological starting point helped Sauer justify a new type of cultural geography, one that focused on cultural groups and their geographical spread as ends in themselves, rather than as means to generate theoretical posturing, empire-building, or nationalistic fervour.

Geographies of the cultural landscape

Sauer thus forwarded a cultural geography that mixed the chorological tradition of Europe with the anthropological interest of North America. This formed a branch of the discipline that differentiated cultural groups by the material artefacts they produced in, with and from the landscape around them:

To put this in Sauerian terms, the task was to describe the *morphology* – that is, the shape, form and structure – of a given landscape, and in so doing to reveal the characteristics, trace, distribution and effectivity of the human cultures that had inhabited and moulded it.

(Wylie, 2007: 23)

Sauer argued that by studying the landscape itself the keen geographer could read off the cultures that inhabited it. For Sauer, therefore, the relationship between the 'environment' and 'culture' was better put in terms of 'landscape' and 'culture' (perhaps reflecting his anthropological rather than natural science inspiration). The interconnection between landscape and culture was far from being the environmentally determinist interpretation of previous branches of the discipline, in fact for Sauer the onus of agency was the other way around. No longer was 'man [sic] a product of the earth's surface' (Semple, 1911: 1), rather cultural geography was 'the story of how human beings have transformed the earth' (Wallach, 2005: 2). As Sauer (1925: 52, my emphasis) put it:

The natural landscape is being subjected to transformation at the hands of man, the *last and for us the most important morphologic factor.* By his cultures he makes use of the natural forms, in many cases alters them, in some destroys them.

So, for Sauer the onus of agency in the relationship between culture and landscape was reversed. For him human cultures were the most important factor influencing the metamorphosis of any place. It was human culture that moulded the natural landscape, and through this interaction the natural landscape was transmogrified into a cultural product.

Echoing the European regionalists, Sauer felt that in order to identify cultures through landscapes it was necessary to document the material things laid down onto them by the humans involved. For Sauer, 'the geographer, therefore, is properly engaged in *charting the distribution over the earth of the arts and artefacts of man* [sic],

to learn whence they came and how they spread' (Sauer, 1969: 1, my emphasis). This interest in the material inscriptions on a landscape resonates with the notion of trace-making that forms the basis for this book (see Figure 2.2).

This branch of the cultural geography family tree focused on artefacts in the landscape, such as barn and housing types in different areas of America, or on how indigenous tribes were distributed in different locations.

Our approach to the cultural geography of place also shares with Sauer the recognition of the dynamic nature of geographical sites. Sauer identified the changing nature of landscapes through drawing attention to the ways inscriptions on the landscape build up over time. He noted that every generation lays down their products onto the landscape, 'superimposing' their cultural artefacts on the 'remnants of older' ones (1963: 343). In this way, Sauer postulated that landscapes should be understood as *palimpsests* – as surfaces with multiple inscriptions that build up over time and mark the presence and passing of different cultural groups (see Box 2.1).

Of course, by reversing the onus in culture-landscape relations, Sauer drew attention to the agency humans can have in the physical world. Sauer himself was disquieted by the increasingly technological, rational, and environmentally destructive tendencies of Western industrial cultures and gravitated towards landscapes in which a more 'harmonious' (Parsons, 1987: 156) relationship between humans and the land could be ascertained. Such a predisposition had two notable effects. It not only tied Sauer into the Romantic tradition that predated his work, but also led him and his colleagues to focus their attentions on more historical landscapes where, they argued, these sort of relations could more easily be found. For the Berkeley School, therefore, cultural geography came to focus on landscapes found away from cities and developed hubs, and on folk or antiquated cultures. As Wallach (2005: 2) puts it:

Barn 1

Barn 2

Barn 3

Barn 4

Barn 5

Barn 6

House A

House B

House C

House D

House E

House F

Figure 2.2 Artefacts in the cultural landscape

Barns:
1 www.flickr.com/photos/ddnlj/2680683753/ Southern State, ddnlj
2 www.flickr.com/photos/geliza/3152160835/ Upstate New York, Geliza
3 www.flickr.com/photos/20314119@ N08/2705276126/ Wisconsin, Boodie 131
4 www.flickr.com/photos/wwworks/686880246/ Oklahoma, Woodley wonderworks
5 www.flickr.com/photos/atelier79033/2873132621/, Upstate New York, Ed Brodsinksy
6 www.flickr.com/photos/lensman45/2712751644/ Neffsville, Lensman 45

Houses:
A www.flickr.com/photos/mcmorr/654271199/ Iowa, McMorr
B www.flickr.com/photos/sunsurfr/457613520/ Alabama, sunsurfr
C Cannery Row, Monterrey Bay, Author
D Alamo Square, San Francisco, Author
E www.flickr.com/photos/silvery/3138621906/ Bryan, Texas, Si1very
F www.flickr.com/photos/21066820@ N06/2084344682/New Mexico Pueblo, jefg99

BOX 2.1 PLACES AS PALIMPSESTS

'Rome Palimpsest'
Source: Author

'Westbury Palimpsest'
Source: Neil France

The idea of landscape as palimpsest suggests that inscriptions are left in a cultural landscape by different generations over time. The inscriptions of different historical cultures can be seen in the above examples of Rome, Italy, and Westbury, England. In Rome built forms can be identified in the foreground that remain from Ancient times (now protected by modern-day fences from modern-day tourists), to eighteenth-century churches and nineteenth-century civic buildings (centre and back left), as well as twentieth-century monuments (back centre) with twenty-first-century scaffolding. In Westbury, Iron Age ramparts are written over with an eighteenth-century white horse. Modern-day farming systems complete the ongoing composition of place as palimpsest.

his aversion to the modern world explains why Sauer's own courses in . . . cultural geography stopped somewhere about 1850 . . . Sauer believed – this is hardly an exaggeration – that human history had peaked sometime before 1900.

Summary: cultural geography and the cultural landscape

We can see that Sauer's version of cultural geography investigated the transformation of natural landscapes into cultural ones (after Price and Lewis, 1993: 1). This investigation focused primarily on empiricist fieldwork description rather than positivist law-making, and sought to document the material artefacts produced by the specific culture in question and how these products changed the appearance and function of the landscape. It documented the distribution and scope of these landscapes, and traced their historical progression. The empirical interest centred more on rural and historical landscapes, rather than the commercial modern landscapes of contemporary America. This branch of the cultural geography family tree therefore focused more on the products of cultures, rather than the

Table 2.2 Geographies of the cultural landscape

Source of key ideas	Anthropology
Methodology	Empiricist fieldwork
Cultural agent	Culture
Geographical scale	Landscape
Social values	Romantic/ environmentalist
Political ends	Anti-technology/ pro-tradition

processes that generated them; it centred on the shape, rather than the shaping of the landscape. As Solot (1986: 508) puts it:

> [Sauer] believed that the proper objects of study in cultural geography are material manifestations of culture – land use, settlement patterns, technology, and other artefacts . . . it encouraged the study of physical changes in particular traits or in the landscape itself rather than the processes of change affecting culture itself.

Such an approach to cultural geography therefore focused on questions concerning the nature of objects and their distribution, morphological questions of 'what?' and 'where?', and the connections between the two. The key characteristics of this branch of cultural geography are outlined in Table 2.2.

Geographies of the cultural landscape and the thorny question of culture

This American branch of the discipline thus focused primarily on the study of cultural artefacts. Whilst this was a project of note, interest became focused on the *cause* of the variations in cultural products. If the agent causing variation was 'culture', what exactly *was* this thing called 'culture'? Sauer was silent on this thorny question, as Price and Lewis (1993: 11) critique:

How then did Sauer conceptualise culture? Most evidence would suggest that he devoted little effort to the question . . . Sauer loosely employed the term to describe and distinguish the range of human diversity . . . it was an heuristic tool, not an explanatory concept.

Sauer's silence on the thorny question of culture led to internal debates within the discipline of geography, particularly in the UK. Scholars such as Jackson (1989), Duncan (1980), and Cosgrove (1989) all questioned Sauer's heuristic use of the term culture, and the lack of critical analysis of the processes and practices that contributed to its production. If, in Sauer's words, 'the shaping force [that transformed the landscape] . . . lies in . . . culture itself' (1963: 343), they considered it logical to investigate what this thing called culture actually constituted. Sauer's preference for field study over theoretical reflection meant he devoted little effort to this question and its associated analysis. In light of this, many commentators turned to Sauer's own anthropological inspiration in an attempt to gain insight into what this thing called culture could be.

What is this thing called 'culture'?

For Sauer's anthropological antecedents, culture was 'superorganic' in nature. This 'superorganic' view, held by scholars like Kroeber and Boas, supposed that culture was something that existed independently of human beings; it was an entity that operated under laws of its own, with the agency to govern human behaviour from a distance. As Zelinsky (one of Sauer's students) puts it,

> We are describing a culture, not the individuals who participate in it. Obviously, a culture cannot exist without bodies and minds to flesh it out; but culture is also something both of and beyond the participating members. Its

totality is palpably greater than the sum of its parts, for it is superorganic and supraindividual in nature, an entity with a structure, set of processes, and momentum of its own, though clearly not untouched by historical events and socioeconomic conditions.

(Zelinsky, 1973: 40–41,
cited in Jackson, 1989: 18)

From this perspective then a culture was something that operated beyond human power struggles, it was a reified entity above society that controlled humans as if they were marionettes, remotely dictating their actions and reactions. It can be argued, therefore, that such a superorganic conception of culture had the effect of simply replacing the deterministic role of the environment that Sauer disliked so strongly with the determining role of this unexplained and apparently inexplicable thing called culture. As a result, many scholars became frustrated at the lack of theoretical or practical purchase such an approach gave them to understand and explain many of the pressing issues of the latter twentieth century. It was argued that the social unrest of modern societies required a cultural geography to not simply question the 'what' and 'where', but also the 'how' and 'why'. In short, it was argued that a new branch of cultural geography was required.

Conclusion

In this chapter, we have seen the development of the family tree of cultural geography up to the end of the twentieth century. From its roots in the natural science of Darwin and the empiricist geographies of environmental determinism, we have seen it branch off into environmental possibilism, and graft in anthropological ideas to bloom into the cultural geographies of landscape. Each of these generations has formed through responding to the contemporary external context in which it operated, but also by learning from the strengths and weaknesses of those branches that predated it. It is important to recognise that each branch had its time when it was the dominant way in which cultural geography was studied, and although each generation has been taken over by new growth, old branches have not entirely died away but continue to develop and influence the discipline as a whole.

In the next chapter we continue to map the growth of cultural geography into the twenty-first century. We will see the development of the discipline from the 'representational' to the 'more than representational'. From these shoots we then come to (re-)establish this book's approach to the cultural geography of place.

Suggested reading

The following texts are useful to explore the issues raised by this chapter.

Books

In terms of books, there are a number of excellent overviews and anthologies covering the family history of human geography.

Johnston, R. and Sidaway, J. (2004) *Geography and Geographers: Anglo-American Human Geography since 1945*, 6th edn. Arnold: London.

Johnston and Sidaway provide a major survey of the key debates, thinkers, and schools of thought in human geography. An ideal text to explore further the foundations of cultural geography, especially environmental determinism and environmental possibilism.

Livingstone, D. (1992) *The Geographical Tradition*. Blackwell: Oxford.

David Livingstone offers a critical account of the development and revolutions through which cultural geography has emerged, in particular the role of Darwin in the founding of the discipline, and role of race and empire.

Driver, F. (2000) *Geography Militant: Cultures of Exploration and Empire*. Blackwell: Oxford.

Driver traces the emergence of the modern culture of exploration and how it produced and disseminated geographical knowledge in the age of empire.

Agnew, J., Livingstone, D. and Rogers, A. (eds) (1996) *Human Geography: An Essential Anthology*. Blackwell: Oxford.

This anthology gives concise insight into some of the key writings of the discipline. It includes Donna Haraway's illusion of the 'god trick', what Peter Kropotkin thinks 'Geography ought to be', and excerpts from the work of Boas, Vidal de la Blache, Semple, and, of course, Carl Sauer.

An extended introduction to Sauer can be found in the 'Morphology of Landscape'; Sauer, C. O. (1925) *The Morphology of Landscape*. University of California Press: Berkeley.

Journals

In terms of journals, Solot usefully reprises the work of Carl Sauer: Solot, M. (1986) Carl Sauer and cultural evolution. *Annals of the Association of American Geographers*, 76 (4), 508–520.

Branching out: twenty-first-century developments in the family tree of cultural geography

Introduction

In the last chapter, the genealogy of cultural geography was outlined up to the last decades of the twentieth century. We saw how the family tree of the discipline developed from a focus on environmental determinism, through ideas of environmental possibilism, and on to the cultural geographies of landscape. This chapter takes us from the twentieth into the twenty-first century and charts the contemporary developments within the discipline. More specifically it takes us from cultural geographies of representation, and the types of understanding this is seen to privilege, and towards a cultural geography that identifies gaps and silences within this study. The chapter then outlines how cultural geographies have moved beyond a focus on representation and towards 'non-representational' as well as 'more-than representational' geographies. The chapter charts this development, the implications it has, and how these bring us to the approach to cultural geography that this book suggests.

Historical background to representational cultural geography

As we have seen, new generations within the family history of cultural geography are conceived due to external pressures and internal debates. As the American geography of Carl Sauer engaged with the nineteenth- and early twentieth-century landscapes of rural and Latin America, so new generations of cultural geographers, particularly in the UK, wanted their cultural geography to engage with the urban society of the late twentieth century. From the 1960s onwards, social and economic changes had prompted a range of new social movements based on race and gender. Anti-war and student protests questioned governments and states, and environmental groups campaigned against industrial pollution. Cities and rural areas underwent substantial change through economic restructuring, whilst different areas of the globe became at once independent from colonial rule and increasingly interdependent within a corporate global economy. A cultural geography that was oriented towards rural and relic landscapes did not have the empirical capacity to usefully engage with these external changes. Similarly, a cultural geography that had no sense of culture as brokered through struggle between competing social actors had no political capacity to engage with these contemporary circumstances. Sauer's cultural geography was beginning to seem 'antiquarian' (see Agnew and Duncan, 1989; Cosgrove and Jackson, 1987; Jackson, 1989) and 'sterile' (Cosgrove, 1989) and in order to remain politically and academically relevant, a range of scholars set about re-orientating cultural geography away from the relic and towards the 'representational'.

Representational cultural geography: from anthropology to eclecticism

In order to do so, representational cultural geographers branched away from a solely anthropological bias, and turned towards a range of social science disciplines for inspiration. These included humanism (including the works of Relph, 1976; Buttimer and Seamon, 1980; Tuan, 1974, 1977; Lowenthal, 1967), feminism (e.g. Haraway, 1988, 1991), social and cultural theory (e.g. Foucault, 1973, 1980; Lefebvre, 1968, 1971), and increasingly to 'post-' theories (including post-modernism (Harvey, 1989), post-structuralism (Doel, 1999; Murdoch, 2006), and post-colonialism (Loomba, 1998; Jackson and Penrose, 1993)). Grafting cultural geography to this eclectic array of social science ideas facilitated a number of changes to how culture was understood.

What is this thing called 'culture'? Culture is human practice

Due to this shift in philosophical lineage, cultural geography was no longer tied to the superorganic notions of Sauer's antecedents. For representational cultural geographers culture was no longer 'the agent' creating landscapes (Sauer, 1963: 343), rather agency was accredited to humans themselves; *we* were the active producers of cultural products and processes. As Anderson and Gale (1999: 3) outline, culture was no longer a separate entity acting above and on human beings, rather it became a 'process in which people are actively engaged . . . [it was] a dynamic mix of . . . practices that people create not a fixed thing or entity governing humans'.

In this approach culture remained about what people make, how they make it, and the effects

that these products have. But crucially, the theoretical source of this cultural production was not some outside force, but the people themselves. Here, culture is not what is done to people; rather it is what people do. Different cultures are produced when individuals get together to live their lives differently.

What is this thing called 'culture'? Culture is symbol and significance

Representational cultural geography also considered the material manifestations of culture in a different way to Sauer. Instead of his focus simply on the nature and effect of cultural products – an 'antiquarian object fetishism' (Price and Lewis, 1993: 3) – it was argued that the definition of culture centred on the *significance and value* given to these practices and products by different social groups. Influenced by work in semiotics and discourse analysis (see de Saussure, 1967; Barthes, 1973; as well as Foucault, 1973) geographers recognised it was not simply what groups produce that was important to forming culture (the institutions, commodities, rituals, artistic forms, etc.) but also the meanings given to these entities. Geographers recognised that cultural products are not just important in terms of the form and function of the things themselves. To paraphrase a famous Marx Brothers line, for them 'a cigar is *not just* a cigar', but also represents something else: an idea, a value system, even a joke. In this sense, cultural products and processes function as signs, and it was up to geographers to decode their meanings.

So this 'new' cultural geography did not want to limit itself to investigating the material entities of a culture, it wanted to explore and understand the mental associations that places, things, and processes had for different cultural groups. Where Sauer would have just looked at landscape and artefacts, this branch of cultural geography considered places, materialities, *and* mental associations; Cresswell explains this 'combination' approach to cultural objects with reference to the example of a church. From this new perspective:

> Places . . . are combinations of the material and mental and cannot be reduced to either. A church, for instance, is a place. *It is neither just a particular material artefact, nor just a set of religious ideas; it is always both.* Places are duplicitous in that they cannot be reduced to the concrete or the 'merely ideological'; rather they display an uneasy and fluid tension between them.
>
> (Cresswell, 1996: 13, my emphasis)

Thus this new cultural geography considered both the substantive and symbolic – summed up by the title of Jackson's seminal book *Maps of Meaning* (1989). What this cultural geography sought to do was investigate the meanings, the symbolic values and cultural ideas that were inscribed into different entities and processes by various cultural groups, and the spatial ramifications these had.

What is this thing called 'culture'? Culture is politics, culture is power

Through focusing on materialities and meanings, representational cultural geography illustrated how understanding cultural struggle was crucial to understanding the landscapes around us. By focusing on the 'combination' of the mental and material, it was suggested that cultural products are not passive; rather they actively influence behaviour by creating and perpetuating particular meanings. It was how different groups celebrate, resist, or disapprove of various products and practices (and their meanings) that was vital to understand. Landscapes therefore became the medium of cultural struggle between different groups; geographies were created through the contestation and conflict between various cultures which valued things and activities in

different ways. Recognition arose, therefore, that culture was political – if some meanings became dominant then some cultures could be encouraged whilst others could be repressed. As a consequence, geographers would have to question how particular valuations became the most dominant in any given location. This conceptualisation of culture meant that issues of power became crucial to any analysis of cultural geography (see Chapter 5). Questions regarding who has the ability to convert or oppress what became orthodox were fundamental. The study of culture thus became the study of power and political struggle; representational cultural geographers saw the landscape as 'the materialization of the ongoing struggle to represent the norms, values and meanings that define the community. Thus, reading the landscape involved examining how dominant agents inscribed the world as well as how those inscriptions were regularly undermined' (Rose, 2002: 458).

What is this thing called 'culture'? Culture is 'text'

> We write the culture in which we live.
>
> (Bolter, cited in Thomashow, 1995: 184)

Not only did this representational cultural geography foreground issues of conflict and struggle in its assessment of cultural landscapes, it also began to adopt a more interpretative rather than morphological methodology to studying it. It introduced the idea of cultures and landscapes as 'texts'.

Texts are usually considered to be written material: books, manuscripts, poems, even maps. It is common for us to identify how these texts can often be read, or interpreted, in different ways. In a novel, for example, some readers will argue how the hero is really a villain; others can guess the murderer before the final page. Others enjoy the author's interpretation but question it, whilst others still go along for the ride until the story's conclusion. The possibility of these different readings or interpretations in relation to books, poems, or even other cultural 'texts' like films, is often familiar to us. In relation to the cultural landscape, due to the introduction of new ways of configuring culture – as webs of significance and meaning that different groups can interpret in different ways – many cultural geographers began to use the idea of texts to help us understand how we should configure landscape in this representational cultural geography. As Cosgrove and Jackson (1987: 96) outline:

> Conceptualising landscapes as configurations of symbols and signs leads inevitably towards methodologies which are more interpretative than strictly morphological . . . This interpretative strand in recent cultural geography develops the metaphor of landscape as a 'text' to be read or interpreted as a social document in the same way that Clifford Geertz (1973) describes anthropology as the interpretation of cultural texts.

As Aitkin (in Flowerdew and Martin, 1997: 198) has identified, these cultural geographers 'read' landscapes, along with the other cultural processes and institutions that go together to constitute them, as 'texts'. Instead of seeing the cultural landscape as simply the entity that is produced by culture from a 'natural medium' (following Sauer); the landscape in this new cultural geography is something that is inscribed with a range of meanings by various social groups. Landscapes therefore became signs or symbols that we read every day – they tell us how to behave, whether to walk or not walk, whether to keep off the grass, where to smoke; they tell us where we are allowed to go and what we are allowed to do there. It was up to cultural geographers to study the meanings of these 'texts', work out the meanings attached to them, the politics underpinning them, and how these were (re)produced.

Representational cultural geography and the language of place

Thus not only did this new cultural geography look at the same materialities in new ways, be they cultural artefacts as meaningful entities, or landscape as texts, it also led to the construction of completely different objects of study. As Jackson put it, there was a move away from geographies, 'that focus[ed] exclusively on landscape' (1989: 9), and a turn towards looking at the various *places* that become contested by different cultural groups. The notion of *place* was introduced to usefully bind together the various new ideas that this branch of cultural geography focused upon. In the first instance, place allowed geographers to make tangible the active role of humans in creating culture, and the importance of power in controlling and making rules that define geographical sites. However, it also allowed geographers to move away from looking at landscapes as a whole, perhaps as more unified, monolithic, or even static entities, and to focus instead on sites of different types at different scales. These places could be conventional sites such as towns, plazas, or parks, but they could also be at scales both beneath and beyond – for example, they could be rooms or nation states (after Sack, 2004: 243). Places could also be located within films or other cultural products (unifying the idea of place as a 'text' that could be read), or indeed as literal or metaphorical sites, e.g. places of the body, of wild nature, or the imagination. Thus, as Duncan identifies, new vocabularies led to new objects of study, the notion of place came to be the site around which the study of the social and the spatial could be effectively combined. The age of place had arrived in representational cultural geography.

Summary: representational cultural geography

Representational cultural geography therefore broadened the array of legitimate subjects and objects that were within the geographers' compass. As Cosgrove put it, now 'cultural geography is everywhere' (1989), as Shurmer-Smith cleverly has it, cultural geography is 'all over the place' (1996). At the end of the twentieth century these changes led to cultural geography becoming:

> one of the most exciting areas of geographic work at the moment. Ranging from analyses of everyday objects, views of nature in art or film to studies of the meaning of landscapes and the social construction of place-based identities, it covers numerous issues. Its focus includes the investigation of material culture, social practices and symbolic meanings approached from a number of different perspectives.
>
> (McDowell, 1994b: 146)

The representational branch of cultural geography thus introduced a departure of sorts from the forms of study that preceded it. It instigated a turn away from a purely anthropological heritage towards a more eclectic array of humanist, feminist, and critical theory influences. It argued for a culture that was produced by humans themselves, acting to take and make place in line with their own political values and preferences. It shunned a purely empiricist epistemology to position researchers within their studies, and combined empirical exploration with theoretical analysis to get a better understanding of the world 'out there' (see Chapter 14). The key facets of representational cultural geography are outlined in Table 3.1.

Cosgrove and Jackson (1987: 95) sum up this representational cultural geography as follows:

> If we were to define this 'new' cultural geography it would be contemporary as well as

Table 3.1 Representational cultural geography

Source of key ideas	Humanism, feminism, critical theory, cultural studies
Methodology	Theory and practice
Cultural agent	Cultural groups
Geographical scale	Place
Social values	Importance of difference
Political ends	Ambivalent?

historical (but always contextual and theoretically informed); social as well as spatial (but not confined exclusively to narrowly-defined landscape issues); urban as well as rural; and interested in the contingent nature of culture, in dominant ideologies and in forms of resistance to them. It would, moreover, assert the centrality of culture in human affairs. Culture is not a residual category, the surface variation left unaccounted for by more powerful economic analyses; it is the very medium through which social change is experienced, contested, and constituted.

Critique of representational cultural geography: theories and things

Despite its many advances, there was a growing critique of this combinational approach to cultural geography. The critique centred on how a focus on representation privileged certain aspects of human and cultural life, and ignored other important dimensions. The root of this critique was that representational cultural geography had become preoccupied with two interrelated elements: *theories* and *things*.

Firstly, it was argued that representational cultural geography had become distracted by *theories*. Through adopting a range of social science theorisation, it was argued that this branch of cultural geography had become obsessed with the languages and discourses used to interpret the world around us. As a consequence, geographers were said to inhabit their own 'wordy worlds' (Thrift, 1996), and through doing so had become detached from the actual world itself. As Duncan (1994: 404) put it at the time: 'The new [theoretical] vocabulary is sometimes dismissed as simply constituting the private, idiosyncratic language of a small group of geographers.'

It was suggested therefore that an interest in theory had led many cultural geographers to be more concerned with decoding ideas than engaging with practical actions (after Curry, cited in Price and Lewis, 1993: 12). An obsession with theory was argued to put these geographers at risk of being abstract and distant from people's everyday lives.

The accusation of elitism and abstraction from everyday life was also central to the second element of this critique. It was argued that representational cultural geography had become preoccupied not only with *theories* but also with *things*. Through focusing on interpretation and decoding, geographers had centred their attention on intentionally created and durable cultural artefacts – on 'things'. These 'things', such as Nelson's Column, were static and symbolic enough to be interpreted with a cultural geographer's expert theoretical armoury. Yet what about other aspects of existence? What about less durable and more temporary events? What about improvised practices or processes? What about activities social actors get involved in that are not as 'thought through' as the creation of an architectural edifice?

This critique centred, therefore, on how interest in the explicit representation of things and their subsequent theoretical interpretation had led to the silencing of other languages and intelligences we use to engage with the world. As Whatmore suggested, this preoccupation led to 'a sense in which the life seems to have been sucked out of the worlds we variously study' (Whatmore, 1999: 260, also cited in Greenhough, 2004: 255), or as Greenhough

put it, we have, 'become [so] fixated by the language of the *word*, with the politics and practice of representation, that [our] accounts run the risk of forgetting the *language of the world*' (Greenhough's emphasis, 2004: 255).

The language of the world: scowls and goosebumps

'Had enough?'

'Yeah.'

'I just wanna sit out here for a minute.'

'Look at you.'

'What?'

'Usually you have this intense sort of scowl of concentration on your face, like you're doing all this for a school project or something . . . Look – it's gone. If I didn't know better, I'd say you looked almost happy.'

'I can't describe what I'm feeling.'

'You don't have to.'

'Goose bumps.'

> (Tyler to Johnny Utah, after his first moonlit surf, in *Point Break*, 1991, director: Kathryn Bigelow)

Refocusing the 'geographical sensibility' (Anderson and Smith, 2001: 9) towards the 'language of the world' thus formed new growth on the cultural geography family tree. But what do these geographers refer to when they speak of the 'language of the world'? Like the Johnny Utah character in *Point Break* (above), we have all been in situations where we can't find the words to express ourselves, to talk about how we feel, or why we do things. Our bodies may betray our experiences – we may blush, shiver, get goosebumps, but we can't adequately translate our experiences into words. How do we try to communicate what a kiss feels like, how falling in love affects us? How it feels like to be underwater? To surf? When the ball hits the back of their net, or worse, ours? How do we communicate loss? On other levels, how do we discuss why we do mundane, everyday things, e.g. why we shop where we do, why we doodle, why our habits have taken hold of us the way they have? Sometimes these issues not only elude our capacity to 'think things through', but also our capacity to verbalise them. In this situation we resort to other types of language to express ourselves – we may create art, cry, begin a new society, or start a war. This focus on experiences, practices, and feelings that are *before or beyond* conventional linguistic representation is what geographers refer to when they speak of the 'language of the world'. It is these hard to pinpoint but essential moments that the critique of representation sought to focus on.

The language of the world and the 'non-representational'

As we have seen, representational geography was critiqued due to its overemphasis on theories and things. The scholarly impulse to draw attention back to experiences, practices, and feelings that are beyond or before representation – the 'languages of the world' – coalesced under the banner of 'non-representational' geographies. Non-representational geography defined itself as different to – as before or beyond – representation. It began from the assumption that representation is partial and does not necessarily include all aspects of cultural life in our studies. As Nigel Thrift, one of the principal instigators of this new branch of cultural geography, put it: 'the varieties of stability we call "representation" . . . can only cover so much of the world' (Thrift, 2004: 89).

A non-representational geography therefore began with a focus on practices, on the *experiences* rather than the *things* that constitute our world. In the words of Lorimer, it centred on the 'everyday routines, fleeting encounters, embodied movements, precognitive triggers, practical skills, affective intensities, enduring urges, unexceptional interactions and sensuous

dispositions . . . [that make] a critical difference to our experiences of . . . place' (2005: 83). It was argued that this range of experiences are not simply interesting asides to the fundamental geographies of theories and things, rather they are 'at the root of the geographies that humans make every day' (Smith, 2002: 68).

Thus non-representational cultural geography centred on aspects of life that we may engage in without really having worked out why. It was often experimental, working in areas its proponents liked to consider as a 'brave new world of fringe science' (Lorimer, 2005: 88). In this world, cultural geographers explored dance (McCormack, 2003), musical performances (Wood and Smith, 2004), engagements with nature (Szersynski et al., 2003), sensory affects (Rodaway, 1994), and artistic expression (Kaye, 2000). It was hoped that by engaging with experiences beyond and before representation parts of our engagements with place that are often obscured could be accounted for. For example, why do some experiences make our hairs stand on end? What do these experiences move us to do? How do they define us in relation to our world? As Dewsbury (2003: 1910) anticipated, refocusing the geographical sensibility aimed to answer these questions:

> If representation at its present limits misses the reality that we seek to capture through it, maybe art [for example] . . . 'stands as both the medium of this failure and the agency for the hope of its restitution . . .' (Heathfield, 2000, page 21). For me this is about getting back to a moment of prediscursive experience; to recommence everything, all the categories by which we understand . . . the world . . . I think this means moving against the general rub of representation.

The implications of the non-representational: methodologies and research practices

As we have seen in previous chapters, we can define the substance of each branch of the cultural geography family tree not simply by what it studies, but how it studies it. 'Moving against the general rub of representation' had important implications for the methodological dimensions of cultural geography. Researchers turned away from methodology that relied overwhelmingly on texts and other representational devices to make sense of the world. These methodologies, it was argued, tended to focus on linguistic forms of communication – i.e. verbalised and thought-through 'talk' – that respondents use to represent their experiences, rather than focusing on the experiences themselves. As Hinchcliffe (2000: 578) outlined,

> the turn to performativity [had] important effects on research practices. Interviews, focus groups, surveys, discourse analysis [see Chapter 14] – all of these have a tendency (by no means inherent) to concentrate on 'what' is being said. In doing so, they sometimes miss what Thrift neatly captures as language's

performative function as often 'simply to set up the intersubjective spaces of common actions, rather than represent them as such' (1996: 39).

Thus it was argued that these 'spoken' communications often failed to capture the 'language of the world'. This problem was compounded when such communications were cut, pasted, and rendered context-less in academic papers and books; in these circumstances even the bodily expressions and (un)spoken nuances that give life to the words uttered were lost (see Latham, 2003). In response to these critiques, some researchers turned away from any preponderance with methodologies that attempted to solely make sense of the world post-event. To chart new 'non-representational' maps, geographers experimented with methodologies that engaged with the experience of the world in their moments of creation. Interventions were now timed in-event (or in-situ, in the ethnographic tradition) with geographers participating, observing, and filming phenomena, as well as requesting participants themselves to document events (through photographs, film, narration, etc.) to create real-time recordings. These documentaries could then be used as records in themselves, but also

---

BOX 3.1 PLACES AS PERFORMANCES

This branch of the cultural geography family tree thus emphasises how a place should no longer be seen as simply a 'text' to be read – as a fixed entity that requires its meaning to be decoded by a cultural geography expert – but also as a process that is here and now, existing but uncertain in this very moment. In other words, place is not simply considered as a closed text (like a book), but rather as a text that is performed (like a book read aloud). Coupled to this, if you and I and other actors are performing this place (i.e. we are all reading lines from this book out loud), it is possible for us to add our own lines to the text, improvising and planning new directions for the 'ongoing composition'. In this sense, place is a process that is enacted through performances. Place is a fixed noun no longer, but an active verb (see Chapter 4). In this view, as Denzin notes: 'context replaces text, verbs replace nouns, structures become processes . . . the emphasis is on change, contingency, locality, motion, improvisation, struggle, situationally specific practices and articulations, and the performance of con/texts' (2003: x).

used as cues to generate further post-event perspectives through conventional interviews and focus group discussion (see Chapter 14).

Summary: non-representational cultural geography

Moving against the general rub of representation and into the world before or beyond the rationally thought through has thus generated a new 'non-representational' branch of the cultural geography family tree. It was a response to the preoccupation with theories and things. This preoccupation, it was argued, had led to a partial view of the world, with many of the unintentional, unanticipated, and emotional moments that define our cultural lives being unaccounted for. The key facets of non-representational cultural geography are outlined in Table 3.2.

Despite the advances offered by the 'non-representational', this book argues that this branch of the cultural geography family tree is not without its weaknesses. As we will see, the problem with focusing exclusively on experiences before or beyond the 'representational' is that it is impossible to escape the language of representation itself.

Table 3.2 Non-representational cultural geography

Source of key ideas	Performance theory, phenomenology, passionate sociology
Methodology	Everyday practice
Cultural agent	Cultural groups and individuals
Geographical focus	Intimate scales: bodies, actions
Social values	Importance of emotional and everyday
Political ends	Ambivalent?

The problem of non-representation

How can we represent that which lies beyond the scope of representation?
(Davidson et al., 2005: 11)

Although there is a world beyond words (e.g. bodily functions (reddening, sweating), movements (trudging, dancing, falling), and expressions (smiling, wincing, shrugging)), this world does not exist in isolation, it is inevitably tied to processes of representation. The problem with non-representational cultural geography is the assumption that we can escape the world of representation. As soon as the language of the world is uttered or then heard by others, the forces of representation have taken hold.

The idea that we cannot escape the world of representation may seem abstract, so let's explore an example. When we think quietly to ourselves – just like you might be thinking now how you are hungry, or when will this chapter finish – this may involve an instinctive feeling that we understand as hunger, boredom, or confusion. We register this feeling in words in our heads. We have the impulse, we give a word to it, we thus have a way of understanding the impulse in our world; we have a meaning to understand our physical and emotional state. Languages therefore help us to understand our personal world, and also to communicate this world to others. We must remember, of course, that these languages aren't the feelings themselves, but stand in for them, like a lawyer for a defendant. So whenever we use language, we are involved in representing. In this way of considering the world, all 'language' is representational. Dance, architecture, art, *may* be non-linguistic, but they are still languages of communication. They may be no more adequate than spoken or written English, Arabic, or Paraguayan at conveying to another person what your first kiss felt like, but we nevertheless use them to try. Similarly, these languages may be used to communicate non-thought-through things – they may be used impetuously, impulsively, or 'instinctively', to exclaim or profane an experience. Knowledge is therefore practical and it can be non-cognitive, but it is nevertheless representational in the broadest sense.

Thus as soon as we 'contemplate' the language of the world (Dewsbury, 2003) we transform it into something different from the original impulse. In short, we morph it into a form of representation. In this situation, how can we write, talk about, and research the language of the world in a non-representational way? How do you say – in whatever form – the unsayable? Bondi (2005a: 438) recognises this problem:

> non-representational geography has been characterized by a wariness of forms of meaning-making that might somehow sequester those elusive qualities of quick and lively geographies that always exceed representability. Herein lies the paradox of non-representational geography: how can our own texts ever honour that which lies beyond the scope of discourse?

The paradox of the non-representational is therefore that whatever new methodologies or ways of writing are employed, the products of research can never be the experiences or practices themselves. Our written reports, papers, and books, even our multimedia and film interventions (especially at conferences or lectures), remain re-presentations. Even though scholars want to 'get at' the worlds beyond words, they are confined within the world of representation to communicate it.

Standing up for . . . re(-)presentation: the importance of the 'so what?' question

We have seen therefore that engaging with the non-representational was a significant step in the development of cultural geography, but it cannot occur without at once also engaging with the representational. Although some cultural geographers have begun to 'draw back from advancing interpretative claims about the practices with which they engage' (Bondi, 2005a: 438), this book takes the position that it is not a 'bad thing' to engage with representation. It is not only inevitable, simply through the choice of words, images, or practices used to share experiences, but it is also a fundamentally important thing to do. It is through giving value to our improvised, unexpected, and uncanny experiences of the world that we generate our cultural ideals. Perhaps it is even more crucial to value these experiences and practices (such as growing in love, being intoxicated with passion, falling in fear, etc.) as it is these that move us to action, that form our life decisions, and come to define us as individuals (as Dewsbury notes, 'these are not light matters for they forge the weight of our meaningful relation with the world' (2003: 1907)).

Furthermore, it is not just 'expert' cultural geographers that decode meaning in both 'things' and 'practices', but all humans. Indeed, engaging with both 'things' and 'practices', and then giving meaning to them, helps us all to answer perhaps the most important question of all: the '*so what?*' question. As both students (in essays, dissertations, or papers), and as citizens making real decisions about the 'good life', this '*so what?*' question helps to define ourselves in relation to these practices, things, and places. Our work as cultural geographers and our lives as people then become more than simply experiencing and describing events. They become about sharing what these events mean to us on an individual level, finding out what others say they mean to them, and allowing our audience to take in these positions and work out their own impulsive and thought-through opinions on them.

We have come to a position therefore where the study of the representational and the non-representational within cultural geography is not an either/or choice. The tension between the newest branches of the cultural geography

DEVELOPMENTS IN CULTURAL GEOGRAPHY

family tree exists, but this book argues it can be harnessed to progress cultural geography in new and innovative ways. Adopting a third option, that breaks down the closed logic of *either* doing representational *or* non-representational study, favours 'a different, more flexible and expansive logic of the "both-and-also"' (Soja, 1999: 258). We are moving towards our approach to the cultural geography of place: place as an ongoing composition of traces.

A 'both and also' approach: place as an ongoing composition of traces

The approach of this book, of looking at cultural geography through places and traces, endeav-ours to combine not simply the material and its meaning, not only theories and things, but also emotions and experiences. It is not simply a cultural geography of representation or non-representation, rather it is both and also. It does not seek to rely on old divisions between thought and practice (see Nash, 2000: 657), nor of empiricism and theory; the point is to consider 'how contemporary [culture] is based as much upon the elaboration and reproduction of particular affective capacities as it is upon the manipulation of representations' (Thein, 2005: 14).

Such an approach has a 'commitment to . . . understand . . . practice and performance that refuses to privilege mental representations' (Hinchcliffe, 2000: 567), but at the same time, *refuses to exclude* these representations too. In

---

BOX 3.2 NEW GROWTH ON THE CULTURAL GEOGRAPHY FAMILY TREE

As we have seen, cultural geography is always changing. Who knows what new branches of the cultural geography family tree may grow in the future? Early shoots suggest that the discipline may be moving into new empirical 'media'. For example, cultural geographers are exploring 'atmospheric' spaces (McCormack, 2009; Adey, 2014) and water worlds (Anderson and Peters, 2014), even beginning to consider the geographies of outer space itself (MacDonald, 2007; Dunnett, 2012; Beery, 2012). Interest in these new media point towards cultural geographers interrogating the 'volume' of places as seriously in the future as they have explored the eye-level perspective in the past (see Elden, 2013). Cultural geography is also changing in terms of the disciplines it seeks to work alongside. In the past, cultural geography has cross-fertilised with natural science, anthropology, the humanities, and social science, and perhaps in the future it will seek to strengthen its growth in the arena of the arts. As Hawkins suggests, cultural geography shares 'increasingly conceptual common ground and a growing set of shared practices [with] artists [and] art theorists' (2014: 1), whilst Cresswell notes how 'performance theorists, historians, artists, people who study music, architects, literary theorists, sociologists and others' (Cresswell, 2010c: 19) may invigorate cultural geography in new and unexpected ways.

Through such cross-fertilisation, the means through which cultural geographers communicate and disseminate their understanding of the world may also change. It is possible to discern signs of a move towards more 'creative geographies' (after Hawkins, 2014), with scholars engaging with their publics not simply through academic critique, but also through art installations, films, and creative writing (see also Chapter 14). As Wylie states, he can 'envisage cultural geographical genres of place-writing and landscape-writing, segueing with travelogue, memoir and spatial history, [in other words] a more performative and literary cultural geography' (2010: 212, see also Anderson, 2015). Such moves raise fascinating opportunities to redefine what cultural geography is, broaden its capacity to engage multiple publics, and persuade us to engage positively in the cultural world.

the words of Bondi, it seeks to 'incorporate rather than bracket off representational, narrative or substantive meanings' (2005a: 444). This, therefore, is a more entangled and combinatorial cultural geography that is not non- or simply representational, it is, to borrow Lorimer's useful phrase, a cultural geography that is 'more-than representational' (2005: 84).

Conclusion

The approach to the cultural geography of place as detailed in the remainder of this book thus builds on the many strengths of the cultural geography family tree that have been outlined in Part Two. This approach involves analysing and interrogating all the agents, activities, ideas, and contexts that combine together to leave traces in places. The traces left in this range of contexts may be material *or* non-material, durable *or* temporary, left by humans *or* non-humans. They may produce the intentions desired by the trace-makers, or have unintentional effects on the agents and their audiences.

Taking this approach to the cultural geography of place has a number of important implications. Firstly, it suggests that places can be claimed and counter-claimed, in other words, places can be fought over. As places are taken and made by different cultural groups, claims are staked for 'ownership' of that area. Staking these claims may produce a range of outcomes: in some cases 'other' groups will no longer be able to use that area – they are *displaced* both geographically and culturally; in some cases the place may become a battleground – a 'no man's land' as 'ownership' of it becomes an issue of disagreement; in other cases the area may become shared by a number of groups, producing a 'hybrid' home-place for groups with differing identities and ideas.

Secondly, this approach suggests that battles for place are never over – places are *ongoing* compositions of traces. However fixed or substantial a place, its identity, and its activities appears it is *always* up for grabs and open to new traces. Boundaries can always be crossed, so places can always be changed. As a consequence of this, the claims cultural groups make on places have to be perpetually guarded from challenges and resistance. Places have to be constantly maintained. Thirdly, therefore, part of cultural geography's approach to place is to interrogate how such places are ordered and bordered (see Chapter 4), and how particular cultural geographies are maintained or changed. Through analysing these ongoing compositions we thus interrogate who ultimately decides on the nature of our places, and, by extension, who decides on our culture, and our world? We ask, whose places are these anyway?

Finally, mapping our cultural geography approach to place also involves positioning ourselves in the world we study. Through doing so we have to critically reflect on our own actions. We begin to consider our own cultural views, the actions we take, and the traces they produce. We are encouraged to ask ourselves what cultural views do we (and do you) want to endorse, and want to resist? We ask what do we (and you) want the world to be like? What is *your* sense of place? What could it be? What should it be? In the sections that follow we will look at different cultures and geographical contexts and interrogate the ways in which traces come together to define places and peoples. In each chapter we will introduce both empirical examples and theoretical ideas which facilitate understanding about how the world around us is generated, and our role within it. The next section begins by devoting a chapter to a key category in our mapping of cultural geography – the notion of 'place'.

Suggested reading

The following texts are useful to explore the issues raised by this chapter.

Books

Jackson, P. (1989) *Maps of Meaning*. Unwin Hyman: London.

Anderson, B. and Harrison, P. (eds) (2010) *Taking-Place: Non-Representational Theories and Geography*. Ashgate: Farnham.

Peter Jackson's 1989 book *Maps of Meaning* is seminal to the representational branch of cultural geography. It marks out its significant departure to the traditional approach of Sauer, focusing on the role of meaning, value, and symbolic significance in our cultural places.

Anderson and Harrison's edited collection is useful to gain an overview of non-representational approaches to the discipline.

Journals

A range of journal articles discuss the relative merits of the representational branch of the cultural geography family tree. These include the following:

Cosgrove, D. and Jackson, P. (1987) New directions in cultural geography. *Area*, 19 (2), 95–101.

Cosgrove and Jackson sum up the critique of traditional cultural geography offered by the representational branch of the family tree.

Duncan, J. and Duncan, N. (2004) Culture unbound. *Environment and Planning A*, 36, 391–403.

Mitchell, D. (1995) There's no such thing as culture: towards a reconceptualisation of the idea of culture in geography. *Transactions of the Institute of British Geographers*, 20, 102–116.

Price, M. and Lewis. M. (1993) The reinvention of cultural geography. *Annals of the Association of American Geographers*, 83 (1), 1–17.

Cosgrove, D. and Duncan, J. (1993) On 'The reinvention of cultural geography' by Price and Lewis. *Annals of the Association of American Geographers*, 83 (3), 515–519.

Nash, C. (2002) Cultural geography in crisis. *Antipode*, 34 (2), 321–325.

The above authors debate the relative merits of this branch of the cultural geography family tree, assessing whether this 'reinvention' of the discipline sparks a 'brave new world' or a 'crisis' for the discipline.

Non- or more-than-representational approaches to cultural geography are usefully discussed in the following papers. Of these, perhaps Bondi is the most accessible.

Bondi, L. (2005a) Making connections and thinking through emotions: between geography and psychotherapy. *Transactions of the Institute of British Geographers*, 30, 433–448.

Thrift, N. (2004) Intensities of feeling: towards a spatial politics of affect. *Geografiska Annaler*, 86 (1), 57–78.

Lorimer, H. (2005) The busyness of 'more-than-representational'. *Progress in Human Geography*, 29, 83–94.

New growth on the cultural geography family tree is tended by the following authors:

Cresswell, T. (2010b) New cultural geography – an unfinished project? *Cultural Geographies*, 17, 169–174.

Wylie, J. (2010) Cultural geographies of the future, or looking rosy and feeling blue. *Cultural Geographies*, 17, 211–217.

PLACE AND POWER

Knowing (your) place

Introduction

As the cliché used in Chapter 1 tells us: 'cultural life does not take place in a vacuum'. As we have seen, a key component of this statement is 'take place', and it is this component we shall explore in detail in the following two chapters. Chapter 5 focuses on how groups take place through the exercise of power. It investigates how power can be understood and how it affects both culture and geography. In this chapter, we focus on the different ways we can understand the notion of 'place' itself.

> Place is one of the trickiest words in the English language, a suitcase so overfilled that one can never shut the lid.
>
> (Hayden, 1997: 112)

In the lived world we cannot escape places; they are all around us, making up the fabric of our cultural life. As Rodman (2003: 204), points out, 'we are as situated in place as we are in time or culture'; just as we cannot step out of time, we cannot extricate ourselves away from the vast array of materialities and meanings associated with place.

The difference that place makes . . .

Places are at once the medium and the message of cultural life. They are where cultures, com-munities, and people root themselves and give themselves definition. Places then are saturated with cultural meanings. They are our 'home', our 'backyard', our 'turf'. Similarly we can feel 'out of place', *dis*placed, and outsiders in particular places. Places then are crucial for understanding who we are and where we fit in to the culture and geography of our lives. Seeing the world as a world of places is, as Agnew tells us, 'a political as much as an intellectual move' (2004: 91). For example, seeing the world as a world of *places* is fundamentally different to seeing it as a world of *spaces*. Although we may use the terms place and space interchangeably in everyday conversation, when used in the field of cultural geography we should be sensitive to the difference that *place* makes.

> if *space* is where culture is lived, then *place* is a result of their union.
>
> (Lippard, 1997: 10, my emphasis)

Places and spaces could not be more dif-ferent. Spaces, as Lefebvre tell us, are 'empty abstractions' (1991: 12), whilst places are 'drenched in cultural meaning' (Preston, 2003: 74). Spaces are scientific, open, and detached; places are intimate, peopled, and emotive. You may travel through spaces (perhaps isolated in a car or train), but you will live your everyday life in places. Place then is the counterpoint of space: places are politicised and cultured; they are humanised versions of space. It is from the empty abstraction of space that different cultures take and make place. As Kaltenborn puts it, 'geographical space becomes place when human beings imbue it with meaning' (1997: 176).

Places then are 'carved' out of space by cultures. In contrast to space places are meaningful, they root people both geographically and socially, and are fashioned by culture from context. Places are understood as being created predominantly by humans, as elaborated further in the following definition:

> Place refers here to something we humans make. A place is made when we take an area of space and intentionally bound it and attempt to control what happens within it through the use of (implicit and/or explicit) rules about what may or may not take place. This bounding of rule-making leads to the creation of place at any scale, from a room to a nation-state.
>
> (Sack, 2004: 243)

Here Sack emphasises the idea that place is a construction 'carved' out of space by human culture. However, Sack's definition requires minor qualification. Firstly, it is important to note that although places are human constructions, we should be open to the influence that *non-human actors* have in taking and making place. As outlined in Chapter 1, birds and animals can influence places, whilst the non-human environment also plays a role in effecting place. Think, for example, of the traces left by beavers in a place – their felling of trees, construction of dams, and the subsequent pooling of river systems all have a dominant effect on their chosen habitat. Another example of non-human actors taking and making place could involve mosquitoes – the traces that they leave in and on people, animals, and harvests can have significant effects, particularly in some regions of Africa and Asia. Think too of the devastating effects of hurricanes, typhoons, and tsunamis. Even whilst walking my dog I am aware of the trivial traces

that he leaves on every lamp-post he passes. In other words, the wealth of actors – or place-makers – that we need to consider in authoring place does not solely involve the human, but also animals, insects, and other non-human species and processes.

Secondly, Sack suggests that places are made when they are *intentionally* bounded. This idea perhaps characterises the majority of places; it is often the case that cultural groups intentionally take and make place in line with their cultural values to not only define themselves but also an area and its use in a certain way. However, it is also possible for cultural activity to *unintentionally* influence the meaning of a place. This may happen if we don't really think through our actions in a place, or are ignorant of the traces they may leave (perhaps in terms of pollution or environmental health). It is important for cultural geographers to be sensitive to the degree of intentionality actors exhibit in places, and diagnose how objectives (or lack of them) help us define and judge the cultural activity in question.

Thirdly, Sack considers the importance of scale when engaging with place (see also Chapter 6). Places can indeed be considered at a range of scales, as Pocock suggests, 'place may refer to one's favourite chair, a room or building, increasing to one's country or even continent' (Pocock, 1981: 17). In addition, we can also consider place more figuratively, we can study places of the body (for example, our skin, stomachs, or physical form) (see, for example, Kenworthy Teather, 1999; Bell et al., 2001; and also Chapter 13), but also consider places of the mind or imagination. The importance of scale, and how places combine consistently or in tension at every level, is a key dynamic for the study of cultural geography (see, for example, the case of trace-chains, below).

With these minor qualifications in mind, place then refers to something that both humans and non-humans make. A place is taken and made when traces are left with intent or by accident, and these traces form borders that attempt to control the ongoing composition of activities. Place can occur on any scale, and it can be both material and imaginary. Place locates and defines the cultural geography of our lives.

Location, location, location

We can build on this brief introduction to place by turning to Agnew and Duncan (1989: 2). According to Agnew and Duncan, place has three constituent parts: *location*, *locale*, and *sense of place*. *Location* refers to place as an 'objective' point in space, a node, for example, which is 'so-and-so' far from another node. Location can therefore be defined by grid co-ordinates, or lines of latitude or longitude. However two-dimensional this version of place appears, we can detect the groundswell of vibrancy inherent in the term: we can sense that place matters. For example, real estate agents identify the three most important aspects when buying property are 'location', 'location', and 'location'. Here, it is not so much the property itself that is crucial, but *where* it is, the crucial where of location one might say (to paraphrase Cresswell, 1996, see Box 4.1). Hotel chains also use the idea of 'location' to advertise the key positioning of their properties for ease of use of their customers. Here it is not just that these amenities are located carefully at specific nodes or points, but also that they offer a place of safety and familiarity in perhaps an unknown or abstract area.

No locale like home

The second notion that Agnew and Duncan identify is that of *locale*. Locale can be understood as the built, natural, and social environment generated by cultural relations – it is the composite of all the traces that come together in one place. Locale therefore provides the setting for everyday routine and social contact – in the words of O'Loughlin and Anselin (1992: 16), 'locale is the setting of interaction and the contextuality of

social life'. As such, locale comes close to identifying the character and feel of a place, a point usefully outlined by Massey and Jess (1995: 46):

> Very often, when we think of what we mean by a place, we picture a settled community, a locality with a distinct character – physical, economic and cultural. It is a vision which has entered the English language in phrases such as . . . 'no place like home'. . . Places are unique, different from each other; they have singular characteristics, their own traditions, local cultures and festivals, accents and uses of language; they perhaps differ from each other in their economic character too: the financial activities of the City of London mould the nature of that part of the capital; the wide open fields of East Anglia give a particular feel that 'it couldn't be anywhere else'.

The notion of locale therefore invites us to consider how traces contribute to a place's unique character. What traces define the locale of our own places, for example? What are the material artefacts that define your area (for example, what clothes, products, or services define it?), and what non-material practices are represented in your place (e.g. what languages, dialects, belief systems, rituals, and emotions can be associated with your place?)?

Sense of place: 'place must be felt to make sense' (Davidson and Milligan, 2004: 524)

The third constituent part of the notion of place, according to Agnew and Duncan, is *sense of place*. Sense of place refers to the emotional, experiential, and affective traces that tie humans into particular environments. As touched upon in Chapter 3, this aspect of place has not always been central to social science study, marginalised in the general antipathy to studying emotional connections. However, sense of place is the key way in which humans, culture, and environment

BOX 4.1 THE CULTURAL GEOGRAPHIES OF LOCATION

SA1

Source: Author

5102

Source: Author

Location

Source: Author

The image on the left is apartment '5102' in Bristol, England. Its name locates this building purely in terms of its co-ordinates of longitude and latitude. In a similar way, SA1 (top right) represents a postcode location in Swansea, South Wales. In the 1990s, the SA1 location was deprived and, for many, undesirable. In the twenty-first century it is the focus of regeneration, with its postcode the centre of marketing campaigns and prestigious marina-related development. Even in less prestigious areas of Cardiff, Wales, hotel chains use the idea of location to market the geographical proximity and desirability of their services (image bottom right).

Locations can often influence key decisions about where people live and the sort of lifestyles they aspire to. As the musician Moby outlines, in North America having a New York zipcode is an important element in locating your life and identity, even if this comes at a cost:

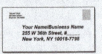

Get a Real Midtown Manhattan Street Address

Included with your subscription is a real New York City street address for your use. Instead of an address in some obscure town that no one's heard of, you'll have a prestigious address in the finance, commerce and arts capital of the world, just as any local business or resident might have

I don't know why [people] aren't moving to Newark. It's only fifteen minutes away from Manhattan and remarkably cheap . . . People would rather move to the worst part of Brooklyn and still have the magical 'NY' in their address. That single consonant on their mail – 'Y' as opposed to 'J' – seems to keep people from making that [move].

(Moby, 2014)

We will see in Chapter 6 how zip- and postcodes can often have strong meanings for individuals (and thus resonate with issues of 'locale' and 'sense of place' that we will discuss in detail below). It seems as if location is where it's at. How much would you pay to have the 'right' zipcode for your apartment?

BOX 4.2 A NEW YORK STATE OF MIND: THE LOCALE OF NYC

The idea of locale suggests that each individual place has its own unique traces and character. We have seen in Box 4.1 how the *location* and zipcode of New York is important to some people, yet what may be unique about New York's *locale* that means people wish to live in this city? What makes New York different from anywhere else? What are the core cultural traces of the Big Apple?

It may seem like a world away from New York City, but inhabitants of Newport, south Wales, are equally proud of the locale of their town. In a genuine but humorous parody of the Alicia Key's *Empire State of Mind* (found here www.youtube.com/watch?v=Eijc2tGe-zM) Newport's locale is described as celebrating the following cultural traces:

Big up to Plaid Cymru [a nationalist political party], and the Welsh Assembly [the devolved Welsh Government],

Big up to the Millennium [the national sports stadium in nearby Cardiff], we don't need no Wembley;

Core of the Big Apple
Source: Kimberley Peters

> Keep your Big Apple, we'll have a tangerine, ____ off Alicia, Shirley Bassey is our queen . . .
> We're from the small Welsh town of Newport: concrete jungle nothing in order, not far from the
> border; when you're in Newport: chips, cheese, curry makes you feel brand new, washed down with
> a Special Brew; repeat the word Newport, Newport, Newport . . .

are united together. We can perhaps imagine our own places and the ways in which we might be tied to, or defined to some extent by, our connection with place. Perhaps we have genealogical linkage to the land through our ancestors owning a house or a farm; perhaps we are connected due to the destruction of a favoured place or haunt. Perhaps we are connected to place due to our religious beliefs, due to everyday rituals, or to the geography of key events in our life. These place attachments, or senses of place, are often individual, but sometimes collective. Perhaps culturally important senses of place are commemorated or protected (through National Parks or memorial gardens); perhaps more individual senses of place are ritualised through holiday excursions, special visits, or pilgrimages. Some cultures value and celebrate this often intangible sense through words that point to this special connection between humans and place at different scales. The Welsh, for example, have the word 'hiraeth' to point towards their national sense of place, suitably explained by Elmes (2005: 98):

> 'hiraeth' [he-rye-th] is that sense of belonging or homesickness that Welsh people feel in their hearts when a long way from the Valleys of the south or the mountains of the north. Now it may be that this special sense of emotional attachment to a place, of longing and belonging, is much the same the world over . . . and maybe this sense of magical connection to a country and culture is simply yet another of those myths of 'localness'. Yet I'd bet there's scarcely a single Welsh man or woman who doesn't recognise . . . 'hiraeth'

as representing something for them that's very special and very, well . . . Welsh.

This sense of place is thus fundamentally important in defining our connection to geographical areas, and why they become significant to our own sense of who we are. As Seamon notes, 'it has become clear that we are emotionally related to our environment and that we must give scholarly attention to human attachment to place' (Seamon, 1984: 757). When focusing on these attachments we can begin to see that our own identity and that of our cultures are closely aligned to the identity of place. This close alignment is worth commenting on further.

The co-ingredience of people and place

> Approaches to 'place' have suggested the vital importance of a sense of belonging to human beings. The basic geography of life is not encapsulated in a series of map grid references. It extends beyond the idea of location . . . Crucially, people do not simply locate themselves, *they define themselves through a sense of place*.
>
> (Crang, 1998: 2, my emphasis)

As Crang outlines, the notion of a *sense of place* leads us to consider how *who we are* is fundamentally connected to *where we are*. From this perspective, our identity – our sense of selfhood – is a geographical thing, it is characterised to some extent by our geographical and cultural context. In short, our identity is defined by place. We can consider this in our own lives: if you had to think of, let's say three, ways to describe yourself, perhaps many of these will be 'geographical' in nature: I'm a student, a female, from Washington State, USA; I'm black, a swimmer, from Sydney; straight, Tower Hamlets, London. As we will see later in this book, all these identity positions are to some extent geographical, but we can see at this stage how in some instances we explicitly define ourselves in relation to place (e.g. Washington State, Sydney, London). This connection between the identity of *self* and the identity of *place* has been explored further by Casey (2001). Casey argues that the relations between self and place are not simply connected, but they are more than this – they are *conflated*. As Casey (2001: 684) explains:

> we can no longer distinguish neatly between physical and personal identity . . . place is regarded as constitutive of one's sense of self . . . The relationship between self and place is not just once of reciprocal influence . . . but also, more radically, of constitutive co-ingredience: each is essential to the being of the other. In effect, there is no place without self and no self without place.

What Casey is arguing here is that geographical context is integral to human's identity. He posits that through inhabiting places, involving a range of everyday mundane practices as well as special one-off experiences, a person-place relationship is inevitably developed. This relationship, he argues, affects both the identity of place – it involves humans taking and making place through leaving a range of traces in different sites, but also affects who we are – traces and their meaning come to influence how we feel and respond, as well as coming to define how we think about the world and where we belong. As a result, as Casey points out, 'places [begin to] possess us . . . insinuating themselves into our lives' (Casey 2000: 199).

It is important to recognise, therefore, how co-ingredient relationships are always developed between human selves and their geographies, at a range of scales. This relationship often prompts political action when these sites are threatened (see Anderson, 2004a), and perhaps grief and mourning when they are lost. For example, I still feel a sense of outrage and loss over the council selling my local park to housing developers. (How could they?! That's where I played football as a kid! It's part of me, gone!) Low (1992: 179) illustrates a range of rather more evocative examples, including an elderly man who wept when two giant palm trees in a local plaza were felled: 'he had spent his entire life under those palm trees – they were like friends and made his bench a special place'. Box 4.3 outlines the case of Mostar Bridge, Bosnia-Herzegovina, and the role that this geographical site had in defining the identity of those from this area of the world.

BOX 4.3 A BRIDGE BETWEEN PEOPLE AND PLACE: THE CASE OF MOSTAR

Mostar Bridge in Mostar, Bosnia-Herzegovina was destroyed during ethnic conflict between Muslims and Croats in November 1993. This is how Mostarians described their loss:

> On the first day the shelling started at about 2 o'clock, it lasted all day. We heard a mortar whistling and we threw ourselves down. When I looked up I saw a cloud of dust, yellow like the sun, I knew then our bridge had gone. I felt like all true Mostarians, as if part of my body had been torn off'.

I remember the river was red as if it had turned to blood. Fifty-year-old men were crying like babies. I know my family are only mortal, that we'll all die sooner or later, so I was ready for anyone's death. But not the Old Bridge.

I can somehow accept that my mother and my husband had been killed in the war. People die, they disappear, but not the Old Bridge, it had been here since 1566. I had a feeling that life had stopped, that we would all disappear with the Old Bridge. When they can destroy eternity, what has man become?

Men, women, and children were killed and raped in the war. But to consciously destroy the Old Bridge that was sacred to all Mostarians hurt us to the core.

(all cited in BBC, 2001).

The bridge was so fundamental to Mostarian identity that it was rebuilt, reopening in July 2004 (see below). The redevelopment was seen as symbolic of the healing of divisions between the Muslims and the Croats, intentionally seeking to re-establish the connections between people and place. As Sulejam Tihic, Bosnian Presidency Head stated, 'The reconstruction of the Old Bridge is a victory for Bosnia-Hercegovina, ethnic coexistence and tolerance' (BBC World News, 2004).

The re-opened Mostar Bridge
Source: Erkan Black

BOX 4.4 EXPRESSIONS OF A SENSE OF PLACE

Despite the importance of sense of place to our identity this dimension of life remains difficult to articulate and express. In popular culture, constitutive co-ingredience is expressed in the following phrase 'you can take the boy/girl out of [insert place], but you can't take [place] out of the boy/girl'.

More artistically, constitutive co-ingredience could be articulated in the following way, with a city 'mapped' into the very heart of one's being (see images below).

In Spanish, the concept of the pueblo points towards the constitutive co-ingredience of people and place. As Hancox tells us,

pueblo is a wonderful word which means village, town, or even city, [but] simultaneously also a people – and in that dual meaning lies the key to its magic. A village is its people. You might travel far away from

National co-ingredience

Source: Author

Heart of the City/Nation

Source: Author

its boundaries, never to return, and subsume yourself into the heady multi-cultures of city life, but you are still, in your essence, a son or daughter of the *pueblo*, and you will never lose that.

(Hancox, 2013: 26, my emphasis)

Although Español may have a word that expresses a form of constitutive co-ingredience between people and place, in terms of written and spoken English, it is often a difficult concept to express. As Huxley notes, 'touch the pure lyrics of [place] experience and they turn into the verbal equivalents of tripe and hogwash' (cited in Burgess and Gold, 1982: 3). Without resorting to the worst excesses of tripe and hogwash – there follow some examples of how a sense of place can be 'rooted in the core of one's being' (Tuan, 2004: 48). The first example is from American nature writer Annie Dillard, recording her relations to an island in her book *Pilgrim at Tinker Creek*, the second is from a Cardiff poet Dannie Abse, writing of his connections to his particular city:

I come to this island every month of the year. I walk around it, stopping and staring, or I straddle the sycamore log over the creek, curling my legs out of the water in winter, trying to read. Today I sit on dry grass at the end of the island by the slower side of the creek. I'm drawn to this spot. I come to it as to an oracle; I return to it as a man years later will seek out the battlefield where he lost a leg or an arm.

(Dillard, 1975: 18)

Cardiff was my home town, one that I felt attached to as if by an invisible umbilical cord.
    Home town; well most admit affection for a city: grey tangled streets I cycled on to school, my first cigarette in the back lane, and, fool, my first botched love affair.
    First everything . . .
    So many people I valued lived in Cardiff, so many occasions remembered, so many locations I cared for . . . I thought [this place] could console even a man condemned.

(Abse, 2003: 4)

The final example is from Benioff's story about a man condemned. Before beginning a prison sentence Monty walks with his dog (Doyle) through the neighbourhood they've grown up in:

Monty and Doyle walk west, pausing behind a fence to watch a basketball game, the teenage players taking advantage of the warm air, one last game before school. Doyle sniffs posts that stink of yesterday's piss. Monty assesses the ballers quickly, accurately, and disdainfully. The point can't make an entry pass to save his life; the two guard has no left; the big man down low telegraphs his every shot. Monty remembers a Saturday when he and four friends owned this court, won game after game after game until the losers stumbled away in frustration, an August afternoon when every jump shot was automatic, when he could locate his teammates with his eyes closed and slip them the ball as easy as kissing the bride.
    . . . This is his favorite spot in the city. This is what he wants to see when he closes his eyes in the place he's going: the green river, the steel bridges, the red tugboats, the stone lighthouse, the smokestacks and warehouses of Queens. This is what he wants to see when his eyes are closed, tomorrow night and every night after for seven years; this is what he wants to see when the electronically controlled gates have slammed shut, when the fluorescent lights go down and the dim red security lights go up, during the nighttime chorus of whispered jokes and threats, the grunts of masturbation, the low thump of heavy bass

from radios played after hours against rules. Twenty-five hundred nights in Otisville, lying on a sweat-stained mattress among a thousand sleeping convicts, the closest friend ninety miles away. Green river, steel bridges, red tugboats, stone lighthouse.

(Benioff, 2003: 12–17).

Thus we can see that sense of place not only involves emotional connections – or tracings – to particular geographical areas, these tracings come to define who we are. In a way these emotional connections come to define our own sense of place, our identity in relation to our culture and geography. We can see this in relation to the positive connections expressed by the examples above, but these emotional and identity connections can also be negative too. Tuan (1974) and Relph (1976), for example, have identified how affection for place (topophilia) can be counterposed by dislike, aversion, or hate for place too (topophobia) (see also Chapter 8). Horror films, for example, play on cultural topophobia, perhaps locating their main protagonists in woods at night, or in other 'haunting' locations. Can you think of other examples?

Green river

Source: www.flickr.com/photos/noise64/2880106776/ Noise 64.

Smokestacks

Source: www.flickr.com/photos/gordongerard/3017009123/ Gordon Gerard McLean.

Stone light

Source: www.flickr.com/photos/wallyg/1108617755/in/set-72157600947831176/. Wally G

Steel bridges

Source: www.flickr.com/photos/wallyg/1033802020/ Wally G

Tugboat

Source www.flickr.com/photos/11606523@N06/2079412059/ Tom Turne

Tinker Creek

Source: www.flickr.com/photos/49859104@N00/1429145066/in/set-72157601106791030/ star shooter supreme.

Gaining a sense of belonging: cultural ordering and geographical bordering

Following Agnew and Duncan (1989), place has multiple dimensions. It can be considered as a straightforward location, in terms of its cultural locale, as well as the sense of attachment it provides. The combination of these multiple dimensions gives us an impression of not only where we are, but also who we are; it tells us whether we belong in a particular place. In some places we may feel comfortable or have a sense of home, in others we may feel as if we are a fish out of water, its cultural traces may be strange, making us feel scared or dislocated. Traces of location, locale, and sense of place therefore give us a feeling of belonging, or not belonging, to a particular geographical site.

Senses of belonging to particular places are thus created by a variety of traces. In many cases, the intention of these traces is to regulate who enjoys this sense of belonging and who doesn't. Traces can designate the 'home' place for material things, non-material attitudes, performative practices, and groups of people. When such traces are successful it becomes clear some activities, ways of life, actions, and languages are welcome in some places, but others are not. This regulation is explained further in Box 4.5.

BOX 4.5 'IF I WANTED TO BE DISTURBED, I'D GO AND STAY IN A NIGHTCLUB'

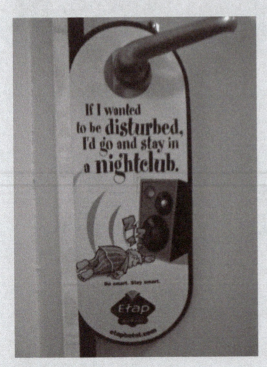

Door label

Source: Author

The image to the left depicts a sign found on a hotel room door. It reads, 'If I wanted to be disturbed I'd go and stay in a nightclub'. Placing this trace on a hotel room door has the effect of suggesting how the occupant views the nature of that place: it is not like a nightclub and so loud nightclub-like activities (e.g. dancing, shouting, or partying) are not appropriate here. By implication, the hotel room is a place of quiet, a place of privacy, a place to sleep. By publicising this view the occupant seeks to influence the behaviour of other hotel users and persuade them to conform to the occupant's preferred use of place. Thus through this trace, its expectations of behaviour, and its positioning, particular preferred uses of place are revealed. How we are expected to behave is publicised, and we should conform to these expectations if we wish to belong. If these expectations of behaviour are followed, the place of the hotel corridor will become a place of quiet. Its identity will be established.

The sign is thus one example of how traces can seek to influence the nature of places. It illustrates how there can be expected ways of behaving in particular places, and that certain people or actions are deemed appropriate in particular places but not others. However, many places do not need ordering signs to maintain their identity. In these cases, cultural uses of place are engrained into the population, and are learned and inherited through daily practice. These uses of place are often only revealed when incorrect place use is undertaken. We can use the statement above to illustrate this:

If I wanted to _____, I'd go to the _____.

Fill in the blanks in a range of ways. Put in activities in places that make sense to you, but also ones that don't. For example, 'If I wanted to *play football*, I'd go to *the library*'. In my culture, I would never do this – playing football in the library is not considered an appropriate use of place. There are no signs in the library telling me not to play football; there is no need for them: the majority know what a library is for, and generally conform to its expected use. Those who may not know look at others and pick up cues of behaviour as they pick up their short loans. Thus from this juxtaposition, and my reaction to it, it becomes clear to me how certain activities are expected in certain places, they are woven into the very fabric of those places, whilst others are not. Let's look at another statement:

If I wanted to be_____, I'd go to the _____.

Fill in these blanks as you see fit. For example: If I wanted to be educated, I'd go to the university. Fair enough? If I wanted to be entertained, I'd go to the university. Fair enough?! Both statements may be true some of the time, but maybe not others. Perhaps this statement throws into relief a grey area, the ambiguity and ambivalence inherent in some activities and states, in some places. In this grey area, ordering and bordering processes are not complete, they are open to change and there is a possibility for a range of positions to be taken. You could be educated or entertained at university, perhaps vice versa, or perhaps even both. This position raises a number of questions about universities, but also other places in which grey areas exist. What does this ambiguity say about the nature of these places? What does it say about the regulating processes that go to define them, and the other trace-making activity in them?

The effect of traces like those examined in Box 4.5 is that some activities and people are welcome and accepted in some places, but others are not. Places then are *culturally ordered* by traces. Signs, regulations, and laws seek to create cultural order by ascribing some people and activities a sense of belonging, but disregarding or ignoring the claims of others. Importantly, however, this cultural ordering goes hand in hand with a *geographical bordering*. As traces order our sense of belonging, barriers, frontiers, walls, and wire reinforce this order through physically imposing limits to movement, people, and places. This dual process of *cultural ordering* and *geographical bordering* occurs in all places at all scales, from prisons, to cities, to national frontiers. For example, I'm sure we have put up signs, even barriers, to our own bedrooms at some stage in our lives (me, aged 12!). Through undertaking this practice we are effectively saying 'Keep out! This place is mine!' In the case of my room, these signs impose my will over the cultural order of this place – I decide what goes up on the walls, and what series I download. I decide who can come in, and who is unwelcome here. Through establishing these ordering traces, this place comes to define me – there is reciprocity between its posters, music, and smells

(!), and *me*. However these signs do not function on their own. They work hand in hand with the material border of my bedroom walls and door to geographically classify my place (this material border marks where my room begins and ends). The process of laying down traces therefore attempts to both *culturally order* and *geographically border* places. All traces perform this dual function, within any geographical border there is always a cultural order, and every cultural order requires and brings into being a geographical border. Thus this *(b)ordering* process attempts to define the organisation and limits of a place, as well as the cultural group who claims it (see Box 4.6).

Which side are you on?

Groups therefore seek to use their traces to establish a cultural and geographical system of order in a location. Traces are used to control

---

BOX 4.6 (B)ORDERED PLACES: DIVE-INS AND DRIVE-INS

Cultural ordering and geographical bordering occurs in all places to different degrees. In the UK, for example, it is common for swimming pools to have a range of signs indicating appropriate codes of behaviour, whilst in France it is not unknown for attire to be included in swimming pool (b)orders.

(B)ordering swimming
Source: Author

Let's think too about the place of the cinema. Are there cultural orders and geographical borders here? How should you behave in the cinema, in your view? The BBC Radio Five Live film review show has derived its own 'Code of Conduct' for appropriate cultural practices in the cinema; which ones you do agree with and why? What is your response to those who regularly cross these (b)orders?

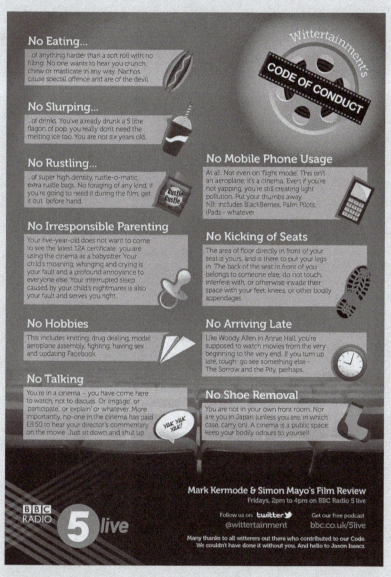

(B)ordering cinemas

Source: BBC

and regulate belonging and identity. In the words of Cresswell (2004: 103), they create a system where:

> Someone can be 'put in her place' or is supposed to 'know his place'. There is, we are told, 'a place for everything and everything in its place'. Such use of [traces] suggests a tight connection between geographical place and assumptions about normative behaviour.

By ordering and bordering place in this way, those who establish control are not only saying who and what belongs in a place, but are also saying who or what does not belong. In this process groups are included as part of a culture, but groups and activities are also excluded. 'Purifying' (Sibley, 1988, 1995b) places in this way has a number of effects. Firstly, it gives us a strong sense of place – this place is ours to defend, a place to live, and maybe even die for. But at the same time such bordering traces also function to exclude, clearly defining who is *not* part of our gang, our tribe, and our place. As Newman and Paasi (1998: 191) suggest, the border is 'the point of contact or separation', and 'usually creates an "us" and an "Other" identity'. In this sense therefore, which side of the bordering trace you are on defines who you are. This sense is what is referred to when we say someone is from 'the wrong side of the tracks'. Of course, if bordering traces are unclear, there exists a grey area where the identity and ownership of a place is not explicitly ordered. In these cases, places exist as a 'no-man's land'.

All places have (b)orders, but their nature varies in terms of the substance and strength. Some (b)orders can be crossed, whilst other are less permeable. The de-militarised zone between North Korea and South Korea is one example of a (b)order that is very difficult to cross both on the ground and in the air (see Hunter, 2013), whilst *The Economist* (2011) is certain that the (b)order between India and Pakistan is the 'world's most dangerous' (Figure 4.1).

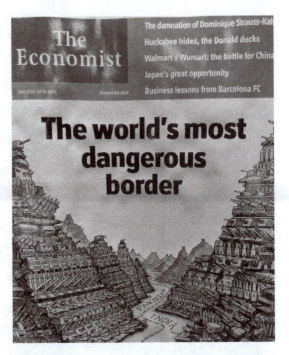

Figure 4.1 The world's most dangerous border?
Source: Author

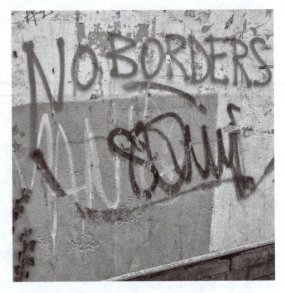

Figure 4.2 No borders?
Source: Author

Think about how you may be able to cross borders – at whatever scale. What do you need to do? How do you need to amend your behaviour and practices? Are there some borders you can't cross – due to gender, race, or age, for example? In many cases, borders are considered as fixed and locatable, and the places that they surround as static and unchanging. As Tuan argues, when borders are considered like this, place 'is essentially a static concept [and we can easily] develop a . . . sense of place' (1977: 179). If (b)orders function in this way, places come to be 'identified with what does not change; their "sense of place", "character" or "identity" is seen as relatively stable' (Dovey, 2010: 3). In short, static borders create still places.

However, as Price argues, with increasing globalisation and mobility the suggestion that places are timeless and unchanging may be a 'romantic notion' (2013: 122). In the contemporary cultural world, crossing (b)orders of all kinds may be getting easier. As a consequence, places – and our co-ingredience with them – may be changing in their nature too. Are we becoming defined by our mobility (see Chapter 8)? Are we now defined by more places in different locations across the globe? (B)orders *can* therefore fix our understandings of place as static and stabilised, but – in line with the broader argument of this book – we must also remember how places can be understood as an ongoing process. How (b)orders function therefore directly influences the nature of places – in what ways do they structure a place, and in what ways do they encourage change?

Crossing (b)orders: places and their trace-chains

Despite the presence of orders and borders, places often remain influenced by the connections they have to other sites, at other scales. Due to the ability of some traces to cross (b)orders, a whole host of traces from a range of scales come together to influence local places. When we think about defining places we therefore have to consider the traces that cross borders and contribute to the ongoing composition of geographical sites.

As noted above, processes associated with 'globalisation' are one key example of traces that cross (b)orders. The spread of capital and media communication across the globe, the relocation of jobs and industries, the threat of environmental catastrophe, and the migration and mobility of people from their traditional locales, are all examples of traces that emanate in one place, but have repercussions in others. These traces cross geographical and cultural (b)orders and set in motion new effects in other local places. One example of how such traces can affect place is given by Doreen Massey (1994). Massey illustrates how border-crossing traces affect her local high street in London. Simply by walking up Kilburn High Road she identifies 'the world in her city' (after Wrights and Sites, 2006: 107). I undertook a similar exercise in my town of Cardiff; these are the images of shop fronts and posters on City Road (see Box 4.7).

The ongoing composition of traces that define any place are therefore never purely local in nature. Despite the existence of cultural and geographical borders, it is impossible to insulate a place from networks of trace-chains. Even those places that seek to cut themselves off from a wider world (for example nations like Cuba or North Korea) are still defined by their (lack of) relations and sense of threat from 'other' places (and vice versa). Places, then, are never wholly defined in isolation, but also by the trace-chains that connect them to the broader world. As Massey (1994: 153) puts it:

Instead . . . of thinking of places as areas with boundaries around [them], they can be imagined as articulated moments in networks of social relations and understandings, but where a large proportion of those relations, experiences and understandings are constructed on

BOX 4.7 TYPICAL CARDIFF? THE WORLD IN ONE CITY ROAD

He claimed to be typical Cardiff: part Portuguese, Irish, Italian and something slightly more exotic, from a further flung soil that he hadn't yet managed to trace.

(O'Reilly, 2003: 30)

'Let's go and eat it in the castle grounds,' I suggest, so we make our way to the market to buy ham and tomato cobs. I imagine I'm a vampire looking at the various people in the market, wondering how they would taste. So many flavours for a Cardiff vampire to choose from: Welsh of course, Indian, Chinese, so many East Europeans, redheads, Aryan blondes, dark little Celts, Somali, Sikh, Swansea tastes. A boy couldn't make his mind up which to choose first. I would probably go for a redhead, pale skin and blue/green eyes, golden hair dusted onto icing white skin. I suppress a shudder and pay for my cob and diet coke.

(Williams, 2005: 208)

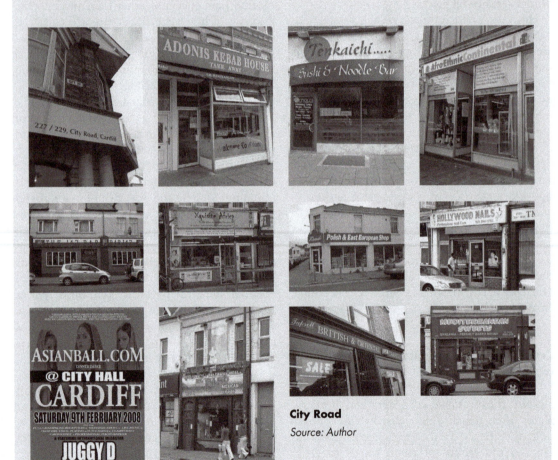

City Road
Source: Author

The world in my City Road consists of a range of traces from a range of geographical places. Traces are inscribed from Wales and Britain, from Western and Eastern Europe, and from the Americas, Asia and Africa. The businesses pictured in the above images are the traces set in motion by globalising processes, but have real effects on my local place. We could see other border-crossing traces in City Road if we considered the origin of residents in this place (from semestering students, transient East Europeans, second generation immigrants, and lifelong locals), or if we investigated the environmental, political, and economic influences on this place. The traces found in my City Road highlight how 'typical Cardiff' is defined not simply by its local, or Welsh traces, but also by these border-crossing traces. Cardiff, then, is not wholly defined by its isolation from other places – it is not defined just by its borders – it is also defined by the 'trace-chains' that tie it into places all over the world. Trace-chains are those connections that cross borders, tying one place to another in different and interesting ways. Following Massey, you can walk down your own street and identify the range of local traces that come to define your place, as well as how these combine with trace-chains emanating from locals elsewhere.

Here are some examples you could consider from streets in Santa Cruz and San Francisco, California, as well as video from the film *25th Hour*, based on New York City (see www.youtube.com/watch?v=TgL_5QcZCMo). What do these traces tell us about the culture in these places, particularly perhaps in terms of nation and race (see Chapters 6 and 10)?

Noah's New York Bagel House, Santa Cruz
Source: Author

Chinatown, San Francisco
Source: Author

Chinatown, San Francisco
Source: Author

Chinatown, San Francisco
Source: Author

Chinatown, San Francisco
Source: Author

Beefeater Gin (London) advert, San Francisco
Source: Author

a far larger scale than we happen to define for that moment as the place itself, whether that be a street, or a region or even a continent.

There is, therefore, a 'constitutive interdependence' (Massey, 2007: 21) between places. Every place is an ongoing mix of traces that are purely local and non-local. Places are therefore connected together, but each remains a unique composition within these trace-chains. No place will ever be the same as another. This may be due to the individual combination of local and global traces in one location, or due to the influence of traces from the past.

Trace-chains from the past

Places then are produced by an ongoing composition of traces emanating from both the local and the global. But new traces do not act on their own, rather they combine with previously constructed traces to order and border place identity. As we saw in Chapter 1, places can be defined by the stirrings of past cultural traces. I am sure we all have reminisced about events and meanings that, although in the past, continue to affect a place's meaning in some way. Richard from East 13th Street gives us one example:

> When people from this block come into this gallery, they may seem to be looking at that painting over there or whatever-it-is over here, but what they are really doing is remembering. That's where we had the bar that Junior sold drinks for fifty cents. That's where we had the pool table. That's where I met my girl friend. That's where that dude got shot. That's where little Elizabeth was crowned Princess of East 13th Street and did her victory dance . . .
> (Richard of East 13th Street, quoted in Lippard, 1997: 261)

Perhaps you can think of other examples from your own experience. I still 'see' my local park where I played football as a kid, even though it is now covered by suburban sprawl. These traces from my past still influence the nature of that place for me. In other sites, perhaps these past-traces are commemorated; perhaps they remain in built form, or only stir in our memories when we re-trace our steps in places. However they remain, it is clear that although places change, traces of the past do not necessarily leave them (see Box 4.8). Places then do not always correspond to a straightforward time line, moving forward from the past to the future, but meanings linger, re-erupt, or disappear, as human cultural activity affects them. As Battista et al. (2005: 439) comment, places are

> not made up of simple linear histories, which stretch from the past in smooth lines towards a vanishing point in the far-off future. Histories scrumple up, times past pop into the present, modifying them for all time. [Places] in the present wobble as stories from the future and the past shimmer in front of them.

Traces assemble

Places are thus formed by traces from the local, from the global, from the present and the past. As Solnit puts it: 'Places . . . always refer beyond themselves . . . What we call places are stable locations with unstable converging forces that cannot be delineated either by fences on the ground or by boundaries in the imagination' (2010: vii). The importance of traces, their origin, and their effect, have led to some cultural geographers talking about place in ways that emphasise the enduring and active nature of these traces. For Massey, traces are turned into trajectories, and place becomes 'a constellation of on-going trajectories' (2006: 92). For Delanda (2006), cultural traces are 'assembled', with place becoming a 'territorialized assemblage' of traces (Dovey, 2010: 17). As Dovey (2010: 16) explains,

BOX 4.8 DREAMING PAST PLACES: DETECTING VICTORIAN LONDON TODAY

'It's as if I'm in a dream', Sally told me. We sat together in a pub just off the Strand at the end of her guided walk, advertised as following 'In the footsteps of Sherlock Holmes'. But here she spoke of a private enthusiasm, her fascination with the River Fleet, an ancient waterway now buried beneath the modern city. 'I'm well aware', she went on, 'that as I walk around London I'm in a dream, imagining it as it was. If I walk down Farringdon Road, which is the dreariest road imaginable, I'm not looking at what's there now, I'm looking, you know, at what it was like before'.

Sally told me that she saw . . . past lives of the river . . . She saw open pasture and distant hillocks . . . At other times the river was dark and filthy, cast in shadows by a canyon of tenements that overhung the banks and led down to its mouth at the Thames. Sally remembered seeing coarse cloaked women emptying their slop-buckets into the water, butchers with greasy leather aprons hacking at cow carcasses and allowing the discarded entrails to float away, or coal barges with their sails down losing their loads overboard as they tried to navigate a narrow course. Sometimes the river appeared completely dried up, buried under rubbish and built over with tumbledown yards and alleys of slum dwellings. Down there, she said, everything was pitched in darkness, but the sky was hazy and bright, a reflection of the gas-lit world of respectable streets on the ground above. Then with a start Sally would find herself back in Farringdon Road, once again confronted with the roar of modern traffic.

Like the other London enthusiasts whom I met, Sally claimed to be able to perceive the pasts of the city in the present. I was told that this ability was a solitary gift, not easily communicated to others. On walking tours, for instance, she reported that customers often failed to share her vision. 'Sometimes you get somebody, who says, "I can't see that. Where is medieval London?" You know, "You've not shown me anything." And then I realize that a lot of it is in my head. I can see things mentally which have gone. You explain it to people and some of them can do it, but not everybody does. They can't seem to imagine how it has changed.'

(Reed, 2002: 129)

a street is not a thing nor is it just . . . a collection of discrete things. The buildings, trees, cars, sidewalks, goods, people, signs, etc. all come together to become the street, but it is the connections between them that makes it an assemblage or a place. It is the relations of buildings-sidewalk-roadway; the flows of traffic, people and goods; the interconnections of public to private space, and of this street to the city, that make it a 'street' and distinguish it from other place assemblages such as parks, plazas, freeways, shopping malls and marketplaces. Within this assemblage the sidewalk is nothing more than a further assemblage of connections between things and practices. The assemblage is also dynamic – trees and people grow and die, buildings are constructed and demolished. It is the flows of life, traffic, goods and money that give the street its intensity and its sense of place. All places are assemblages.

Resonating strongly with the notion of places as ongoing compositions of traces we can see here how places can be considered not as discrete, essential things in themselves but, rather, are formed by the coming together of more or fewer random component parts. It is *how* these component parts relate, connect, and interact that forms particular places. These relations are not permanent and, therefore, places may change, decay, and transform. Similarly, component parts can become detached or removed and the place become something else. Places are constituted not simply by 'things' (or material traces) but also by non-material traces (e.g. the building of houses, the movement of people, the 'flows of life'). As a consequence these places are also constituted by the experiences of those involved in these practices, the meanings they bring to these places, and the intensities of affect produced by their interactions with the other connecting parts. As Dovey suggests, 'change any of these and [the street] would still be a place, but not the same place' (2010: 24).

The notion of ordering and bordering place thus creates a sense of identity for both places and their peoples. Cultural groups may feel loyalty, ownership, and responsibility for their places, and often seek to protect them from changes that are unwanted or unwelcome. It is therefore important to study the nature of new traces in any place and the trace-chains they set in motion. What changes do new traces seek to make? Which groups seek to encourage them, and which groups seek to retain the status quo? Where do these traces come from? The intimacy of our own places often leads us to consider that the origins of new traces should be local in nature. However, particular locations are often affected by traces originating in other areas. The ways in which such new traces affect places, and how we think about them as a consequence, are important issues to consider.

Summary: places and trace-chains

The cultural orders and geographical borders of any place are thus influenced not just by traces at the local level, or traces in the present, but by trace-chains that affect place from across the globe and places-past. With these points in mind the important thing to focus on is the diagnosis of trace-chains that make each individual place. What traces come together to take place? What do these traces stand for, and where do they come from? Do they come from the local or the global, from the present or the past? From this diagnosis we can begin to understand the nature of a place. How does each trace affect our sense of belonging? Are some traces defended, and some resisted? How, and why? These are questions that we will return to in the next chapter.

BOX 4.9 TRACE-CHAINS: 'CASTLES 641 STARBUCKS 6'

Castles 461
Source: Wales Tourist Board

In this Welsh Tourist Board advert for Wales, a view from the inside of Cardiff Castle is shown. The castle itself has clear geographical borders. It has walled fortifications that mark its perimeter, keeping the unwanted on the outside, and protecting those accepted and desired. With this image of the castle in mind, you'd be forgiven for thinking that the Wales of today retains its strong borders, resolutely ordering its places and culture from those 'foreign' and 'other'. The tagline to the advert: *'Castles 641, Starbucks 6'* also plays to this sense of place. With its proud heritage and traces-of-the-past so far outscoring the modern and global traces symbolised by the coffee chain Starbucks, (in)famous for its omnipresence on many high streets, the Tourist Board is highlighting how 'pure' the culture and places of Wales are.

Although Welsh borders may be strong, trace-chains still influence the location, locale, and sense of Wales. Starbucks may not be dominant (see Chapter 5), but its traces do register in this place. The borders of Wales are therefore not absolute and can be crossed. This fact is demonstrated by the possibility (and desire) for tourism in the first instance. The presence of these 'foreign' traces amongst those of the past therefore gives a hint of the cosmopolitan nature of Wales. This combination of the local and the global, the past and the present, generate the unique composition of traces that is Wales.

That is one reading of the above advert. You will surely be able to come up with others. Similarly, there will be many possible readings of your local place. How do trace-chains combine to influence its identity? What are the affects of local and global traces on your sense of place? Do trace-chains change your sense of home? Do they threaten it, make it more cosmopolitan, or enhance it? Do people seek to defend their borders from trace-chains? Which ones, and why?

Conclusion. Placing traces

every person leaves some residue of their passing and of their actions on their physical surroundings.

(Battista et al., 2005: 448)

Place is vital to human life. As Relph puts it, 'to be human is to live in a world that is filled with significant places; to be human is to have and know your place' (1976: 1). Through our every-day actions, whether intentional or otherwise, we leave traces that take and make place. As traces combine, cultural orders and geographical borders are established that attempt to define and control the identity of a place. These orders and borders give us our sense of whom and where we are, they give us our sense of belonging. However, as traces combine from distant as well as local places, as well as from the present and the past, borders are crossed and orders change. Place, and our belonging to it, is therefore not fixed in a real sense, it is dynamic and evolving: places are ongoing compositions of traces.

Places then are not 'nouns' – they are not fixed and solid things, rather they are 'verbs', they are doings, and they are always active. Cultural geographers therefore analyse and interrogate all the agents, activities, ideas, and contexts that combine together to produce these places. Attempts can be made to order and border it, and define the connections it has to the wider world. The nature of these actions means that places can be divided and segregated, or related and united. Taking and making place is therefore a political and cultural action. It involves and produces relations of power between different cultural groups in different areas. The exercise and agency of power, and how it affects place, is to be explored in the next chapter.

Suggested reading

The following texts are useful to explore the issues raised by this chapter.

Books

Cresswell, T. (2004) *Place: A Short Introduction.* Blackwell: Oxford.

Cresswell, T. (1996) *In Place/Out of Place.* University of Minnesota Press: Minneapolis.

Tim Cresswell offers an insightful introduction to the notion of 'place', as well as an exploration of how people and practices are accepted or marginalised within different geographies and cultures.

Massey, D. (1994) *Space, Place and Gender*. Polity: Cambridge.

Massey, D. (2007) *World City*. Polity: Cambridge.

Oakes, T. and Price, P. (eds) (2008) *The Cultural Geography Reader*. Routledge: London and New York.

Doreen Massey introduces us to globalised senses of place, both in terms of her walk down Kilburn High Road (1994, reproduced in Oakes and Price, 2008), but also in terms of global cities, such as London (2007).

Relph, E. (1976) *Place and Placelessness*. Pion: London.

Dovey, K. (2010) *Becoming Places. Urbanism/ Architecture/Identity/Power*. Routledge: Abingdon.

Here Relph outlines different definitions of place, and how these relate to 'insider' and 'outsider' identity positions, and notions of 'placelessness', whilst Dovey introduces ideas of place as a process.

Journals

Casey, E. (2001) Between geography and philosophy: what does it mean to be in the place-world? *Annals of the Association of American Geographers*, 91 (4), 683–693.

Casey draws on humanist and phenomenological positions to outline how place and personal identity are closely related.

Taking and making place: the stuff of power

Introduction

> The study of culture is . . . closely connected with the study of power. A dominant group will seek to establish its own experience of the world, its own taken-for-granted assumptions, as the objective and valid culture of all people. Power is expressed and sustained in the reproduction of culture.
>
> (Cresswell, 2004: 124)

We have seen how vital place is to cultures. In taking and making place different cultural groups generate an array of traces that have the effect, intentionally or otherwise, of arranging, managing, shaping, and transforming places in line with their belief systems and political values. As we all leave traces through our everyday actions, we all contribute to taking and making place. As Maxey (1999: 201) outlines, the cultural world

> is produced through the acts each of us engages with every day. Everything we do, every thought we have, contributes to the production of the social world . . . We are in a sense all activists, as we are all engaged in producing the world.

Of course, sometimes we are not aware of the traces we leave in places, sometimes they are temporary and lack significance – perhaps footsteps in the grass as we walk across the park. But sometimes they are more thought through

and attempt to have a wider political and cultural meaning; perhaps we decide to buy a free trade product, or go to a victory parade, perhaps we stand for our national anthem, or engage in some form of demonstration or rally. When we engage in these more explicitly political and cultural activities perhaps we begin to ask ourselves some questions. Who decided that these products are 'fair', or that we should celebrate some victories and not others? Who decided on the national anthem, and how we should respond to it? What happens if we didn't respond in the accepted way? Why are political demonstrations necessary? In short, we perhaps begin to ask: if I am a trace-maker, then why isn't the world produced in my image?! It seems that some trace-makers are more important than others in making the cultural world. How do some trace-makers acquire this capacity, and what do people do to perpetuate or change this situation? All these questions are connected to issues of power.

Taking and making: the stuff of power

As Maxey has told us, the cultural world is produced through the acts each of us engages with every day. Each of these acts is a demonstration of power. In a basic sense, therefore, power can be defined as the ability to act. As a consequence, we all have a certain level of power; we can all leave traces in places. Traditionally, however, power has been considered as more than simply the power to act. Power has traditionally been considered as the *power to influence others*, to change what they do and where they do them (see Weber, 1994; Marx and Engels, 1984; Marx, 2003). To exercise power in this sense, your acts must *transform* places. Foucault (1980, 1984) considered power in this way. To him, power is a *transformative capacity*; it is the ability to transform the traces of others in order to achieve certain strategic goals (see also Cresswell, 2000: 264). In a basic sense, therefore, we all have a degree of power, yet the power to transform how we act and how we think is of paramount importance. This is the power to create culture and place.

When individuals and groups act, there are always geographical consequences. Traces are made, orders issued and borders constructed. It is through geography – through taking and making place – that power is exercised, made visible, and has effects. The struggle for place is therefore both the manifestation of cultural struggle and the medium of that struggle, it is in place that power is constituted and played out. Studying place allows us to read power struggles but also to identify who has power to transform them. As we have already noted, the power to transform place isn't exercised by all people equally within a society. Certain groups have more opportunity to exercise this power than others. This power may be exercised by our parents in the family home, bullies or teachers at school, or the police in broader society. The groups who already control culture and places can be considered as having this power to transform: they have 'dominating power'.

Dominating power

Dominating power is that power which is successful in controlling or coercing the action of others (see Lukes, 1974). This is the power that can stop individuals doing exactly what they want, where they want, and sometimes even make individuals act against their own interests. Through imposing a range of ordering and bordering traces, dominating power manipulates, encourages, or enforces people to act in certain ways in certain places. Such traces could be the simple construction of signs that request good behaviour, appealing to the spirit of individuals of 'fair play' or cultural collegiality. They could also involve the threat of cultural exile or 'othering' through failing to conform to conventional codes. Often the pull of belonging – wishing to remain tied in to the identity of a people and a place – is powerful enough to ensure conformity and conventional behaviour.

Other means of cultural control are also used. Groups may, for example, impose their will on others through threat and the exercise of punishment (see Box 5.2). This punishment may involve the removal of cultural 'goods' (e.g. money through fines, freedom through imprisonment or curfew, or identity through designation as 'criminal'), or the imposition of physical injury (through corporal or ultimately capital punishment). In other cases control is achieved when individuals 'go along' with dominating traces simply because it's the most straightforward thing to do. In many cases, as Gramsci has noted, '"spontaneous" consent [is] given by great masses of the population to the general direction imposed on social life by the dominant fundamental group' (Gramsci 1972: 12). Sometimes we just go along with (b)ordering traces because it makes life easier. Thus cultural control can be exercised through threat, through manipulation, or simply through acquiescence. Dominating traces then can be material and non-material, affective in physical, symbolic, and psychological ways.

Dominating power and the implementation of ideology

Dominating power is thus power which is successful at making people obey. It makes people

BOX 5.1 REGULATING CULTURAL PLACES: ACTS OF DOMINATING POWER

Signs

Source: Author

The images above are examples of dominating power transforming the actions of people. The signs above tell people where to walk and not walk, which sides of the tracks to stay on, how to act in a particular area, where to stay away from, or how to dress. They are at once orders and borders – they attempt to culturally order a particular area, as well as geographically border people and actions. You can do this there, but not here. This is the power that takes and makes place.

BOX 5.2 ORDERING PLACES: THE ALL-SEEING EYE

Dominating forces need a variety of support infrastructure in order to transform behaviours and places. A credible threat of being caught failing to obey (b)orders is necessary, as is a credible threat of punishment for this disobedience. If we think that a dominating power is watching, and that they will show no mercy to inappropriate behaviour, we are more likely to conform to their (b)orders. Can you think of ways in which our behaviours are monitored in

Or what?

Source: Author

Are you being watched?

Source: Author

contemporary society? What mechanisms are available for a dominant power to enforce its will on places in our cultural world?

One example we might consider is technological surveillance. This is one support infrastructure that allows governments to monitor the behaviour of citizens in a range of places. When considering surveillance systems such as closed circuit television (CCTV), satellite imagery, or telecommunication monitoring, relating them to the notion of the 'Panoptican' is insightful. The Panoptican is an idea originally conceived by Jeremy Bentham (1843), but extended by the French philosopher Michel Foucault (1979). Bentham's original idea was a design for a circular prison with a central observation tower. This tower would be surrounded by prisoner cells. Due to the design of the building, it was possible that the guards could monitor prisoner activities at all times. This threat of perpetual observation greatly improved the possibility that prisoners would obey the dominant power: the risk that disobedient behaviour would be seen and punished was near absolute. In this situation, it was argued that obedience to (b)orders would become automatic.

Big Brother

Source: Anonymous

Foucault developed this idea to discuss the ways in which cultural groups can control all sorts of places – that public places could become panoptical in nature. If a group could turn the cultural world into a Panoptican, they could ensure there would be little 'danger of plot, [or] attempt at collective escape, [or] planning of new crimes for the future' (Foucault, 1979: 195). Perhaps realising a panoptical world would mean there would be 'no disorder' (ibid.)?

Although it may be difficult to imagine a material Panoptican in contemporary culture, it is possible to gain an all-seeing eye through technology and surveillance. Closed circuit television and satellite imagery enable many governments and intelligence services to monitor our movements (see MacAskill,

2013). As Jenkins (2013) reports, 'now the vast databases of Microsoft, Google, YouTube and Facebook are open to government . . . [this form of dominating power] intru[des] in the private lives and communications of every citizen . . . As police states have said down the ages, the innocent have nothing to fear. [But] for innocent, eventually read obedient.'

conform to another group's vision of what the world should be like, whether it is in their interests or not. Dominating power is thus successful in imposing its *ideology* on particular places. Ideology, as Ryan outlines, 'describes the beliefs, attitudes, and habits of feeling which a society inculcates in order to generate an automatic reproduction of its structuring premises' (cited in Wallace, 2003: 239). In other words, ideology comprises the particular ideals and beliefs of one cultural group and presents them as the 'common sense' for all. This ideology – this 'common sense' – may not be fair or equitable, indeed it often seeks to control place for the dominating group's benefit. (As Tabb notes, 'the rich and powerful set the rules and the rules, not surprisingly, favour the rich and powerful' (cited in Smith, 2000: 1)). Ideology thus often produces forms of prejudice and repression, as Sharp et al. (2000: 2) identify: 'patriarchy, racism, and homophobia are all faces of dominating power . . . [This] power engenders inequality and asserts the interests of a particular class, caste, race or political configuration at the expense of others.'

As cultural life always has a geographical dimension – as all orders create borders – when dominating power is based on prejudice the places around us become unequal and defined by repression. As Watson states, 'people build their tastes and prejudices, their ideals and adorations into the land itself . . . This should be an axiom in geography. People generate prejudice; prejudice governs place' (1983: 388). We can see, therefore, how relations of inequality – differences of wealth, privilege, and status – are tied directly to the outcomes of dominating power.

Through the imposition of dominating traces, groups create the orthodox 'common sense' way that cultural life should be. Thus in many ways dominating power is about the power to define. Groups with dominating power define the cultural order, they get to construct conformity. Dominating power is thus the ability to define what a culture considers to be normal and appropriate behaviour, the definition of basic notions such as right and wrong, what is acceptable or improper, what should be tolerated, and what should not. Dominating power creates systems of 'normality' that we all *should* conform with to be 'good citizens'. Through this conformity, these particular cultural preferences become less political and increasingly normal. As time goes on, these values don't seem like one point of view anymore, they come to seem like the natural course of things, the best way to be. This creation of cultural orthodoxy reaches its zenith when individuals can no longer imagine any alternative to the cultural values they are conforming to. Bourdieu (1977) calls this situation the arrival of cultural 'doxa', when the cultural world is made to seem as if it were not only normal, but as 'natural' as motherhood and apple pie (see Box 5.3).

Dominating power and defining traces

Dominating power is thus the power to define. If one cultural group can achieve this power it can authorise its own version of reality as the 'official' one; it can create a common sense view that the majority of people subscribe to, and ultimately ensure the reproduction of its own

BOX 5.3 THE CREATION OF CULTURAL ORTHODOXY: ORDERING AND BORDERING THE TRAINS OF SOUTH WALES

Arriva loud

Source: Author

Arriva twerp

Source: Author

Arriva wander

Source: Arrive Trains Wales

The above images are part of a series that seeks to establish ideologies of behaviour on the trains and railway platforms of South Wales. To the train company, some things are 'obviously' too loud (see the image on the left) – be it overly floral wallpaper, shouted conversations, or amplified MP3 players. By describing this behaviour as 'obviously' out of place the train company seeks to inculcate the passenger into their point of view, increasing the possibility they will agree and acquiesce to this ordering of behaviour. Similarly, the centre and right images attempt to impose an ideology of action. By satirising those who are 'anti-social' (be he the 'Supertwerp' who vomits on platforms, or the 'Wander Woman' who lets her child run loose on trains) passengers are encouraged to identify with good behaviour and be a 'good' citizen. Identifying someone who behaves 'badly' and satirising their behaviour as anti-social attempts to reproduce the ordering and bordering premises desired by the train company. Surely only a 'supertwerp' would act anti-socially on a train?

culture. As cultural geographers we can be sensitive to the range of definitions given by the dominant. Which traces are seen as appropriate in a given location? Which conform and which are controversial? Which traces are seen as '*natural*', '*normal*', or '*novel*'? A typology of these constructions and examples of the cultural definitions ascribed to them is given in Table 5.1.

As Table 5.1 outlines, traces can be judged as 'natural', 'normal', or 'novel'. Each of these categories includes a range of pejorative descriptions which seek to apply to them moral and

Table 5.1 Defining traces

'Natural' traces ⟷	'Normal' traces ⟷	'Novel' traces
Doxa	Orthodox	Unorthodox
Essential	Popular	Different from the 'norm'
The only way to be	Mainstream	'Other' than expected
Everyone	Us	Them
Conforming	Appropriate	Inappropriate
In place	In place	Out of place
Common sense	Conventional	Controversial
Natural	Clean	Polluting
True	Right	Wrong
Ordinary	Good	Evil

political values. For example, a 'natural' trace will be described by a dominating group as 'essential', as 'right', as 'good' or 'appropriate' in a particular place. Can you think of an example of these 'natural traces' in a place of your choice? Other traces may be described as 'normal'. These traces may be popular, mainstream, or orthodox (and if we think about the music charts, 'pop' music is a good example of a 'normal' trace). Although such traces are mainstream they have yet to attain the status of a 'natural' trace; as a consequence other traces can be accepted as equally valid, if a little unconventional (e.g. jazz, heavy metal, or emo music). 'Novel' traces in turn are seen as dangerous, risky, and inappropriate in particular places. In many cases they are seen as 'dirty' (see Box 5.4), in some cases even 'evil' (Box 5.5). If dominating power is successful, definitions of traces as 'natural', 'normal', or 'novel' will be unquestioned by the general population. The moral and political values given to these traces will then influence our behaviour. The more 'natural' a trace seems to be, the more likely individuals will conform to it, unable to think outside the box and dream up alternatives. The more 'novel' (or unorthodox, different, strange, dangerous, etc.) a trace appears, the easier it will be to marginalise and outlaw.

BOX 5.4 TO DOMINATE IS TO DEFINE

Dominating power seeks to influence how we think and how we act in different places. It does so by giving positive meanings to acts of conformity, and negative meanings to acts of deviance (or non-conformity). One example of this pejorative framing of particular actions, peoples, or places is to claim they are 'dirty' or 'polluting'.

For Douglas (1991), notions of order and disorder are often allied to ideas of cleanliness on one hand, and dirt on the other. As Douglas outlines, dirt is 'the by-product of a systemic ordering and classification of matter, in so far as ordering involves rejecting inappropriate elements . . . [dirt is] that which cannot be included if a pattern is to be maintained' (Douglas, 1991: 35/40). Dirt therefore is matter out of place, as Freud suggests, 'dirt seems to us incompatible with civilisation' (1930: 56/7, cited in Sibley, 1999: 135).

Who would want to be dirty, with all its cultural connotations of lack of hygiene, disease, smell, and squalor? Even the generation of this question reflects the success of the cultural orthodoxy of clean/good, dirt/bad. There are many examples of the ways in which places and the activities and people in them are deemed to be dirty and polluting, and thus matter 'out of place'. Papayanis (2000) for example has researched the dominant perceptions of the sex industry in New York, and how, he states, 'sex in public is, quite simply dirt' (2000: 350), whilst Sibley's work on gypsies in Hull (1999) graphically depicts how this cultural group is seen to be 'polluting' mainstream places (see also Chapter 10). Describing these practices, peoples, and places as 'dirty' generates a 'common sense' opinion about these traces. It seeks to perpetuate the ordering of culture and the bordering of place, keeping some behaviours and people in one place, and out of another.

BOX 5.5 WHAT'S IN A NAME?

Dominating power seeks to influence how we think and how we act by defining traces in particular ways. When an action, person, or place is defined, it shows how one particular group values that entity; the definition demonstrates its significance to that culture. For example, a culture is unlikely to give its state buildings comedic or satirical names; similarly a road is more likely to be named after a cultural icon, rather than a pariah. Whole countries can even be named in particular ways, as the following example illustrates.

The 'Axis Of Evil'

Following the attacks on the World Trade Center on 11 September 2001, the USA made attempts to track down those who instigated them. This process has not been straightforward since the battle is not against an enemy which can be physically located within a nation state, but rather it is a battle between cultural ideologies – between 'free market capitalism and democracy' on one hand, and 'terror' on the other (at least from the point of view of the USA at the time). The then President of the United States, G.W. Bush attempted to geographically (b)order his enemy through exercising his dominating power. He did so in a speech on 29 January 2002:

> [Our goal] is to prevent regimes that sponsor terror from threatening America or our friends and allies with weapons of mass destruction. Some of these regimes have been pretty quiet since September 11th. But we know their true nature. North Korea is a regime arming with missiles and weapons of mass destruction, while starving its citizens. Iran aggressively pursues these weapons and exports terror, while an unelected few repress the Iranian people's hope for freedom. Iraq continues to flaunt its hostility toward America and to support terror . . . States like these, and their terrorist allies, constitute an axis of evil, arming to threaten the peace of the world.
>
> (Democratic Central, 2008)

In this speech Bush defines the actions and nature of his ideological enemies in two ways. Firstly, he geographically borders the enemy as he sees it. He locates the focus for his ire on North Korea, Iran, and Iraq.

Secondly, Bush attributes a value or moral judgement on those within these borders. To him, they are 'evil'. Bush has exercised his power to define to create a morally divided cultural world. To Bush, the world is also now geographically divided between the axis of evil on one hand and the allies of the USA on the other. He is attempting to convince as many as possible that these countries are actually evil, and this should be the 'common sense' view. The extent to which we accept Bush's attempt to transform the world in this way depends in no small part on whether we uncritically accept his dominating power or seek to question it. It is worthwhile to note not everyone shares Bush's vision. Hugo Chavez, President of Venezuela, describes the United States and its allies as an 'axis of evil' (BBC, 2006a), whereas on the other side of the world the following graffiti from a Cardiff street makes its views clear:

USA evil?
Source: Author

BOX 5.6 DEFINING PRACTICES IN PLACES: THE DEFINITION OF 'TERRORISM'

In 2009, an attack was carried out on a military base in Fort Hood, Texas. Nidal Hasan killed 13 people, and injured a further 32. In 2013 on a London street, Michael Adebolajo and Michael Adebowale stabbed to death off-duty soldier Lee Rigby. At first glance, perhaps the least important element in these unspeakably tragic attacks is how they should be defined. However, defining these practices accurately is an important endeavour. Some officials and media outlets reported the attacks as 'terrorism' (Fox News, 2009; New York Post, 2013). Yet should these attacks be defined in such a way? According to Greenwald (2013), 'in order for an act of violence to be "terrorism" . . . it must deliberately target civilians. That's the most common means used by those who try to distinguish the violence engaged in by western nations from that used by the "terrorists": *sure, we kill civilians sometimes, but we don't deliberately target them the way the "terrorists" do.'* In relation to the attack in Fort Hood, Greenwald (2009) argues that: 'Hasan's attack was carried out on a military base, with his clear target being American soldiers, not civilians . . . Can an attack on soldiers – particularly ones in the process of deploying for a war – fall within any legitimate definition of "terrorism"?'

So how would you define these attacks? Similarly, how would you define Western activities, such as drone or missile strikes, in overseas territories? Is the 'who' (in terms of 'aggressor' and 'victim') and the 'where' the most important element in defining 'terrorism'? What is at stake when activities are defined as 'terrorist', 'criminal', 'civil', or 'acceptable'? At second glance, therefore, the power to define activities in the cultural world is critical. It positions individuals in relation to particular versions of 'right' and 'wrong', and legitimises particular responses as a consequence.

Dominating power and hegemony

Dominating power thus works to instil its own cultural values on the traces of society. These values are not 'true' or 'natural', but groups attempt to persuade us that they are. As Jordan and Weedon (1995: 5) explain:

> It is through these [dominating traces] that we learn what is right and wrong, good and bad, normal and abnormal, beautiful and ugly. It is through them that we come to accept that men are better leaders than women, that Black people are less intelligent than Whites, that rich people are rich because they have worked hard and the poor are poor because they are lazy.

If those with dominating power convince us that their cultural values are 'natural' they can be said to have achieved '*hegemony*'. Hegemony is the dominance of one cultural group over another, acquired through a range of processes including coercion, regulation, as well as the willing acquiescence of citizens, even though this acquiescence may be against their interests. Hegemony thus refers to a situation where it has become accepted that this cultural group will dominate over all others, it is ordinary and natural that this hegemomic group 'rules okay'. In these cases, a hegemonic group has been successful in transforming its own values and beliefs into the ideological order of its culture. Do you live in a cultural hegemony? Can you imagine an alternative to the current system of cultural control? Do you always do what you are told to do by those with dominating power (see Box 5.7)?

Disobedience and transgression: stepping out of line

As we all have the ability to act, we all have a degree of power. Although dominating groups use persuasion and punishment to ensure conformity with cultural (b)orders, some people don't always do what they are told and use their power to step out of line. This process of stepping out of the cultural and geographical (b)orders constructed by a dominating group is known as 'transgression'.

You committed a transgression when, intentionally or otherwise, you didn't keep off the grass, you overtook across chevrons, or you jumped the queue. On all these occasions, you literally and metaphorically crossed the line, the cultural and geographical (b)order. By doing so, whether you did it intentionally or by accident, you raised the possibility of another way of being. By crossing the line to unorthodoxy you made doxa and hegemony visible for what it was – not the *only* way of being, but simply *one* way of being. You put a question mark in the minds of all those who witnessed your actions – why did they do that? How can they get away with that? If they can, why can't I? Through the act of transgression the 'natural' (b)order became reduced to a political order rather than the only way of life.

BOX 5.7 DO YOU ALWAYS DO AS YOU ARE TOLD?

Dis/obey

Source: Anonymous

From early childhood we are taught to obey, to do the 'right' thing. We assume this is for our own safety and the well-being of society at large. Even so, we'd like to believe that we wouldn't obey just any order. Wouldn't we? The following excerpt is taken from a Channel 4 documentary called *Five Steps to Tyranny* which shows we often obey, sometimes even without thinking.

> *Vox Pops:*
> 'Do you always do what you're asked?'
> 'Nah!'
> 'Hmmm, no, not always.'
> 'No.'

As the narrator of the documentary comments: 'A famous psychology experiment in the 1960s by Stanley Milgram demonstrates how wrong these people are. We often obey with little questioning.'

The experiment ran as follows. A man on a train asked fellow passengers whether they would mind giving up their seat so he could sit down. This was a novel request as there were many spare seats nearby, and the man was in good health. All passengers queried the oddness of the request, but one in two passengers (50%) did what they were asked and gave up their seat. The experiment was conducted for a second time, but on this occasion the man was accompanied by an actor in a police-style uniform. When this 'authority figure' was present, every single passenger did what they were asked (100% obedience). Professor Philip Zimbardo (Dept of Psychology, Stanford University) commented on the experiment as follows:

> Many of us have the tendency to act without thinking. That's when we have danger, when we can get people to act on these impulses, on these conditioned habits, without stopping to say, '*Well, why? Why should I do this?*' In the course of human history there have been more crimes committed in the name of obedience that in the name of disobedience. So it's not the disobedient, it's not the rebel, it's not the unu-sual kind of deviant person who is the threat to society, the *real* threat to all societies are the mindlessly, blindlessly obedient people who follow any authority.

> (in McIntyre, 2001)

If any cultural group achieves hegemony then dominating power can become a dangerous thing. Although the above example is trivial in nature, when citizens begin to act without thinking, they may begin to impose and perpetuate injustice, even if they assume they are doing the 'right' thing.

One example of transgression was a sticker campaign mounted on the Underground metro system in London. These stickers parodied the (b)ordering signs of London Transport and introduced alternative codes for behaviour (see BBC, 2012b).

The act of posting these stickers was itself transgressive – it is illegal to add traces to Underground trains without authorisation. However, by suggesting alternative forms of behaviour these stickers also offered commuters the opportunity to think about the signs which (b)order their existing behaviour on the Underground, and remind them that other forms of practice are possible.

Turning borders into orders, or lines into question marks, thus occurs through acts of transgression. Transgression raises questions about who is in control; it questions authority and 'represents an opportunity for thinking differently about something we might otherwise take for granted' (Sargisson, 2000: 104). Transgression opens up places to alternative futures, enabling the possibility of thinking and acting differently. Transgressive acts are therefore dangerous for those with dominating power. Transgression bends, blurs, or breaks the (b)orders holding their hegemony together, with the power to transform no longer clearly held by one group alone. Transgression can therefore set in motion significant changes to how places are taken and made (see Box 5.8). As a consequence, its transformative potential is often intentionally harnessed by many groups within society. These intentional acts are known as resistance.

Figure 5.1 Re(b)ordering the London Underground

Source: Author

BOX 5.8 ACTS OF RESISTANCE

Fuck authority
Source: Author

Road closure
Source: SchNEWS

Bodily resistance
Source: SchNEWS

Saddam statue
*Source: Brainyskeeta www.flickr.com/photos/
brainyone/343784711/*

In each of these images, places are taken and made to resist the impositions of dominating power. Be it through graffiti, street parties, physical resistance, or monument destruction, the authority of dominating power to order and border place is contested. In the first image, graffiti on a Cardiff monument commemorating military victory is defaced with a slogan that says 'Fuck Authority!' This slogan, written illegally on this monument, openly flouts the conventional (b)orders of this place. Who is in control here now? (The resisting power of graffiti is outlined in more detail in Chapter 11.) The second photo highlights a street party in the UK where the normal use of place – for cars and mobility – is resisted by campaigners who shut the street down to traffic and chose to dance there instead. In the third picture a single protester physically resists police at an anti-nuclear demonstration in Germany, using her body to inconvenience dominating power and the way it seeks to use place. Fourthly, US soldiers destroy a statue of Saddam Hussein in Baghdad. Here the dominating power of Hussein's regime is symbolically destroyed by the resisting power of the US military.

Resisting power

Resisting power is that which seeks to intentionally oppose, challenge, and dispute acts of domination. Resisting power seeks to transform the traces of dominating groups and dismantle their cultural orthodoxy. Acts that reform, reconfigure, or reject the traces and places of cultural orthodoxy come in many different forms. Resisting power can involve

> very small, subtle and some might say trivial moments, such as breaking wind when the king goes by, but it can also involve more developed moments which discontent translates into a form of social organisation which actively co-ordinates people, materials and practices in pursuit of specifiable transformative goals.
>
> (Sharp et al., 2000: 3)

Power as resistance therefore comes in small, subtle forms, such as not stopping at a red light or not clearing up after your dog. Drinking in class, putting your feet on the chairs, even sitting on tables, can be seen as intentionally breaking cultural and geographical (b)orders. Even though the absolute, concrete effects of these activities may be marginal, they should not be overlooked as forms of symbolic and direct confrontation with dominating power. They all seek to take and make place in a different way than that intended by the dominant.

Resistance can, of course, be more organised than this. The value and meanings of places can be subverted and challenged through resisting groups taking and making place through demonstrations, rallies, coups, or revolutions. There are a range of examples illustrated above.

Resistance and domination are therefore both acts of power. How we define these acts is, on the face of it, a straightforward process. Dominating power has the capacity to transform our behaviour; resisting power seeks to prevent this transformative potential. However, in many ways any act of power contains both these elements: it is *both* transformative *and* restrictive – it not only seeks to stop the imposition of one cultural (b)order, but also impose its own. The same act of power can dominate some groups and places, but at the same time act to resist the ideas and manoeuvres of a competing group. Acts of power are thus both progressive and repressive, they seek to liberate but also impose. It is important, therefore, to diagnose the ways in which one act of power dominates, and the ways in which it resists. What cultural (b)orders does it seek to challenge, and which ones does it seek to impose? Let's undertake this diagnosis for some of the photos depicted in Box 5.8.

In the case of the street party image above, the campaigners shown have exercised their power to take and make the street for a carnival. Through so doing they are resisting the dominant use of place – for cars and traffic – and seek to impose their own (b)orders which make manifest their own cultural ideology: streets should be for people not for cars. In the fourth image, the US soldiers dismantle a statue of Saddam Hussein. Through this exercise of power the soldiers are thus at once resisting the cultural ideology and dictatorial regime of Hussein, but at the same time imposing a new ideology which

BOX 5.9 FREEDOM OF SPEECH AS BOTH DOMINATION AND RESISTANCE

Right to speak

Source: Author

As Article 10 in the Federal Constitution of Malaysia outlines, 'the right to freedom of speech, assembly and association' is guaranteed. The First Amendment to the United States Constitution has a similar guarantee, prohibiting the making of any law which abridges freedom of speech, interferes with the right to peaceable assembly, or prohibits the petitioning for governmental redress of grievances. Freedom of speech can thus be seen as a basic tenet of democracy, allowing all citizens the right to voice their opinion, whether or not this challenges or resists those with dominating power.

Zucotti Park, New York, 2011. In this place many citizens exercised their right to freedom of speech and assembly. These 'Occupy Wall Street' protesters sought to draw attention to the economic inequality produced by capitalism, and the role of the banking sector in influencing the nature and (in)equity of the economic system. From the images, identify the different types of power expressed in these practices. Is it straightforward to identify acts of domination, and acts of resistance? What is the role of place in these practices?

In some senses, the 'Occupy' activists are expressing their rights of assembly and speech. However, what happens when one person's rights infringe another's 'right' to access and mobility? As the then Mayor of New York, Michael Bloomberg stated: 'People have the right to protest, [but] what they don't have a right to do is to deny other people free access and the right to "not protest" – the first amendment works both ways' (Reuters, 2011). On these occasions, an act of resistance can also have an element of domination within it – it seeks to impose its will and ideology on the actions of others. In the same way, the actions of the police could be seen as dominating in nature, seeking to minimise and dispel resistance protests (often through violent means). At the same time, the actions of the police also resist the uprising in order to maintain the orders and borders for which citizens have voted through electoral democracy. Diagnosing cultural acts of power is thus not a straightforward issue of 'good' versus 'bad', or 'right' versus 'wrong', but a complex exercise in the relations between domination and resistance.

Occupy

Source: Anonymous/Author

in time will construct new (b)orders in line with Western values.

How individuals and groups use their power to take and make place is the very stuff of cultural geography. How different cultural groups exercise their power, what (b)orders they seek to impose and resist reveals much about their own ideology and the places they seek to make. Despite attempts to impose cultural ideology and conformity within a place, the nature of power means that some groups will always cross the line and practice transgression and resistance. Further questions thus arise concerning how actors deal with those who openly resist cultural ideology. Is it possible, for example, to maintain cultural (b)orders and allow free speech? What trace-chains are set in motion by transgression and resistance, and what is the consequence for the ongoing composition of place?

The struggle for place: domination, resistance, and trace-chains

The ongoing competition between acts of power are vital to consider in cultural geography. The tactics used by both sides – tactical traces if you will – often reveal much about the cultural ideologies of the groups involved, and contribute significantly to their respective success and failure. For example, in some cases one group may use oppressive acts to deal with opposition (for example, the use of martial law, violence, or prison). For them, these tactical traces intend to re-establish order in the face of a danger-ous threat. However, such traces may have the unintended effect of generating sympathy and support for their opposition who are seen as sub-ject to draconian responses from a heavy-handed authority. (Think for example of the emotions stirred by any bully picking on an underdog, sol-diers attacking children, or police using violence against peaceful protesters).

Alternatively, groups may choose not to retaliate in the face of intentionally transgressive activity and allow resistance to take and make place in unorthodox ways. Ignoring resisting activity may seem a sign of weakness on behalf of the dominating group; however in some cases such disregard can pacify its challenging threat. Such a tactic sees resistance activity in line with the notion of the '*carnivalesque*'. The carni-valesque involves the, 'temporary suspension, both ideal and real' of the conventional (b)orders of cultural society (Bakhtin, 1964/1984: 10). In the carnival dominating power *licenses* dissent, therefore effectively turning a blind eye to some deviancy, in order that control is maintained in the long term. Employing such a strategy there-fore says something about how the dominating group frames the nature of resistance, but also their own power. Perhaps resistance is seen as a necessary homeostasis mechanism for a healthy society – far from it amounting to a wholesale challenge to the dominant (b)orders it may simply be a way of letting off steam before its instigators return to life-as-normal. However, adopting such a strategy involves clear risks for the dominating group. As Shields (1991: 91) points out, carnivals have the potential to be far more than 'steam release' valves, they offer the 'the enactment of alternative, utopian social arrangements' which may sow the seed for further transgressive traces elsewhere and when.

Diagnosing the ongoing competition between acts of power is therefore a complex but crucial exercise in studying the struggle for place. Trace-chains can be set in motion by acts of domination and acts of resistance that are difficult to predict, with unintended and unforeseen outcomes often occurring. The unpredictability of trace-chains is omnipresent, as Foucault famously states, 'People know what they do; they frequently know why they do what they do; but what they don't know is what what they do does' (1984: 95). Who could have predicted, for example, the trace-chains set in motion by the actions of the Arab Spring and Occupy Wall Street? (See Box 5.10.)

BOX 5.10 THE TRACE-CHAINS MOBILISED BY AN ACT OF RESISTANCE: CONNECTING THE ARAB SPRING TO OCCUPY WALL STREET

The Arab Spring was a popular uprising that occurred across North Africa and the Middle East, beginning in 2010. This uprising entailed populations of countries from Algeria to Yemen seeking to overthrow their dominating rulers. The Arab Spring had a catalysing effect on many people in the Middle East, prompting them to realise that another future was possible other than that offered by the dominating powers of their nations. As Palestinian writer Raja Shehadeh states, the Arab Spring caused 'the psychological barriers [in the population] to fall, and when the psychological barriers fall, *everything* is possible' (2012). The possibility that (b)orders could be questioned and doxa overthrown prompted a wave of 'trace-chains' that could be seen to mobilise a range of protests, including the Occupy movement in Europe and America. As Addley (2011) argues, 'Occupy Wall Street [fully] acknowledges its debt to the Arab Spring'. In turn, the Occupy movement itself continues to prompt further resistance activities; as Pickerill and Krinsky reflect: 'to many [Occupy Wall Street] was the moment when resistance to the inequalities of capitalism finally emerged: a tipping point in which the unfairness of bank bailouts juxtaposed against rising personal poverty triggered a moment of clarity of the absurdity of the current economic and political system' (2012: 279). At the time of these cultural practices it wasn't only nations that 'occupied' territories (like Israel in Palestine, for example, see Chapter 6), it was people (this is *our* street, see Box 5.9), or to be more precise, it was a different group of people than normal determining the present. (In the terminology of the Occupy movement, it was the 99%, not the 1%.). However, are the Arab Spring and Occupy examples of revolution or simply the carnivalesque? What further trace-chains will they set in motion, and how will they influence the ongoing composition of places in our cultural world?

Conclusion. Power in and through place

We have seen in this chapter how crucial the practice of power is to cultural geography. It is through cultural acts – the exercise of power – that places are taken and made. Geographical places and cultural acts are the tangible expressions of power, and through these trace-making exercises, and the meanings attributed to them, cultures take both shape and place. Power can be exercised by all individuals, yet the power to transform place is of paramount importance. In some cases power can dominate – transforming the actions and places around us, in others power can transgress or resist, attempting to challenge the ideology of those controlling our cultural world. Through acts of domination and resistance, traces, places, and meanings are fought

over. Traces that were once 'natural' can be questioned and contested. Similarly, traces that were once deemed as 'novel' can become part of the mainstream. Power thus is at the root of cultural geography, creating, stabilising, and destroying orders and borders. Through so doing, power endlessly shapes the ongoing composition of traces that places come to be.

This book proceeds to interrogate traces and places through a number of examples. In each chapter we will look at different cultures and geographical contexts and interrogate the ways in which traces come together to define places and peoples. In the next chapter we explore the different scales at which we act and through which we form our identity and belonging. In short we explore how our sense of place can be formed by allegiances to different geographical scales.

Suggested reading

The following texts are useful to explore the issues raised by this chapter.

Books

Gramsci, A. (1971) *The Prison Notebooks*. Lawrence and Wishart: London.

Gramsci offers an advanced but important introduction to ideological power, change and critical theory.

Pile, S. and Keith, M. (eds) (1997) *Geographies of Resistance*. Routledge: London.

Sharp, J., Routledge, P., Philo, C., and Paddison, R. (eds) (2000) *Entanglements of Power: Geographies of Domination/Resistance*. Routledge: London and New York.

More accessibly, Pile and Keith provide an excellent introduction into notions of power related to resistance, whilst Joanne Sharp et al. offer a sophisticated account of how power is 'entangled' through acts of domination and resistance.

Scott, J. (1985) *Weapons of the Weak: Everyday Forms of Peasant Resistance*. Yale University Press: New Haven and London.

Jordan, T. (2002) *Activism! Direct Action, Hacktivism and the Future of Society*. Reaktion Books: London.

Scott's 1985 book gives interesting empirical examples of how resistance can be practised in everyday, mundane ways, whilst Jordan outlines more spectacular forms of resistant practice and the potential utopia these promise.

6

Senses of place: scales and beliefs

Introduction

As we know, the identity of self and the identity of place are closely connected. In this chapter we explore how our sense of place can be formed by allegiances to different geographical scales. As we have seen in Chapter 4, attachment can be generated at a range of geographical scales, from the local to the national. In this chapter we will examine the range of scales we may identify with, how that identification is generated, and to what end. We will see how our sense of place is generated at each scale, and how these senses in some way coincide, but in some ways conflict with one another. In the second half of this chapter we examine senses of place relating to belief systems. We explore how these identity positions synergise or compete with our more 'scalar' senses of place. The combination of religious and scalar affiliation has profound affects for how we consider our sense of place both culturally and geographically. Can we have multiple senses of place that sometimes contradict one another? Is our 'place' fixed and solid, or dynamic and fluid? We look for answers to these questions by exploring ideas of multiple identities and individualisation. We conclude by considering how these ideas can affect how we think about our senses of place, and what influence they have on our actions in the world around us.

A sense of place: my (local) backyard

We form a sense of our-selves and a sense of our-places at a range of scales. The local scale is perhaps the most immediate and the most obvious, and it is at this level that we begin. At the local scale there are a range of possible (b)ordering mechanisms that define both our-selves and our-places. Do you, for example, identify with your electoral constituency? Perhaps with your post- or zipcode? These are types of geographical (b)ordering that give important meaning to our lives. Electoral constituencies determine who we vote for, what political party represents us at the local level, and, by extension, the national scale. Post- or zipcodes determine our place within a national house ordering system, my street gets its own postcode, demarcating my 'manor', and those of my neighbours, from others, so our post arrives in the correct place. However, I would find it difficult to attach myself to these (b)orders in any meaningful way *beyond* these administrative functions. It may be different if I were a postman or a councillor, but these (b)orders leave me cold; there is no 'sense of place' for me, and my heart and mind, here (for an alternative perspective, see Box 6.1). There are, however, types of (b)ordering mechanism that lie at the local level that give me a stronger sense of self and place than those imposed on me by electoral administrators or postcode gurus. These (b)orders are those that I take and make myself through my everyday actions. In my neighbourhood, my work, and my local leisure places I leave traces through my daily involvements and actions. In these places I not only contribute to their definition and meaning, but concomitantly generate my own sense of who I am and where I belong. A short example will explain what I mean.

A personal case: the backyard of my youth

I remember as a child the routes I used to make on my bike, cycling to friends' houses, or walking to school, knowing which alleys I could go down, or whose gardens I could sneak across. Through creating these *routes* I put down *roots* in these places, through my practices I took and made these places as my own, albeit temporarily. These places were my travel lines, my mobility, my dens and haunts, rather than simply an abstract alley or garden. In this way they came to be parts of my life, and still in a way are. For example I remember events and incidents through remembering places in which they happened. When I go to visit my parents now, it is as if I can revisit my old self through going back to these places, even if they have changed in appearance and function during the intervening years. These places, then and now are my manor, my backyard.

Through these actions I therefore trace myself into my backyard, and my backyard into myself. The intimate, everyday interaction with the local generates traces that tie us into this sense of place. Having a coffee in your local café (see Laurier and Philo, 2006; Latham, 2003), going to 'your local' pub, competing in the local leagues, going to the town carnival or village fair: these practices build up a (local) sense of who we are. These are literal senses of a geographical location – they are the smells, sights, textures etc. that give us senses of our own place in a cultural order, as well as contributing to the definition of the place in which the events occurred.

As we saw in Chapter 4, Casey argues that the relations between self and place are not simply connected, but there is a 'constitutive co-ingredience' between our cultural and geographical identity. At the local level, Casey suggests that this co-ingredience is generated through processes of 'incoming' and 'outgoing'. These processes occur when the human body physically encounters our local places and inscribes traces of location on the human self by laying down 'incoming' strata of meaning (see Casey, 2001: 688). According to Casey, over time, this process also influences the meaning of places ('outgoing'). Continual practice sediments meaning onto places, with personal memories meshing with cultural meanings on an individual and (potentially) societal scale.

From family contacts to friendships and gang memberships, from local gossip to the local newspaper, everyday practices build up a sense of local place. This direct, often face-to-face, form of interaction is unique to the local scale. Only at the local scale can people share a personal affinity, only through everyday contact can extended families come to know people and generate a trust and social capital (see Putnam, 2000). At other scales it is harder to generate senses of place through such means. For example, sensing your place at the national scale is a very different process. Even in these times of mobility and mass communication, at the national scale you can never know everyone. National senses of place are therefore not based on face-to-face contact in the 'local' way; alternative (b)ordering mechanisms are needed to re-place us.

National senses of place: patriotism

Patriotism is not commonly considered an example of a sense of place. But why not? Patriotism may indeed be the most fervent sense of place now in existence. It is fervent, it can be passionate, because it can draw on all the inarticulate feelings that we have for

BOX 6.1 ZIPCODES AND GANG IDENTITIES

Everyday activities at the local level build up identification and belonging between people and place. An example of this is the ways in which gangs affiliate with particular local neighbourhoods in cities, often directly connected to post- or zipcodes.

Coding co-ingredience
Source: Author

Across the world, zipcodes or postcodes are a means through which cultural groups identify with a particular location. Gangs are well known in Baton Rouge, Louisiana (WBRZ, no date), and Compton, Los Angeles (Curtis et al., 2014), but in many cities identification with code location is key to belonging and identity. In the above images from Cardiff, UK, we see postcodes traced onto the material environment, claiming the area for gangs from CF11 and CF14. We also see the potential for conflict over territorial location between gangs, as individuals from CF5 have created a palimpsest or stratigraphy of traces through overwriting their own postcode on top of the CF14 tag. As Wright (2007) notes, 'For most of us postcodes mean nothing more than a bunch of numbers that indicate where we are or where we want to go. But to a new generation . . . postcodes have literally become danger-zones, marking out no-go-areas.' As the BBC report, in London teenagers marked as 'E5' or 'E9' (postal districts in the east side of the city) risk being attacked for straying into each other's area. Simply crossing to the other side of a street which borders two postcodes could end in violence (BBC, 2007). In these ways, particular cultural groups not only identify with particular local areas, in turn local areas come to be identified with particular cultural groups.

our house with the picket fence and augment them with articulated images, ritual, rhetoric, and music.

(Tuan, 2004: 47)

We often consider our sense of place at the national level. It is something that, in the vast majority of cases, we are born into, and thus to an extent national identity becomes a 'natural' thing. However, sensing our place at the national scale is anything but natural. According to Benedict Anderson (2006), our sense of place at the national level is 'imagined'. Nations are 'imagined communities' because it is impossible to know, meet, and talk to all the citizens at a national scale in the same way as it is in a village or neighbourhood. National unity and fraternity therefore needs to be imagined, rather than directly lived. Although nations may need to be imagined, they are not in any way fake or false – their (b)orders are materially defended and their idea(l)s lead many to kill or die for them. So how is a nation imagined into being?

> We tend to see national identity as natural almost like the air we breathe, when in fact it is a very recent phenomenon in human history. For most of history people identified with much smaller cultural communities than modern states . . . they identified with their local village, their parish or their city.
>
> (Anderson, 1992)

Specific (b)ordering mechanisms were introduced in the mediaeval period to re-scale our sense of place away from the local and towards the national. Chapter 10 will examine some of the ways in which a national sense of place has been generated in both historical and contemporary times (e.g. through genocide, tight control, or partnership), whilst Chapter 12 will analyse nation building attempts through the use of language. Other (b)ordering mechanisms include inventive re-placements for the extended family into the everyday life of 'subjects' (for example through the establishment of a monarch or head of state on stamps and currencies). Belonging is also established through the creation of flags (directly so in the case of the *Union Flag* uniting Scotland, Wales, England, and Northern Ireland together in one symbol of togetherness, or the *Stars and Stripes* uniting the states of America), as well as through music (e.g. national anthems), invented rituals (such as patriotic days and cultural festivals, e.g. the Last Night of the Proms in the UK), and sporting contests. In some places, patriotism is not simply encouraged through these emblems and rituals, but forcibly imposed. In Thailand, for example, the national anthem

is played twice a day, at 8am and 6pm. At these times it is required that all individuals stop what they are doing and stand still to respect the flag and the community it represents. A 'Patriotism' Bill was even formulated in 2007 that required motorists to cease driving when the anthem was played. As was reported at the time,

> The Bill's supporters say road traffic should stop nationwide when the anthem is played during the raising and lowering of the flag 'to preserve tradition and instil patriotism in Thais'. 'The national anthem lasts only one minute and eight seconds, so why can't motorists stop their cars for the sake of the country?' retired General and NLA member Pricha Rochanasen said.
>
> (ABC News, 2007)

Thailand is therefore a clear example of how a sense of national place is projected into the everyday lives of its citizens. The dominating power of the state and the police services seek to (b)order the times of 8am and 6pm (for 78 seconds in duration) to make each individual respect the nation and forge a fraternity, equality and sense of collective belonging. Although Tuan (above) considers that the construction of a sense of place at the national scale is the most fervent in existence, it is up to you to consider how successful the (b)ordering mechanisms of patriotism are. Perhaps your sense of place at the national scale is strong, you are a flag waver and feel a fierce loyalty towards to your country? Perhaps your allegiance is mild, experiencing only an occasional sense of place through common humour, musical taste, or sporting competition, for example?

A sense of place at the national scale can thus be imagined into being in a number of ways, but is often most explicit when a nation is under threat. As we will see in forthcoming chapters, threat from the 'other' – from cultural ideas and practices that are not our own – can threaten our sense of place, and even our place itself.

BOX 6.2 RE-IMAGINING NATION

If nations can be 'imagined' into being, imagine you were ruler of a new country: what would you do to generate a sense of place and belonging in your territory? Would you invent everyday practices to insinuate scales together (for example, through flags, a currency, a national costume, a greeting, or a language?) Would you invent rituals and anthems, or a special day of celebration? What would you celebrate on these occasions – a collective past, or a hoped for future? In terms of creating belonging would you attempt to unite individuals against a collective enemy, or for something that binds them in a positive sense? Would you realise this through inventing a competition to pit different locals or nationals against one another (for example in the fictional *Hunger Games* (see Ross, 2012), or the modern Olympics?) Furthermore, would you ask for payment from your citizens to be part of your new 'nation' (through membership fee or tax)? What would the conditions of entry be (birthright? familial connection? biological characteristics? sworn allegiance?) What can we learn from existing countries? What traces have been invented to imagine other nations into being?

In Wales, national days are celebrated in the mainstream media, creating an expectation that individuals should join in and participate in the ongoing composition of their nation by wearing collective emblems (for example, 'get your leek on' see Capital FM, no date). In Peru, a debate occurred to change the national anthem in order to encourage a more positive association with the country. The anthem's words, 'Condemned to cruel servitude . . . [the Peruvian] quietly whimpered' were deemed to be too depressing to instil nationalistic fervour (see Carroll, 2011). What would you do to imagine a positive sense of nationhood into being?

During war-time, or even pseudo-competition through sporting tournaments, feelings of pride, loyalty, and belonging are intensified. Indeed it is at these times that local and national feelings can coalesce to form one 'nested' sense of place. Here a coherent scalar identity is formed *for* a nation and *against* those that would be in direct competition with 'us'. It is at these times – when the (b)orders between competing groups are made explicit – that you come to realise which side you really are on. This is the case in Belfast.

Do good fences make good neighbours? (after poet Robert Frost,[1] in Harris, 2005)

Belfast is a place profoundly shaped by competing nationalistic senses of place. Physically part of the island of Ireland, but politically connected through colonisation to Britain, different communities within Belfast demonstrate their loyalty to one of these competing cultural groups through where they are, who they speak with, and what they do in this city.

> Belfast's . . . geographies are clearly marked visually in the landscape, most obviously through flags, and colour coding of kerbstones and lamp posts[2]. . . For inhabitants of the city, it is almost impossible not to know where you are (or at least make an educated – or bigoted? – guess), with cultural, political and religious markers ranging from the obvious to the extremely subtle.
>
> (Reid, 2005: 488; see Figure 6.1)

Street murals in Belfast are one clear example of traces of belonging. Each seeks to outline historical events and national allegiances that bring the past and the patriotic into these local places. Such traces designate and reinforce a sense of collective belonging or un-belonging, depending on which community you are in.

Figure 6.1 Traces of belonging in Belfast

Source: 1 and 2 – Mike Gregory www.flickr.com/photos/g_travels/2080533976/
3 – Ian Cowe www.flickr.com/photos/iancowe
4 – citysoundslikecolour www.flickr.com/photos/tachismes/264040428/

There are a range of different (b)ordering mechanisms in Belfast which categorise, define, and divide senses of place. Actors leave traces in places to mark their turf as both local and national(ist). As Morrissey and Gaffikin note:

> the deliberate use of flags, emblems, graffiti and wall murals that characterize contested cities serves not only to demarcate against the outsider, but also to affirm the faith of the insider . . . [H]istorical narratives and symbols are summoned to support nationalist projects (Mann, 2000; Comaroff and Comaroff, 2000). In contested space, these serve to simultaneously proselytise to the believer, while inciting the infidel. Thus, 'the political culture intentionally invests cultural landscape with contentious ideological messages – obvious and subtle mnemonic and didactic devices that remind citizens of their loyalties and responsibilities' (Keirsey and Gatrell, 2001: 2).
>
> (Morrissey and Gaffikin, 2006: 875)

Local and nationalist senses of place are thus traced into the landscape of Belfast through murals, flags and graffiti. These traces remind citizens which side of the national divide they are on, they persuade certain forms of action and behaviour, and perpetuate (b)orders that divide the local community.

Perhaps the most obvious of these divides are the security walls that separate the two communities (see Figure 6.2). These borders physically and psychologically divide peoples, creating senses of place for insiders and outsiders. Such divides therefore create both external and internal ghettos for the inhabitants of Belfast (see O'Hara, 2004). A 'psychology of spatial confinement' (after Reid, 2005: 488) has been established that limits physical movement, but also communication and attitude towards difference. Shirlow's research on neighbourhoods in Belfast (see Brown, 2002; Shirlow, 2005) demonstrates that division exists on a number of levels. According to Shirlow (2005):

- 86.3 per cent of housing areas in Belfast are segregated (i.e. with greater than 61 per cent of inhabitants from one cultural community);
- 68 per cent of 18- to 25-year-olds living in Belfast have never had a meaningful conversation with anyone from the other community;

- 72 per cent of all age groups refuse to use health centres located in areas dominated by the other group;
- only 22 per cent will shop in areas dominated by the other group, whilst 58 per cent travel twice as far as they have to to shop, or to go to a leisure or health centre;
- 62 per cent of unemployed people refuse to sign on in their local social security office because it is in an area dominated by the other group.

Such physical and psychological separation leads to an entrenchment of pride and loyalty towards one community, and fear and distrust of the other. This often erupts in violence, even after the Irish Republican Army's (IRA) ceasefire in 1994 (see Shirlow, 2005; Hill, 2002).

Summary

In the case of Belfast it is unclear how far good fences actually make good neighbours. Competing imaginaries of nationhood conflict in Belfast, and physical (b)orders work to compound the political difference and distrust on

Figure 6.2 Belfast's (b)orders
Source: Susan Moore

both sides. Here the imagined communities of nationality are perpetuated and compounded by those of locality to create a strong sense of place that is exclusionary. By imagining then creating a (b)ordered community with a 'common interest', there are those that are 'in place' but also those that are 'out of place'. Such imagined communities often breed the stereotyping of 'others' as strangers (see Wren, 2001). These synergistic scalar senses of place thus create 'parallel lives' within Belfast (see also Chapter 10), where a city is mixed but its population is not.

Thus we can see from this example that a dual sense of place can be experienced; identity can be sensed at both the local and national scale. In the case of Belfast, the sense of place at the local scale reciprocally combines with senses of place at the national scale to produce a synergised and strengthened 'nested' sense of identity. In some cases, however, these dual senses of place can come into conflict with one another. In these situations, loyalty to a locale can be put in tension with loyalty to a national scale, as we will see in the next section.

Local v national: NIMBYism and the 'national interest'

In many cases local and national senses of place are in conflict. The 'local interest' seeks to defend the ongoing composition of traces that is their locality, yet the 'national interest' often means that these places are redefined by traces

---

BOX 6.3 BIRDS NESTS, HOMES, AND OLYMPIC DREAMS

In August 2008, China hosted the Olympic Games. Through television coverage the Games gave China the opportunity to project its national sense of place into the world's living rooms. However this opportunity came at a high local cost. An estimated 1.5 million people were forcibly evicted from their residences for the construction of the main stadium (the Bird's Nest) and its related projects (although Chinese official statistics put the figures at 6000 (see Beck, 2007)). Eviction for the Games and the good of the country did little to pacify the sense of loss locals suffered about their dis*place*ment. This loss moved many to extreme measures, as one resident stated,

> They might demolish the house at any time. I feel very scared in my heart – scared and lonely. If they come by force I will try my best to negotiate with them. If they don't negotiate with me, I have no other way. I can only pour petrol [around me] and be burnt together with the people who have come for us. I can't let them do this – take my things so easily and cheaply.
>
> (Su Xiangyu, in Branigan, 2008)

> 'Developers shouldn't use the Olympics to take our homes' (Ma Xiulan); 'We don't oppose the Olympics. But it's wrong for them to demolish our house. It's wrong.'
>
> (Liu Fumei, in ibid.)

As these quotations illustrate, the connections that tie these people to their local place are strong. Although the sacrifice of their homes is in the national interest, their local connections remain stronger than those which tie them to the imagined community of the nation. So much so, that they would rather resist the dominating power of the state and sacrifice themselves than live without this connection.

imposed for the good of the nation. In such instances, the orthodox approach to Not In My Backyard campaigns (or NIMBYism) is to dismiss or deride them. As Blacker (2008) notes, it is natural to dismiss NIMBYs as 'irresponsible and self-protective, [as] enemies of progress and the national interest'. Yet we can see that this approach seeks to forward the aims of the nation, rather than the attachment to the local that places us in our everyday life. To repress NIMBYism is to progress the cultural values of the nation at local's expense. This was the case at the Beijing Olympics (see Box 6.3).

> Geography . . . is finally knowledge that calls up something in the land we recognise and respond to. It gives us a sense of place and a sense of community. Both are indispensable to a state of well-being . . .
>
> (Lopez, 1998: 143)

So far in this chapter we have seen how senses of place can be generated at different geographical scales. From everyday practices in local places, to invented traditions at the national scale, senses of place effectively tie us into a set of communities and belonging, they make these places ours. The need to be part of these scalar communities is strong (as Lopez suggests, above). However, it is worthwhile considering how these scalar attachments interact. In some cases the local and national levels coincide, strengthening our loyalty to place. In others they conflict to put our identity and loyalty in problematic relations. In such instances we are placed in tension, we become the (b)order zone: how do we decide between our place in a locality and our place in our nation? How do we decide between our place in a locality and our place in our nation? How far would you go to defend your local place from the national interest? Do your local attachments fade in the presence of national goals?

Values and passports

We can see that in some cases our senses of cultural and geographical place are not always coherent. This is especially the case when we add into this mix senses of place that do not come directly from geography as such, but rather from cultural ideas and beliefs. These identities or communities of interest (rather than communities of territory in the strictest sense) come in the form of race and ethnicity (as we will see in Chapter 10), but also through other cultural ideas such as religion or value system. Attachments generated through these cultural tracings often serve to (con)fuse our senses of place even further.

A brief and perhaps trivial example serves to introduce what I mean. The relationship that football clubs in Britain have to their locality is becoming increasingly (con)fused. Through globalisation they have long ceased to fully and exclusively represent their local place. Clubs field players that haven't been born or brought up in their local backyard, rather they buy the best players they can to represent their club and their fans. Recently in the English Premier League, many clubs are not only failing to field players from their local backyards, sometimes they have no players from the national backyard either. This situation occurs frequently at clubs like Arsenal. Traditionally (b)ordering the Woolwich then Holloway area of London, now the club represents something different, as current manager Arsene Wenger outlines, 'We represent a football club which is about values and not about passports' (BBC, 2006b). At Arsenal, individuals are still signing for a club that is rooted in a location, but the associated senses of place that go along with place are changing. In this case senses of place are not bounded by roots or geographic attachments, rather connections are made through shared values rather than scalar origins. The situation raises a key question: how do 'values' combine with 'passports' to give us a cultural and geographical sense of place?

BOX 6.4 CARDIFF CITY: A LOCAL COMMUNITY OF PLACE OR A GLOBAL COMMUNITY OF INTEREST?

As we have seen from the Arsenal example, football clubs have traditionally been one means through which local people can identify with their area. Through local players, home kits, songs, and shared histories – all tied into a particular local place – football ties people into places. Increasingly however, football culture is being integrated into global geographies, and in this process, communities of place are sometimes being turned into communities of interest.

The case of Cardiff City represents an example of this. In 2010, Malaysian businessman Vincent Tan took over the club, which then went on to secure promotion to the English Premier League. In 2012, however, Tan 'rebranded' Cardiff City and changed their home colours from blue to red, and also considered a change in their nickname (from the 'Bluebirds' to the 'Dragons', see BBC, 2013). Through doing so, Tan wanted to make the club more appealing to potential supporters in Asia, where red is seen as a colour of dynamism and success. In opening up the club to global appeal, one can also suggest that Tan wanted Cardiff to become the 'national team' of Wales. As Dato Chan Tien Ghee, the Cardiff City chairman put it, the rebrand created 'a symbolic fusion with Asia which will allow us to fly the Welsh flag on behalf of Cardiff wherever we go' (BBC, 2012a).

However, many local fans resisted the rebranding. To them, cultural history, local pride, all tied into their geographical roots, were vital to the identity of the club. This co-ingredience meant more than appealing to a global cultural community and its favoured colour preferences. In the locals' view, the new owner 'took something perfect and painted it red' (see Kent, 2013).

Blue or Red birds?
Source: Author

Religion and a sense of place

In the current global climate, there are a range of interests, values, and community identifications that are, at first sight at least, not explicitly connected to place. Gender, brand loyalty, class affiliation, and political party support can all be read to be place-less to some degree. Religion is perhaps another example. Religion is first and foremost a belief and value system centred on powers and deities beyond the human. People all over the world believe in religions, you don't necessarily have to be born in a particular place to be a Buddhist or a Christian, to some extent you can choose your god regardless of geography. These are, on the face of it at least, communities of interest, rather than communities of place in the first instance (see MacKinnon, 2002).

This is not to say, however, that place no longer matters for ideas-based communities. Geography is thoroughly imbued in all these group affiliations (gender for example will be more thoroughly discussed in Chapter 13). With regard to religion, all belief systems have sacred sites at the local level, for example churches, synagogues, or mosques, where individuals can congregate to worship. At the supra-local level there are also sacred sites where key events in the development of that religion took place – Bethlehem, Palestine, Medina, Mecca, where people throng to on pilgrimages. These religious affiliations thus engage with place at a variety of levels tying their practitioners and places together into complex webs of cultural significance.

Of interest here is how these different sets of belongings come together in different places. How do national identities cross-cut, compound, or conflict with other identities, for example religious identities? In some cases, geographical place itself is the point of conflict between different religious groups (the Belfast example is a good example of this, here national affiliations elide with religious beliefs – Unionists are Protestants whilst Republicans are Catholics – and these differences seek to further compound difference between communities). Box 6.5 outlines another example of how geographical place is the point of conflict between two religions, in this case between Jews and Muslims and the disputed territory of Israel/Palestine. In other instances, it is our bodies and selves that are the 'place' of tension, as the next section goes on to discuss.

BOX 6.5 A PLACE OF TENSION: GEOGRAPHICAL AND RELIGIOUS DISPUTES IN ISRAEL/PALESTINE

There is insufficient space here to fully explicate the geographical, religious, and cultural entanglements associated with the Israeli/Palestinian conflict. To generalise is to grossly simplify; however, for our purposes here such generalisation is possible. During the Second World War, followers of the Jewish faith were systematically persecuted and killed by Nazi forces and their sympathisers across Europe. Following the cessation of this conflict, a United Nations Special Commission, supported by the United States and the Soviet Union, recommended that a Jewish state should be created, covering the biblically significant lands of Palestine. In this decision territorial (b)orders would broadly coincide with religious ones. However, such a decision overlooked the senses of attachment felt by those already occupying this place. The (b)ordering of Israel onto Palestine involved the dislocation of many Muslims who felt they had overwhelming rights to this particular backyard. Where could (b)orders be drawn to please all parties? Could separate states, coinciding with religious affiliations, ever practically work?

Decades on from the Second World War, the issue is still politically debated and violently fought over. In 2002, Israel began erecting a wall to divide what they see as Palestinian land on one side and Israeli land on the other. The wall, which is three times the height of the Berlin Wall and over 700 km long, is seen as a 'vital security barrier' by those on the Israeli side, an 'apartheid wall' by those on the Palestinian side, and illegal under international law by the United Nations (Jones, 2005). The wall has had the effect of cutting off movement between communities, affecting businesses, destroying families, and further alienating cultures on either side. As one inhabitant of Bethlehem states,

Palestine
Source: Karen Chalk

We are living in a prison. If you are surrounded by a wall, who will come to you? And where can we go? I have five children, and none of them have seen Jerusalem. I have tried to go: I told the soldiers, 'I want to take my children to see the Old City.' But I am not allowed.

(in Harris, 2005)

The images above depict the scale of the wall being constructed, as well as some of the graffiti painted on it. What interpretations do you have of these cultural traces?

In this case, geographical territory is the very unit of tension and conflict, with different senses of national and religious place attempting to (b)order this area in line with their ideals. There seems to be little scope for 'plural' lives to be lived here. If both communities are tied together (as the graffiti of the donkeys, above, suggests) then how can a more multi-cultural place be imagined?

Boxed into a sense of place?

Conflicts between scalar and religious senses of place do not only occur in a traditional territorial sense, they also occur at the geographical scale closest in – the scale of the body (after Rich, 1986). There follows an example about a boxer, Amir Khan (see Figure 6.3). This boxer may not be known to many – a boxing career is short and often over in one punch – but during his moment in the public eye his situation exemplified how geographical and cultural (b)orders can be written on a body, even when they are conflicting, and how they can be used to establish ideas about appropriate behaviour and action (see also Chapter 13).

Can you be a Muslim and a Brit? The Muslim boxer Amir Khan would be forgiven for feeling that his geographical and cultural (b)orders were a place of tension. Radical Islamic cleric Omar Bakri Mohammed believes Khan's pride in his nation is an insult to his religion:

Amir Khan is not a good example for Muslims. He wears shorts with the Union Jack on them. That is a sin. He should not be wearing the flag because sovereignty is for God. His only allegiance should be to the Prophet Mohammed. The ideal situation would be to have a Muslim team not registered to any state so he can represent the Islamic community.

(cited in France, 2008)

Figure 6.3 Boxed in Amir Khan

Source: The Sun

Bakri even goes as far to say that Khan is 'deviant' for his allegiance to the country of his birth. It is clear here, that from Bakri's point of view at least, to defer one's religious identity to that of one's nation is untenable. Indeed why *should* nation states be the representative scale for sporting contests? Why is this imagined community more important than others? Could not local, religious, or ethnic affiliations be used instead? In contrast to Bakri, Islamic journalist Alam (2004) views Khan in a different way:

> There is something very exciting about [Khan;] a young British Muslim representing his country in the Olympics. It defies the stereotype of the Muslim as an indigestible minority or cultural parasite. British Muslims are rarely celebrated as heroes – Khan's stellar performance in [the] Athens [Olympics] could change all that. They have few genuine role models who are proudly British and faithfully Muslim . . . His family has been seen at ringside, hysterically waving the British flag and, at times, the Pakistani one. It's hard to miss his proud dad sporting a Union flag vest, but look closer and there's a Pakistani cricket T-shirt underneath. Khan's dual identity is something he shares with many second and third generation British Muslims . . . Yes, he is Muslim and Asian and all the things that make him so unique and inspiring, but at the end of the day, he is just a British lad who wants to put everything he's got being successful at the sport he loves. Amir Khan is a home-grown champion. And he is ours.

From Alam's perspective, it is possible to unify national identity and religious affiliation, indeed this hybridity is exciting rather than threatening. For Alam, such hybridity helps to publicise, educate, and dispel prejudice against religious minorities, for those who share national identification with Khan, they may become more open-minded to his minority religious belief (in the UK at least). As he is 'ours', in British terms due to his birth, in Pakistani terms due to his family roots, and in Islamic terms due to his religion, all three communities are potentially bound closer together, at least for the duration of his success.

BOX 6.6 SENSES OF PLACE IN TENSION: HYBRIDITY HIDDEN?

The case of the boxer Amir Khan thus demonstrates the tension that can be placed on the sense of who we are. Senses of place that combine the ethnic, religious, and national are, as in Khan's case, often explicit and visible. In some situations, however, such hybridity is purposely hidden. For example, in the French education system legislation has been passed that bans all school pupils from publicly wearing religious symbols (e.g. crucifixes, stars, veils, etc.). In the eyes of French law, the place of the school is not a place for belief systems: religious iconography is 'out of place' here. Although it applies to all religions, this position has not been welcomed by the Muslim community. In response to the ban Muslim Iraqis kidnapped two French journalists and demanded its repeal if the hostages were to be released. In this case many acts of power seek to take and make place: from wearing religious jewellery on the body; to publicising your religion in school; to banning iconography in education; to restricting the freedom, movement, and safety of hostages. How would you respond if one cultural group threatened your sense of place in these ways? Does it matter if people are banned from expressing their religious identity in certain places? Why?

Janus us?

Senses of place can thus be sited in a strictly geographical sense but also culturally written into our bodies. Through these multiple senses of place it is often the case that we are sited in contact zones, being pulled over (b)orders, with different attachments and connections stretching our identity to breaking point. This situation has been identified by Garton Ash (2004). Garton Ash highlights how our geographical and cultural identity increasingly faces the 'Janus dilemma'. Comparing senses of place to the Roman god of bridges and thresholds, he suggests that identity can be thought of as having many 'faces'. Each face looks in a different direction, and our identity forms a bridge between competing and often contradictory positions. How do we manage the contradictions and tensions of this Janus dilemma? Writers such as Bauman suggest that it is our 'responsibility' (2002: xv) to re-(b)order our lives in ways that suit our own Janus locations. Rather than relying on traditional imaginaries, such as the local, national, or even the religious, we need to 'produce, stage and cobble together' (Beck, 1992: 13) new and individual senses of place. For Beck and Beck-Gernsheim (1995, 2002) this process of 'becoming individual' can liberate us from those conventional ties that connect but also limit us. Yet becoming individual may mean that we no longer share a collective sense of geographical self in any real sense. In these circumstances, what new places will be taken and made in the future? What new cultural affiliations will be created, and what new traces generated? Can we imagine a world beyond the nation or beyond religion? What would it look like?

Conclusion

In this chapter we have investigated how practices – such as anthems, cycling tracks, and wall constructions – and things – such as flags and emblems – are employed to generate particular senses of place. It has explored how senses of place can be personal and how they can be collective, how they can be instigated by everyday actions but also by invented rituals. These tracing practices can occur at the local, neighbourhood scale, but also at other scales up to the nation state and beyond. They can combine with other trace connections, working to strengthen particular senses of place, but also problematising and placing in tension who we think we are. Due to this combination of culture and geography in our identity, our senses of place 'are no longer necessarily or purposively . . . scalar, since the social, economic, political and cultural inside and outside are constituted through the topologies of actor networks which are becoming increasingly dynamic and varied in spatial constitution' (Amin, 2004: 33).

In other words, and as this book argues, our senses of place are not only defined by geographical (b)orders, but also by cultural imaginaries that transgress and change them. Our senses of place therefore,

> come with no automatic promise of territorial . . . integrity . . . they must be summoned up as temporary placements of ever moving material and immanent geographies, as 'hauntings' of things that have moved on but left their mark, as situated moments in distanciated networks, as contoured products of the networks that cross a given place. The sum is [senses of places] without prescribed or proscribed boundaries.
>
> (Amin, 2004: 33)

Our senses of place thus 'fold together the culturally plural and the geographically proximate and distant' (Amin, 2004: 37). Through our own practices and choices they are dynamic ongoing compositions that fuse together the cultural and the geographical, often in unique and contradictory ways. How we weave together our 'locals' and our 'nationals', alongside our beliefs and our religions, will come to define the places of our future.

> How could anyone think a culture stands still? Or that we hand it over intact like a book? It's more like the retelling of a story. It changes every time we put our own spin on it and pass it on. We add our own little bits, forget others and get some of the story completely mixed up. [Through considering culture and geography like this] I felt more at ease with my place . . . I had begun to carve out the idea that there might just be a place for me here but I would have to make it . . . I try to piece together the parts of the collective memory, the essence of my generation that may be meaningful to [me]. Is it the sea stories and the stories of movement; is it the story of unbelonging? The story of global connections and yet local attachment, of boundary disputes, of changing the very idea of [nationalism] and religion? What it is that binds us? How will we reinvent this heritage together?
> (Williams, 2002: 182–190)

Notes

1 Ariel Sharon, when President of Israel, also used Frost's words to make the 'Peace Wall' between Israel and Palestine seem 'natural'. It was a good fence to make good neighbours (see Harris, 2005). This issue is returned to later in this chapter.
2 Unionist areas will make use of the colours of the United Kingdom's 'Union' flag (red, white, and blue), while nationalist areas will make use of the colours of the Irish Republic's Tricolour (green, white, and gold).

Suggested reading

The following texts are useful to explore the issues raised by this chapter.

Books

Sibley, D. (1995b) *Geographies of Exclusion*. Routledge: London.

Here Sibley argues that Western culture is based on exclusion. He suggests that powerful cultural groups place 'others' (including marginal ethnicities) as outsiders, 'purifying' their own locations, breeding fear, and excluding marginal voices.

Jackson, P. and Penrose, J. (eds) (1993) *Constructions of Race, Place and Nation*. UCL: London.

Jackson and Penrose offer sustained analyses of how 'race' and 'nation' vary from place to place, how these spatially construct cultural life, and territorially express racist and nationalist ideologies. Particularly useful chapters include those on Aboriginal Australians and Muslim culture.

Bauman, Z. (2002) Foreword. Individually, together. In Beck, U. and Beck-Gernsheim, E. (eds), *Individualization*. Sage: London. xiii–xx.
Bauman, Z. (1995) *Life in Fragments*. Blackwell: Oxford.

Bauman's work illustrates his argument that cultural life is increasingly fragmented, liquid, and negotiated. He gives key insights into the role that individualisation may play in the construction and exclusionary nature of our cultural geographies.

Journals

Phillips, D. (2006) Parallel lives? Challenging discourses of British Muslim self-segregation. *Environment and Planning D, Society and Space*, 24, 25–40.

In this paper, Phillips challenges ideas of ethnic mixing and self-segregation in the UK.

PART

4

MONEY, MOVEMENT AND THE MORE-THAN-HUMAN

Place and capitalism: global, corporate, and anti-capitalisms

Introduction

In this chapter we turn our attention to perhaps the most influential cultural practice of our time: the culture of capitalism. Capitalism takes and makes place around the globe, leaving traces that affect the commodities we buy, the livings we earn, our methods and motives for travel, and the meanings associated with them. Capitalism affects our identity, our senses of place, and how (b)orders connect and distinguish them. Capitalism, therefore, is not just about making money, it comes to affect who and where we are. As a consequence it is an essential focus for geographers interested in culture.

Capitalism rules okay?

Capitalism is the dominant form of economic organisation across the globe. Yet it is not a new phenomenon; the Ancient Greeks used coins as currency to represent a trade in products and services. So why is capitalism so important to culture in the twenty-first century? The answer lies in its scale of operation. Even a generation ago, different systems of economic and cultural organisation were in real competition with one another. Via the 'cold war', Western style capitalism and Soviet-style communism battled to prove which mode of operation would be the best (or indeed the least worst!) for the world. With the fall of the Berlin Wall (see Box 7.1),

perestroika, and the dismantling of the Soviet Union, capitalism has become the dominant system of economic organisation. As Gough (2000: 17) tells us, 'Following decolonisation and the collapse of state socialism at the end of the 1980s, few areas of the world resist the logic of capitalist markets.'

In terms of economic organisation then, as Porritt (2005) outlines, capitalism is the only game in town.

Capitalism: making money

> You [now] have a system in which anyone can be rich. Careful, not *everyone*. *Anyone*. You can't lose when everyone is betting on being that anyone.
>
> (Advertising executive for Pinochet's election campaign, in Larrain, 2013)

In terms of substance, the culture of capitalism is based on trading products, experiences, and services. It uses monetary currencies to represent the exchange value of these entities, with the aim of making profit through the sale of commodities on the market. The means to make money in capitalism are privately owned. These means may include the ownership of factories that make products; the registering of patents, trademarks, and copyrights; or the ownership of resources, entertainment venues, or land. As these means are in private hands, capitalism offers

BOX 7.1 BREAKING (B)ORDERS: THE FALL OF THE BERLIN WALL

The Berlin Wall, erected in 1961, separated West Berlin (in West Germany) and East Berlin (in the German Democratic Republic). This physical border represented a clear geographical and cultural divide between two very different systems of economic and cultural organisation: capitalism to the west and communism to the east. Ronald Reagan (when President of the United States) made it clear that he thought this (b)order should be removed to allow capitalism to spread. Reagan made the following speech at the Brandenburg Gate, West Berlin, on 12 June 1987. His comments could be heard on the eastern side of the Wall:

Behind me stands a wall that encircles the free sectors of this city, part of a vast system of barriers that divides the entire continent of Europe. From the Baltic, south, those barriers cut across Germany in a gash of barbed wire, concrete, dog runs, and guard towers. Farther south, there may be no visible, no obvious wall. But there remain armed guards and checkpoints all the same – still a restriction on the right to travel, still an instrument to impose upon ordinary men and women the will of a totalitarian state. Yet it is here in Berlin where the wall emerges most clearly; here, cutting across your city, where the news photo and the television screen have imprinted this brutal division of a continent upon the mind of the world. Standing before the Brandenburg Gate, every man is a German, separated from his fellow men. Every man is a Berliner, forced to look upon a scar . . . General Secretary Gorbachev [of the then Soviet Union], if you seek peace, if you seek prosperity for the Soviet Union and Eastern Europe, if you seek liberalization: Come here to this gate! Mr. Gorbachev, open this gate! Mr. Gorbachev, tear down this wall!

(Reagan, 1987)

The cultural and geographical border of the Berlin Wall fell on 9 November 1989, and Germany was officially unified on 3 October 1990. Capitalism became the hegemonic mode of economic organisation in the world.

a democratic dream – as we are all private individuals we all have the potential to make products, sell them, and generate currency. In short, capitalism is a system in which *anyone* can be rich. However, in practice the market is competitive, with many traders seeking to sell their products to consumers. As a consequence, not all individuals can be successful. Capitalism is thus often defined by those who successfully trade products and services, and those who are limited to selling their labour to help others manufacture and provide these products. This is the traditional distinction made by Karl Marx between the bourgeoisie (the owners of the means of production) and the proletariat (the workers). According to Marx,

the most fundamental division in society . . . lay between those who owned the means of production (the machinery and buildings of a factory, for instance) and those who had nothing more to sell than their labour power, the ability to perform paid work.

(Cloke et al., 1991: 29)

Within capitalist culture the distinction between those who own the means of production and those who only own their labour power is crucial as it secures different opportunities to make currency. Selling your labour on the market will realise you financial advantage (to the tune of X dollars, or pounds, or euros etc. per hour). Your time, creativity, and effort are translated into cold, hard cash. However, your labour is also translated into added value for your employer's product, so in turn they realise an economic advantage from your labour, often one that is far greater than that achieved by you. Both parties therefore 'win' through the transaction, yet employers always make *more* money. This process of making (more) money is a key defining element of the culture of capitalism. The generation of profit is the difference between the cost of producing the good or service and its retail price.

The secret to making money: minimising and externalising costs

In order to make profit it is necessary for any capitalist to maximise the difference between cost price and retail price when trading on the market. This can be achieved through minimising and externalising costs. One way in which owners can minimise their costs is to drive down the price of the labour they buy. Employers and employees are thus often in dispute with one another – employers will seek to lower wages, whilst employees will seek to raise them.

Due to differences in standards of living and employment opportunity, workers in different places will offer their labour to potential employees at a range of prices. As employees will look for the lowest possible labour cost (with the skill set that can do the job required), owners are prepared to move their productions systems around countries, continents, even across the world in order to minimise cost. The profit motive of the culture of capitalism therefore produces geographical consequences. As firms move their production practices to maximise profit, jobs that have held together communities and families in one place are lost, with significant cultural repercussions. The jobs that are created in new locations also set in motion a range of cultural changes, often with mixed responses (see Box 7.2).

BOX 7.2 MINIMISING LABOUR COSTS – A CASE IN JAKARTA

Many of the brands we wear are made in Jakarta, Indonesia. Perhaps look at the label on your sneakers, jeans, or t-shirt that you are wearing now and see if they are 'Made in Indonesia'. Firms such as Gap, Adidas, and Nike have moved their production systems to South East Asia because the labour costs in these areas are amongst the cheapest in the world. As a consequence of this shift in production, Jakarta is now ringed with vast factory and worker compounds, known as 'economic processing zones'. Although these zones offer jobs, they also offer a completely different way of life to the traditions of Indonesia. As journalist John Pilger (2001) investigated:

> In these factories are thousands of mostly young women working for the equivalent of 72 pence per day. At current exchange rates, this is the official minimum wage in Indonesia, which, says the government, is about half the living wage and here, that means subsistence. Nike workers get about 4% of the retail price of the shoes they make – not enough to buy the laces . . . At a factory I saw, making the famous brands, the young women work, battery-style, in temperatures that climb to 40 degrees centigrade. Most have no choice about the hours they must work, including a notorious 'long shift': 36 hours without going home . . . these are capitalism's unpeople. They live with open, overflowing sewers and unsafe water for many, up to half their wages go on drinkable water . . . The result is an urban environmental disaster

that breeds mosquitoes. Today, a plague of them in the camps has brought a virulent form of dengue fever, known as 'break-back fever'. . . It is a disease of capitalism – the mosquitoes domesticated as the camps grew.

The culture of capitalism has thus re-made the place of Jakarta for the workers in these factories. The environmental, health, and human rights conditions in these economic processing zones are a direct result of the profit-driven motive of capitalists who (b)order this place. The need for income means that many live and work in these conditions, however unappealing they may appear. As Michael Walker (of the Fraser Institute, a marketing think tank) points out:

Let's look at it from the point of view of the rural peasants who are starving to death . . . the only thing that they have to offer to anybody that is worth *anything* is their low cost labour. And in effect . . . they have this big flag that says, 'Come over and hire us, we will work for ten cents an hour. Because ten cents an hour will buy us the rice that we need not to starve . . . and rescue us from our circumstance.' And so when Nike [for example] comes in they are regarded by everybody in the community as an *enormous* godsend.
(Walker, 2006: 8, emphasis in original)

Whether the effects of the spread of capitalism are 'an enormous godsend' or exploitation is open to debate. In what ways are the traces set in motion by capitalist culture progressive and in what ways are they repressive? How are we tied into these traces? Can we do anything to change them? Should we?

The cultural values of capitalism can therefore be made for employers through minimising labour costs, but other costs can be externalised all together. Costs that can be externalised – known as '*externalities*' – are those costs that can be passed on to sectors of society not directly involved in the primary economic transaction. For example, owners of the means of production not only require labour, but also a safe and secure transportation network, a healthy workforce, a repository for waste, and a generally stable political condition for successful operation. Owners take on many of these costs themselves, but it would cost any private individual an exorbitant sum to ensure the establishment and maintenance of the entire infrastructure necessary for profit making. As a consequence, the costs of this infrastructure provision are passed on to others. Nation-states (i.e. taxpayers) pay for transportation networks, Internet and satellite communication, welfare provision, military forces, and even armed conflict to enable capitalist operations to continue. Society also pays a cost by allowing common property resources such as the air, water systems, and the biosphere to be used by private companies to absorb their waste. The costs of these facilities are thus 'external' to those that have to be exclusively met by businesses; they are met instead by broader society and the state.

Driven by the profit-logic of capitalism, companies behave like 'externalising machines' (Monks, 2006: 7). Private individuals transfer as many costs to broader society as possible, as well as minimising the taxes they pay to particular governments, and thereby maximising their profits. Ray Anderson (2006: 7), CEO of Interface, the world's largest commercial carpet manufacturer, puts it this way:

Running a business is a tough proposition, there are costs to be minimized at every turn, and at some point the corporation says, 'you know, let somebody else deal with that. Let's let somebody else supply the military power to the Middle East to protect the oil at its source, let's let somebody else build the roads that we can drive these automobiles on, let's let somebody else have those problems', and that is where externalities come from, that notion of, 'let somebody else deal with that – I got all I can handle myself'. So, the pressure's on the corporation to deliver results *now*, and to externalize any cost that an unwary or uncaring public will allow it to externalize.

The extent to which publics are 'unwary or uncaring' about the external costs of the capitalist system will be explored later on in this chapter.

Capitalism: it's the economy, stupid. But it's culture too!

Although it is traditional to define capitalism simply in terms of its economic organisation, it is difficult to define capitalism as purely economic in nature. As we have seen, capitalism is equally defined by its cultural dimension, and the profit motive is its primary defining logic. Beyond this, however, a whole array of cultural institutions (including the media, government, and the police) are closely connected to the economic base of capitalism. In a Marxist analysis, over the long term the economic base (or substructure) of capitalism will be underwritten by the cultural institutions of its dominating powers (the superstructure). In other words, governments are likely to support and aid capitalism as they see a broader societal benefit in doing so (the example of nation-states paying for transport networks, security, and welfare systems can be cited as one example of this). However, the relationship between the cultural and economic aspects within capitalism can be interrogated in other ways too.

The economic and the cultural within capitalism are closely related. Products bought and sold under capitalism not only have an economic price, but also a cultural value. Products gain their desirability, their necessity, or their kudos, through the multifarious cultural systems in which they are used. It is thus cultures that give meanings and values to particular goods and services, indeed it is sometimes these meanings and values that are more desirable than the actual utility of the products themselves (we may desire a particular product, for example, because it is associated with a film star, or a fashionable image, rather than because it benefits our lives in any functional way). In turn, these products, or material traces, have particular effects on the places and cultures we live in. They work to perpetuate or erode not only material circumstances of society (traces have affects generated through their use, manufacture, or disposal), but also our non-material circumstances as they affect the particular ideas and value systems that take and make our places.

Capitalism thus involves both economic and cultural dimensions. Cultural groups give value and meaning to an economic product, making it more or less desirable. Yet in capitalist culture the reverse is also true – the economy gives a monetary price to cultural practice (thus making it more or less affordable). This process of bringing cultural life into the economic realm (and vice versa) is known as the process of *commodification*. As Scott (2001: 12) outlines, commodification is the process by which 'goods and services . . . are increasingly produced by capitalist firms for a profit under conditions of market exchange'.

The process of commodification is another example of (b)ordering. It is a process that injects into one product or activity, already imbued with culture and geography, a dose of capitalist value. Through so doing, that product or activity is in some sense ordered and bordered by capitalism: it has become a commodity. Capitalism has become so successful at injecting culture with its values that it is increasingly difficult to distinguish where the original culture ends and the capitalist commodity begins (see Barnes, 2005: 62–63). When we go to a cultural event, for example a sports stadium to watch a match, the ground, the team, the ball, the cheerleaders, even the players, are sponsored by capitalist firms and their products. The popular music we hear at half time is used to advertise goods and services; the artists are even paid to place products in their songs (see Box 7.3). Culture and capitalism are thus increasingly interpenetrated, both give value to each other – culture is used to sell product, whilst product is used to create culture.

Capitalism is thus successful not simply because it is an economic system, but also because it is a cultural one. Capitalism has its own cultural logic – the search for profit – yet this logic is most successful when it combines with the preferences and values of *other* cultures in order to make money. The combination of capitalism and other cultures transforms certain commodities into 'goods' rather than 'bads'. When such a transformation occurs, other cultures can be used to make capitalists' money. Capitalism thus puts a price on cultural life, and through recognising and selling desirable traits from particular cultures, capitalists and corporations begin to influence these very traits themselves. Some commentators go so far as to say that commodification has reached 'into every nook and cranny of modern life' (Thrift, 2000: 96); that whilst the purpose of capitalism used to be to 'sponsor culture', it now has the capacity to 'be the culture' (Klein, 2001: 30). Do you agree with these claims? If so, what consequences does the spread of capitalism have for cultural life in different places? Is the ubiquity of capitalism inevitable? Can we think of alternatives to this system? Has capitalism become doxa?

Capitalism has thus been successful in re(b)ordering cultural life in line with its own values and preferences. This process is not just happening in a few isolated locations, it is happening where you live, and where everybody else lives too. Can you think of examples of how capitalism affects culture in your place? The process of the commodification of culture raises a number of questions concerning the effect of capitalism on the identity of places and people, namely: does *what you buy (into)* say something about *who you are*? (Do you, for example, identify with the slogans: 'I am Starbucks', or 'I'm a PC'?; see Box 7.5.) Does intention have anything to do with your purchasing habits? Would your answers change if you considered consumption an act of endorsement rather than simply an act of purchase? Does your place change due to capitalism commodifying its traces?

BOX 7.3 THE NEW STATUS QUO? THE MERGING OF CAPITALISM AND MUSIC CULTURE

Commodification occurs not simply by capitalist firms advertising at cultural events in order to sell their product, but also by merging with these cultures in practice. Clear examples of this occur in music and films. Many artists are paid by firms to mention their products in their songs (see Ayres, 2005), whilst many films have explicit product placement which influences the story-telling narrative (see Spurlock, 2011). On some occasions songs are completely rewritten, re-released, and played on the radio despite them being little more than jingles for supermarkets. This occurred in the case of veteran rockers Status Quo. In 2012, they rewrote their song *Down down* inserting lyrics that advertised the Australian supermarket Coles. This song was played on Australian commercial radio as a piece of cultural music, rather than as an advert. Listen for yourself (ABC, 2012) and see how capitalism merges with music culture to rewrite the status quo.

BOX 7.4 COMMODIFYING THE SUPER BOWL: A CULTURAL AND/OR AN ECONOMIC EVENT?

I write this on the day after Super Bowl XLVIII. In this final showdown of the American Football season, the Seattle Seahawks thumped the Denver Broncos 43–8 in the MetLife Stadium, New Jersey. But how should we understand this event? Is it simply the culmination of a season-long sporting culture, or is the main event now a showcase of capitalism?

Reading the media coverage of the game across the 'pond' in Britain, banner headlines spoke of Bruno Mars entertaining the fans at half time (a gig for which he was not paid, as the exposure gained from the appearance was thought to be priceless), followed swiftly by a summary of the key adverts which attempted to sell product to a record number of television viewers at home (111.5 million in the United States alone (Inside TV, 2014)). A fairweather fan may fail to realise that the main event was actually a game of gridiron, not a marketing opportunity.

In the Super Bowl's half time adverts (see Salon, 2014), capitalism attempted to associate products not simply with the biggest event in American culture, but also with broader idea(l)s which are valued (theoretically at least) within that culture. One beer company attempted to associate its product with the friendship between a puppy and a horse; whilst a deodorant firm sprayed its message all over romances between unlikely couples in all corners of the world. In their view, their drink or smell are integral to the spread of friendship and peace; here capitalism is attempting to realise its cultural aims (i.e. making money) through insinuating them into the ideals of mainstream American culture – on an occasion when most of America is watching. This is Super Bowl as Super Billboard.

BOX 7.5 'THE NEW MILLENNIUM DESCARTES: I SHOP THEREFORE I AM'? (CHERRIER AND MURRAY, 2002: 245)

As we have seen, capitalism has been successful in injecting its idea(l)s into many cultures. However, can it be argued that capitalism has been able to commodify not simply other cultures, but also our own identity? Laclau suggests that the construction of who we are and what we stand for 'is an act of power' (1990). However, due to the success of capitalism in

I consume therefore I am

Source: Author

C.o.l.a Cola
Source: Author

extending into many arenas of culture, our ability to conceive of our identity as anything more than an act of consumption may be limited. Capitalism encourages us to believe that 'the good citizen is a good consumer' (after Aldridge, 1994: 905, cited in Hobson, 2002) and that we must 'buy into' a model of being where we 'voluntarily choose among commodities in the market with a view to creating a distinctive and individual self-identity' (Warde, 1994: 64). This view suggests that our identities are constructed through the commodities we buy. As Giddens has argued, from this perspective everyday consumption choices become 'decisions not only about how to act but who to be' (1991: 81). Capitalism thus seeks to commodify our individual lives; and the following examples go some way to illustrate this.

In perhaps a counter-intuitive marketing move, in 2013 Coca-Cola dropped their name from the labels of their soft drinks. In their stead, the company inserted popular first names of people who lived in each of the countries they sold products in – they may even have used your own name to sell their drink. This move was an attempt to revitalise profit-making by adhering its product to an individual's sense of self. As Chris Deere, Head of 'Brand Activation' at Coca-Cola Great Britain explained, 'We wanted people not just to find bottles with their own names on, but to surprise a friend or someone they love by seeking out a bottle with their name on it' (in Fisher, 2013). Would this marketing campaign work on you? If you found a capitalist product with 'your name on it', would you be more tempted to buy that product? Why?

In Tokyo, some companies have gone a step further. Here they do not simply put individuals' names on products, but put product names on people. Tokyo women can sell the 'advertising space' on their bodies (in particular their thighs) to companies to further the interpenetration of capitalist culture with individual identity (see ITN, 2013). Interestingly, individuals will only be accepted for this advertising if they have more than twenty 'friends' on social media. What do you consider the sale of this personal (advertising) space to be? Is it the final success in the spread of capitalist commodification, or the expression of personal freedom on behalf of the individual? Perhaps it could be both?

It is important to note at this stage, however, that the process of commodification is not just about bringing different cultures into the capitalist world. It is also about how the world itself has been tied into a particular *type* of capitalism. Commodification is defined not simply by capitalists from *particular locales* injecting a dose of capitalism into the culture of their town or country, it is being achieved by *global* capitalists, operating through trans-national corporations, that come to affect both the economies and cultures in almost every place. Thus, as we will see, capitalism has not simply commodified culture, it is commodifying culture on a global scale.

Corporate capitalism

Through communication technology, aviation, shipping, satellite monitoring, media systems, and the Internet, all aspects of trade are becoming increasingly globalised. Through modern communications I can be 'cold called' from sellers in Mumbai, contact my local high street bank via a call centre in the subcontinent, order an American branded sneaker from a warehouse in Yorkshire that was designed in Europe, manufactured in South East Asia, with the profits going to Grand Cayman.

The dominant players in this global capitalist activity are trans-national corporations. Trans-national corporations are, as their collective name suggests, companies that operate beyond the limits of nation states. Up until recently, this has meant that corporations are integrating once geographically disparate sourcing, manufacturing, marketing, and selling processes into one company structure, usually controlled by headquarters in its country of origin. As a consequence, the traces created by corporations no longer have limited local affects. Trans-national corporations create trace-chains through place dependencies, with markets, peoples, and up- and down-stream companies reliant on decisions made in distant places for the sustainability of their livelihoods. They create traces in the environment, with economic valuations of land often being preferred to local cultural valuations, dispossession of tribes becoming common place as traditional homelands are traded for economic expansion, and pollution and trans-boundary waste becoming a significant problem.

Trans-national corporations thus operate beyond the geographical borders of nation states. However, now operating beyond these borders is not simply a geographical issue. When Stiglitz (2005: no page) identifies that globalisation is the 'increasing economic integration and interdependence of countries and companies', he is referring not simply to the *geographical integration of economies and cultures* across the globe, but also identifying that *a new set of power brokers* have emerged into 'global culture'. Corporations, it is argued, are becoming so powerful within the global economy that they are as important as countries (maybe even more so).

In 1998 I was invited to Washington DC to attend this meeting that was being put together by the national security agency called the Critical Thinking Consortium. I remember standing there in this room and looking over on one side of the room and we had CIA [Central Intelligence Agency], NSA [National Security Agency], DIA [Defense Intelligence Agency], FBI [Federal Bureau of Investigation], Customs, Secret Service, and then on the other side of the room we had Coca-Cola, Mobile Oil, GTE and Kodak. And I remember thinking, 'I am in the epicenter of the intelligence industry right now'. I mean, the line is not just blurring, it's just not there anymore. And to me it spoke volumes as to how industry and government were consulting with each other and working with each other.

(Barry, 2006: 32)

According to Monbiot (2013), the role of business corporations is generally interpreted as that of lobbyists, seeking to influence government policy. However, in the era of trans-national corporations, this has changed. As Monbiot states, 'in reality [these corporations are] on the inside. They are part of the nexus of power that creates policy. They face no significant resistance, from either government or opposition, as their interests have now been woven into the fabric of all main political parties' (ibid.). Marx, writing in the nineteenth century, would have argued that capitalism and the political institutions of the state were always closely aligned (with the 'superstructure' of culture and politics working to underwrite and legitimise the 'substructure' of economic organisation in the long term). However, as Stiglitz observes, now 'the problem is that multinational companies have such tremendous reach and power that they can dominate the governments of impoverished nations and muscle their way past regulations' (2005: no page).

Stiglitz's concern is that the rise of globalisation has meant that trans-national corporations may be able to not only operate beyond national borders, but also beyond national orders too. As illustrated in Box 7.6, this means that unelected corporations are able to order and re-border the world in line with the cultural logic of capitalism. Stiglitz's concern is based on the immense economic wealth of trans-national corporations.

Writing in 2003, Kingsnorth identifies that of the 100 biggest economies in the world, 51 are corporations, whilst 49 are nation states. Records show that, in 2011, the Bank of America had a larger economy than Vietnam, General Motors was bigger than Bangladesh, and Chevron had more economic clout than the Czech Republic (see Trivett, 2011). In some instances corporations exercise their growing economic muscle by working hand in hand with nation states (as Barry identifies, above), whilst in others they use their power to dominate governments. In both cases they operate to secure their own capitalist objectives.

BOX 7.6 THE POWER BEHIND TRANS-NATIONALS' POWER: THE BRETTON WOODS ORGANISATIONS AND THE GROWTH OF CORPORATION CULTURE

So how did corporations become so dominant? The situation where corporations are as important as, and perhaps even more important than, countries is not due to natural selection or innate superiority. It has been facilitated by a number of institutions that operate beyond the level of nation states; namely the International Monetary Fund, the World Bank, and the World Trade Organisation.

The World Bank (WB) and International Monetary Fund (IMF) were established following the Bretton Woods conference held in the USA in 1944. Based on free market principles, these organisations were designed to regulate trade and maintain economic stability on a global scale. The World Trade Organisation (WTO) was established with similar principles, but much later, in 1995. As Lamy (2003; writing when Commissioner for Trade at the European Commission) explains, these organisations,

> provide a framework in which member countries negotiate how to regulate trade and investment, and ensure the respect of the rules agreed by common accord. [They] help us move from a Hobbesian world of lawlessness, into a more Kantian world – perhaps not exactly of perpetual peace, but at least one where trade relations are subject to the rule of law. [They are] the best bulwark against unilateralism.

These organisations thus maintain order in the global capitalist system, they create the rules that regulate trade, offer protectionism opportunities for growing economies, loan money to countries that have cash flow problems, and fund development processes in less advanced economies. In theory, therefore, these 'Bretton Woods' organisations function as useful controlling mechanisms for the global economy, giving everyone a 'fair' crack of the whip. However, in practice their role is not so straightforward.

Although the Bretton Woods organisations are supposed to act unilaterally from nation states and corporations, their free and fair operation is often compromised. In the first instance dominant nation states retain power over them. As Kevin Watkins (Oxfam) states, 'the real power in these [WTO and World Bank] meetings and what gives them real authority is that you've got the US Treasury, the European Union, and Japan, all getting together to see how they're going to run the global economy. And of course what they do is run the global economy in their own interests' (BBC, 2000).

Secondly, it is possible for trans-national companies to influence the direction and agenda of meetings. As Monbiot states, in practice the 'democratic structure [of these organizations] has been bypassed by "green room" meetings organized by the rich nations, by corporate lobbying, and by secret and unaccountable committees of corporate lawyers' (Monbiot, 2003). In practice, therefore, it is often the case that the global economy is not overseen by independent authorities, rather by dominant groups that seek to run the economy 'in their own interests'.

One example of this is the General Agreement on Tariffs and Trade (GATT). This General Agreement, first signed in 1947, was designed to provide an international forum that encouraged free trade for corporations by regulating and reducing tariffs on traded goods. However, GATT also confers rights to corporations to take sovereign governments to court if countries impose any national laws that militate against global trade. In effect, as Buttel (2003) outlines, this general agreement 'has essentially been established to permit offshore veto . . . of environmental regulations [and] . . . measures for enhancing social security . . . by unelected and unaccountable trade regimes'. The effects of this General Agreement will be compounded by others (such as the General Agreement on Trade in Services).

In effect, therefore, these global institutions confer a degree of power and authority to corporations over a nation state that has never been experienced before. It is now corporations alongside, and sometimes in control of even major nation states, that commodify culture and re(b)order place. Cultural and geographical power is thus held by a new group of actors operating on a vast scale. In effect these actors are taking and making place through the processes of global commodification.

Through transnational corporations, capitalism has come to connect together disparate locales and cultures, and corporations themselves have thus gained tremendous influence in spreading these cultures, as well as their own. The growing economic and cultural influence of capitalism thus raises questions concerning how local cultures are affected by processes of global commodification. Why do I, for example, wear a New York Yankees baseball cap when I've never played baseball or been to New York? How is global capitalism interacting with different cultures in local places, and how in turn does this affect our culture, geography, and identity?

Capitalism: Disneyfying and McDonaldising local cultures?

Capitalists seek to take and make place in order to intentionally perpetuate their own profit-making culture; through doing so they may (un)intentionally support the values and customs of some cultural groups, or dismantle and extinguish the cultures of others. The spread of capitalist culture brings with it many benefits. It offers jobs, new investment, and the opportunity to buy new products; however, it can also be so overwhelming that it can dislocate people from their local cultures, change value systems, and disenfranchise us from conventional politics.

From one perspective, the spread of corporate capitalism is often seen as the 'McDonaldisation' of global culture, where the 'principles of the fast-food restaurant are coming to dominate more and more sectors . . . of the world' (Ritzer,

2002: 7). This McDonaldisation allows us to get a Big Mac on nearly every street corner, or buy a pair of Levi's on every high street, thus making every town from New York to Nairobi familiar to us (if not like home). Whether this process is progressive is up for debate: to what extent does this promote a homogenisation of culture, where cultures in every-place are similarly defined by corporate values rather than local(e) customs? And to what extent does this matter? Toynbee (2001: 191) argues that the spread of corporate capitalism is not having positive effects:

> Sometimes it seems as if a tidal wave of the worst Western culture is creeping across the globe like a giant strawberry milkshake. How it oozes over the planet, sweet, sickly, homogenous, full of 'E' numbers, stabilisers and monosodium glutamate, tasting the same from Samoa to Siberia to Somalia. Imagine it in satellite pictures, every canyon and crevice pink with it and all of it flowing out from the USA. Just as world maps were once pink with the colonies of the British Empire, now they are pink with US strawberry milkshake, for 'cultural globalisation' is often just a synonym for Americanisation . . . the milkshake of the mind is spilling across frontiers, cultures and languages, Disneyfying everything in its path.

In an example which you can watch here (www.youtube.com/watch?v=p9KiHBLBW3M), the US rapper Jay Z is paid by a US drinks giant to align his musical philosophy with that of their corporation, and by extension the country they are both from (these are 'products' that are both 'Made in America'). The rapper talks of 'all like-minded people gather[ing] together', 'trading off each other's culture', 'no matter what country' they are in. According to Jay Z, 'we're all going to find a way to come together' as 'lines . . . cannot keep us apart'. He may well be referring to the power of his music, or implying that a certain drink (and a certain country's culture) can cross all cultural and geographical (b)orders (as

suggested by Toynbee above). Whatever he is referring to, this advert nevertheless attempts to realise the cultural aims of capitalism, nation, and corporate musician, at a global level. This example highlights, therefore, how capitalism has the power to commodify art, belonging, and maybe even the future, everywhere.

However, Jackson (2004) argues that although corporate capitalism and globalisation are having affects in many places, local cultures often resist the dominating power of trans-nationals. From this perspective, corporate capitalism does not have its own way, rather global commodification is an 'incomplete, uneven and contested process: an unfinished project whose contours are shaped by locally specific social and cultural practices' (2004: 166, see also Appadurai, 1996). A trivial example of how local cultures can 'indigenise' the process of 'McDonaldisation' can, unsurprisingly, be found in the food sector. Spices are added to burgers in the subcontinent, for instance, and the 'Indianization' of Italian-American pizza is seen in the Punjab through the addition of tikka massala sauce (see Jackson, 2004: 166). Can you think of how certain commodities of global capitalism are indigenised in different places? How does capitalism respond to local cultures where you are? (See Box 7.7 for an example.) In some cases, the arrival of corporate capitalism leaves little room to indigenise its dominating effects (as illustrated in Box 7.8). This chapter progresses by outlining how a growing number of cultural groups are seeking to shake up the 'strawberry milk' of corporate capitalism by resisting its spread into their cultural geographies.

Anti-capitalism: 'another world is possible'

Corporations are powerful only because we have allowed them to be. . . Their power is an artefact of our acquiescence. . . The struggle between people and corporations will be

BOX 7.7 WHOSE TOWN IS THIS?

The power of corporations and their success at commodifying culture means that our everyday places are often composed by many capitalist traces. From shops, services, and advertising, these traces are often accepted as 'normal', even 'natural', in our public places (in contrast to traces left by other cultures, for example graffiti, see Chapter 11).

For example, in Union Square, San Francisco, corporations' advertising dominates the skyline. Although such traces are commonplace, many are beginning to question the right of these traces to influence our cultural sites in this way (as the following sections will investigate further).

> Companies . . . have re-arranged the world to put themselves in front of you. They never asked for your permission . . . Companies scrawl their giant slogans across buildings and buses trying to make you feel inadequate unless we buy their stuff. They expect to be able to shout their message in your face from every available surface.
>
> (Banksy, 2005)

The commodified city
Source: Author

BOX 7.8 ROLL OVER HOLLYWOOD, HERE COMES TAFFYWOOD . . .

In my local culture, the website www.ilovesthediff.com indigenises global culture through merging its own vein of South Walian humour with English cultural artefacts and Hollywood films. Ilovesthediff (i.e. I love Cardiff) markets cards and other memorabilia which combine the style and typeface of Penguin book publishers, familiar Hollywood film titles, and a quintessentially Welsh sense of locale and language. Can you decode the origins of traces in the following cards?

Taffywood

Source: Ilovesthediff.com

How are 'global culture' traces indigenised in your area? How might you try indigenising particular cultural traces?

BOX 7.9. CULTURAL AND GEOGRAPHICAL (B)ORDERS OF FREEPORT MINE, INDONESIA

In the Freeport Mine in Indonesia, little effort is made to indigenise its effects on the local culture and geography. The trans-national corporation Freeport owns the biggest gold mine in the world, along with the third largest copper mine, in the Papua province of Indonesia. Company director James Moffett gained a mining license in Papua in 1967 after alleged bribery of the Indonesian dictator General Mohammed Suharto (see Kenny, 2006). In a special investigative report, the *New York Times* noted that:

Freeport has built what amounts to an entirely new society and economy, all of its own making. Where nary a road existed, Freeport, with the help of the San Francisco-based construction company Bechtel, built virtually every stitch of infrastructure over impossible terrain in engineering feats that it boasts are unparalleled on the planet.

(Perlez and Bonner, 2005; see also Kenny, 2006)

The taking and making of Papua by the Freeport trans-national has been dominant in its effects. Freeport has ordered and bordered its mine and the area around it in line with the cultural logic of capitalism. Since 1996, Freeport has paid the Indonesian military to remove indigenous people from the areas surrounding the mine, (b)ordering them away from ancestral homelands. The several thousand Amungme and Kamoro people who lived in the area were relocated into refugee settlements, as well as gravitating to the mining town of Timika, previously home to a small population. As the *New York Times* reported: 'Now Timika is home to more than 100,000 in a Wild West atmosphere of too much alcohol, shootouts, AIDS and prostitution, protected by the military' (see Kenny, 2006). Without access to their traditional land and with little prospect of employment, the local people are losing their social and cultural cohesiveness. Alcohol abuse and drug dependencies are more common. This has led to what some have called 'cultural genocide' (for more on cultural genocide, see Chapter 10).

In May 2000, Australia's Mineral Policy Institute described Freeport's mine as having 'the world's worst record of human rights violations and environmental destruction'. In 2005, CEO James Moffat declared his income to be $64.8 million. Between 1992 and 2004 Freeport contributed $33 billion in direct and indirect benefits to the Indonesian government, approximately 2% of GDP. The dominating power of this trans-national corporation, over and above the national government, raises questions about who really orders and borders the place of Papua, Indonesia. More broadly, is it corporations that now rule the world?

the defining battle of the twenty-first century. If the corporations win, liberal democracy will come to an end . . . We must, in others words, cause trouble. We must put the demo back into democracy.

(Monbiot, 2001: 356, 17, 358)

As we have seen in Chapter 5, dominating power often prompts resisting acts. With the rise of corporate capitalism this potential has been realised through a set of cultural interests that seek to resist the taking and making of place by corporate interests:

another world is possible
(Banner at Anti-G8 protests in Belfast, 2013, see Reuters, 2013)

For every action there is a reaction, and my reaction is: you need to listen to the people, there is a chance for a better world here. The way you're doing it isn't working and it isn't fair.

(Middle aged female protester at Anti-G8 protests, Belfast, 2013, ibid.)

We're here for our kids today, to teach them how to protest peacefully against what we see are greedy fat cats who are taking over this small part of the world and all of the world.

(Male protester, ibid.)

Resistance activity against corporate capitalism has become rife in recent decades. From Seattle to Genoa, from Belfast to Venezuela, and probably somewhere close to where you live, there have been protests against the growing power of corporate capitalism and its effects on cultural geographies. As Buttel (2003) notes, these have occurred not just in developed nations, but in all areas of the globe:

While we in the North almost always presume that the essence of the movement is that of periodic protests by citizen-protesters from OECD countries against institutions located in the North (such as the WTO, World Bank, IMF, or G8) or corporations headquartered in the North, the lion's share of protests have actually occurred in the global South. Protests have been particularly common in Bolivia, Argentina, Thailand, India, Brazil, and Indonesia.

Taking our cultural geography approach to place we can interrogate these resistance activities against corporate capitalism. We can ask, what do these cultural traces stand for, and how should we understand their challenge to corporate capitalism?

Denaturalising capitalism

Anti-capitalist activities combine a range of groups including the environmental movement, peace and human rights groups, the women's movement, development groups, and union representatives. As we have seen, they occur in places across the world, including both the 'north' and 'south'. The diverse and international nature of these resistance activities have led many to identify the paradox of an apparently 'anti-globalisation' movement being itself global in its reach and scope. However, as Klein (2002: xv–xvi) argues, the globalisation of resistance was almost inevitable due to the tangibility of the effects of corporate capitalism on many across the world.

> The irony of the media-imposed label 'anti-globalisation' is that we in this movement have been turning globalisation into a lived reality, perhaps more so than even the most multinational of corporate executives or the most restless jet-setters . . . Like so many others, I have been globalised by this movement.

From Klein's perspective, the naming of this movement as 'anti-globalisation' was something imposed by media actors. As we have seen, the 'power to define' is an important aspect of cultural geography, conferring a particular value and meaning to groups and their activities. In one sense, the 'anti-globalisation' label is accurate; these movements stand against the increasing commodification of local cultures by corporations acting on a global scale. However, the irony identified in an 'anti-globalisation' movement operating on a global scale cannot be escaped. Perhaps this imposition of this moniker by the media is thus an attempt to satirise and render impotent these resistant groups? In any case, from Klein's point of view, the global element of these resistant activities is vital to their cause. From her perspective: 'the face-off is not between globalisers and protectionists but between two

radically different visions of globalisation' (2002: 6). So from this view, protesters are not 'anti-' globalisation, but 'pro' a different form of globalisation. Wallach emphasises this point:

> Our movement is at a turning point . . . They label us 'anti' – we have to shake off the label. We're *for* democracy, *for* diversity, *for* equity, *for* environmental health. They're holding on to a failed status quo; *they're* the 'anti's. They are anti-democracy and anti-people. We must go forward as a movement for global justice.
> (cited in Kingsnorth, 2003: 219)

Thus we can see that these resistance activities have both a progressive and repressive dimension to them (they seek to forward particular cultural ideas and resist others). It seems these activities are 'anti' commodification by corporate capitalists and dominant countries, and for a redistribution and redirection of their power; as the following examples go some way to illustrate:

> We may feel intimidated and overwhelmed by their [corporations' and countries'] sheer size and power, but it is in the essence of capitalism that their power is already ours. We are the ones that create their wealth – we have only to redirect that wealth creation in the interests of all . . . This is the meaning of anti-capitalism – the promise of moving beyond capitalism to a more radically democratic, self-governing, self-sustaining world.
> (Bircham, 2001: 3)

'Capitalism' is a word which emerged in the eighteenth century to describe a system in which a few control production to the exclusion of the many. The crucial point to understand is that capitalism and a true market economy are not the same thing. In a true market economy, many small firms, rooted in real communities, trade and compete with each other. Under capitalism, large, rootless transnational corporations with no responsibility to anyone but their shareholders destroy real market economies and monopolise production. Let's be clear here – the idea is precisely to abolish capitalism, and with it the institutional form of the limited liability corporation.

(Korten, cited in Kingsnorth, 2003: 213)

Anti-capitalism: what does it stand for?

We can see therefore, that this movement seeks to name itself as 'anti-capitalist' and, due to this stance, stands for a significant change to capitalist's cultural (b)orders. As Jordan observes, there is something 'culturally transformative' about this movement (2002: 95), or in the words of Jennifer from the Infernal Noise Brigade (a music ensemble set up to provide the musical accompaniment to protests in Prague), 'I don't just want to edit what already exists, I want an entirely new revolutionary world' (BBC, 2000).

However it remains unclear what this 'entirely new revolutionary world' seeks to be. Some elements continue to emerge (see Box 7.10), but there are few precise intentions and thought-through objectives for this movement. That, to the anti-capitalists themselves, is precisely the point: in the words of Zapatistan 'leader' Subcomandante Marcos, their aims 'don't fit in the present, but are made to fit into a puzzle that is yet to be finished' (cited in Jordan, 2002: 40). Their potency therefore lies in the sense that an alternative is possible, at least when they are involved in action:

The first time I participated in one of these counter summits, I remember having the distinct feeling that some sort of political portal was opening up – a gateway, a window, a 'crack in history', to use Subcomandante Marcos's beautiful phrase. This had little to do with the broken window at the local McDonald's, the image so favoured by television cameras; it was something else: a sense of possibility, a

blast of fresh air, oxygen rushing to the brain. These protests – which are actually week-long marathons of intense education on global politics, late-night strategy sessions in six-way simultaneous translation, festivals of music and street-theatre – are like stepping into a parallel universe. Overnight, the site is transformed into a kind of alternative global city where urgency replaces resignation, corporate logos need armed guard, people usurp cars, art is everywhere, strangers talk to each other, and the prospect of radical change in political course does not seem like an odd and anachronistic idea but the most logical thought in the world.

(Klein, 2002: xxiv)

Through practice therefore activists are taking and making places that were once controlled and given meaning by capitalist interests, and re-imbuing them with alternative meanings and values. As one activist in London put it, 'on 18th June we owned this place for a day. Totally owned it. It was magic' (BBC, 2000).

Summary

We have always been told that what they call globalisation is an inevitable process. What we have done as a movement is to clearly demonstrate that it is not – that it is something manufactured by a specific set of interests and sold to us as something irresistible.

(Lori Wallach, US Trade Lawyer from 'Public Citizen' watchdog, in Kingsnorth, 2003: 220)

We can see that the cultural traces of anti-capitalism transgress the (b)orders established by processes of corporate commodification. In practising resistance, they turn its lines into question marks, challenging the inevitability of its way of taking and making place. Through their action they reveal capitalism to be simply

BOX 7.10 POWER AND DIS/ORGANISATION IN ANTI-CAPITALIST MOVEMENTS

From the perspective of those within it, part of the progressive agenda for this 'anti-capitalist' movement is adopting a different approach to the practice of power. As Kingsnorth (2003: 31) puts it:

> Autonomy is clearly a key part of it. Every [protester] you talk to will tell you that autonomy – real, local control of their community, economically and politically – is a hard-fought-for principle, rather than an expedient political move. They will tell you, too, that autonomy doesn't for them mean independence, dropping out, isolation – it means control of their own destinies. Linked with that is the commitment to community democracy – real control, by all, at community level, however difficult it may be to implement.

This approach to power, for people 'doing it for themselves', is very different to the indirect form of representation many citizens participate within in democratic countries, and far removed from individuals suffering under dictatorships or military juntas. This form of autonomy when devolved down to the individual level involves the re-creation of power structures and decision-making. Instead of 'organisation' this is commonly known as 'disorganisation'. Such disorganisations replace hierarchies of decision-making with 'flat' structures where there are no conventional leaders. As Jordan (2000: 69–70) puts it, these,

> flat hierarchies [enable] co-ordination [so] all who want to participate do so. [They are an] ethical statement. They are a belief that all who participate in a dis/organisation have something (not necessarily the same thing) to contribute and that forms of co-ordination must strive to draw all the worth they can from everyone's contribution . . . as well as leaving space for spontaneous reinterpretation of actions . . . there is no privileged decision-making point.

This disorganisation, alongside the lack of nominated decision-maker or decision-making point makes the actions of disorganisations difficult to predict – and therefore difficult to know where this form of resistant activity will end up. Will it involve 'revolution' or 'recuperation' (see below for an explanation of the latter idea)? Due to this unpredictability, it also makes this movement difficult to control, as we now go on to discuss.

BOX 7.11 INVOKING ALTERNATIVES

> How can communities defend their local economies, when local earnings are siphoned out of an area by big business or distant utilities? The answer . . . create a new kind of money.
>
> (Boyle, 2000: 200)

Example #1: Time dollars

In the United States, Edgar Cahn set up an alternative to mainstream capitalism, and he called it 'time dollars'. Boyle (2000: 3–4) outlines the concept behind time dollars in the following way:

You go along to your local time dollars project, and tell them what kind of work you are prepared to do – anything from roofing to giving people lifts – and the things you need doing in return. All these details are entered into the computer. Then one of your elderly neighbours suddenly needs a lift to the doctor at a time when you are free – or maybe they just need driving to the shops. They phone the office, who call you up. The final judgement about whether the two of you are suited is made by the time dollar organiser, who phones up first to see if you are available and willing. Result: you spend an hour helping your neighbour and you earn one time dollar. You get a statement showing your earnings at the end of the month, and you feel good about yourself.

What can you do with your new found wealth? Well, that depends on the scheme. You could spend it on services from other people in the system. Or you could give it to an elderly relative who might need it more. Or you could keep it for a rainy day. Or you could just forget all about it: only about 15% of time dollars are actually spent.

As Edgar Cahn explains, 'the real function of time dollars is to provide value in a world where the market economy defines the work which the majority of people do – looking after old people or bringing up children – as useless'. From the perspective of corporate capitalism, all such tasks 'will never be adequately valued in a market system which must devalue what is common or universal. To give adequate value you just have to step outside the system' (Cahn, in Boyle, 2000: 11).

Example #2: The Bristol Pound

In Bristol, England, a new currency has been created called the Bristol Pound. This currency has a one-to-one exchange value with Pounds Sterling, but can only be used within the city. This means that value is not 'exported' beyond the city limits and wealth circulates within the area. As its founders state: 'the primary purpose of the scheme is to strengthen the local economy by incentivising people to use independent local businesses. We believe that the people of Bristol have a strong sense of identity and civic pride. Bristol's independent businesses are crucial to that community identity' (see http://bristolpound.org).

Bristol currency
Source: Author

Example #3: Bitcoin

Another alternative to the corporate economy operates on a global scale. 'Bitcoins' are virtual money, invented for the purpose of inhibiting traditional banks' ability to store money and profit in the process. In Kreuzberg, Berlin, Bitcoin has become a favoured medium of exchange in many shops and bars. Joerg Platzer, the owner of bar Room 77, is helping to establish what he believes to be the world's first Bitcoin local economy; he states: 'Kreuzberg is an area in Berlin where people are politically very aware and people are critical towards existing systems and . . . looking for alternatives . . . therefore Kreuzberg is a perfect ground for Bitcoin' (Grandjean and Connelly, 2013). Operating outside the existing financial system, but being able to trade with it (at the time of writing, one Bitcoin is worth approximately US$72), Bitcoin offers new opportunities for traders. Not only does it create new institutions and structures, but also perhaps new inequalities.

one way of organising economic and cultural activity rather than 'the only game in town'. The doxa of corporate capitalism is thus transmogrified through protest into a simple orthodoxy, as 'normal' rather than 'natural', and as a consequence the possibility of other ways of being are facilitated. This process therefore makes doxa seem like orthodoxy, then turns orthodoxy into 'something manufactured by a specific set of interests and sold to us as something irresistible'. Through a combination of action and intuition, activists have 'take[n] down the first fences – on the streets and in their minds', and replaced them with a new set of (b)orders that represent a new way of being.

The consequences of anti-capitalism

Dominating groups within the culture of corporate capitalism respond to anti-capitalist protests in a number of ways. As activists have started using the term 'anti-capitalism' (rather than say 'anti-globalisation'), and begun invoking alternatives to the mainstream system, so the dominant culture has started to change the way it talks about these issues. This is significant in two senses. Firstly, the anti-capitalists are beginning to influence the terms of the debate, thus the 'power to define' has subtly shifted between groups. Secondly, due to this shift in terms and power, so the debate itself has changed. Instead of it being solely about 'terrorism' or tactics, it is about capitalism and the Bretton Woods organisations. Thus the debate is no longer simply about the culture of anti-capitalism, but the culture of capitalism itself.

In this move we can see the first step in the shift in trace meanings from 'natural', to 'normal', not simply by the anti-capitalist movement, but by the dominant capitalist culture too. As Smith suggests, these actions have led to the traces of 'free trade, globalization, and neoliberalism . . . unmasked [within the mainstream] as social and political projects with decisive agendas, masquerading as economic inevitabilities' (2000: 2). This erosion of doxa has not perhaps gone so far as altering the balance of dominating power away from capitalism, but it is significant nonetheless: the doxa is now being debated (see Box 7.12).

Changes in these debates have set in motion further trace-chains in different places. Dominating authorities in some places have sought to resist anti-capitalism by adopting more repressive and violent policing towards it. In protests in Genoa, for example, where 200,000 activists gathered, dozens were hospitalised during clashes with police, whilst one protester was killed (Hooper, 2008). These responses to anti-capitalist activity represent an attempt to redraw the (b)orders between groups and values by re-emphasising who has the dominant ability to determine places. However, these re-tracings also have the effect of backfiring on capitalism and the institutions that defend it. Through underlining their authority through violence and repression, capitalism and the state make themselves seem at once powerful, but also under threat.

A further way in which capitalist culture has responded to this resistant activity is through attempting to 'recuperate' it. Recuperation is a process where groups 'appropriate antagonistic expressions and render them harmless through

BOX 7.12 DEBATING AND DESTABILISING DOXA

Bruno Latour outlines the culturally transformative shift from traces being valued as 'natural' to 'normal', or 'normal' to 'natural' in the following quotation:

> Take some small business owner hesitatingly going after a few modest shares, some conqueror trembling with fever, some poor scientist tinkering in his lab, a lowly engineer piecing together a few more or less favourable relationships or forces, some strutting and fearful politician: turn the critics loose on them, and what do you get? Capitalism, Imperialism, Science, Technology, Domination – all equally absolute, systematic, totalitarian. In the first scenario, the actors were trembling, in the second, they are not. The actors in the first scenario could be defeated; in the second, they no longer can.
>
> (Latour, 1993: 125–126)

Through anti-capitalist protests and the uptake of alternative currencies, we can see how the (b)orders and values of corporate capitalism are being destabilised. Debate is being held over its nature, substance, and power, and thus the scenario is made possible where it could be defeated.

transformations and integration into some form of commodity' (Aufheben, 1996: 34).

Recuperation then is an extension of the process of commodification. Recuperation not only packages up cultural life into a saleable commodity, it also transforms the cultural activity through the process, removing any challenging aspects and rendering the original cultural challenge harmless. Recuperation therefore is effective at taming the threatening aspects of a cultural trace and absorbing it back into the main cultural system. An example of this recuperation process in relation to anti-capitalism is outlined below (see Figure 7.1).

In this example, the 'novelty' of protest activity is rendered harmless by recuperating it back into the cultural mainstream. Here protest is associated with fashion and clothing products, thus reducing it to a trend, or a political expression that can be articulated through consumption. Here, therefore, capitalism has been successful in not only neutering a threat, but also forwarding its own cultural objectives in their place. In this way, recuperation is a form of 'culture jamming' (see Jordan, 2002). Culture jamming is usually used by minority groups to undermine and subvert the meanings and representations of dominant cultures. However, in this case capitalist culture has 'jammed' anti-capitalist culture, commodifying it for its own ends.

Despite the relative successes of anti-capitalist practice, the example of recuperation illustrates how pervasive capitalist culture can be. Its durability and inventiveness allows many practices and products, however challenging and transgressive, to be absorbed within the system and thus commodified. This commodification retains traces of the transgressive meanings of the original, but with their absolute challenge disarmed. This may occur with alternative currencies such as the Bristol Pound and Bitcoin (see Box 7.11). The Canadian Royal Mint, for example, is using Bitcoin as a template for the generation of its own virtual currency. Will these alternatives end up as an anti-capitalist version of the carnivalesque? Massey ponders this conundrum when she states, 'it is sometimes hard to see how resistance can be mounted without it subsequently, and so quickly, being recuperated' (2000: 281).

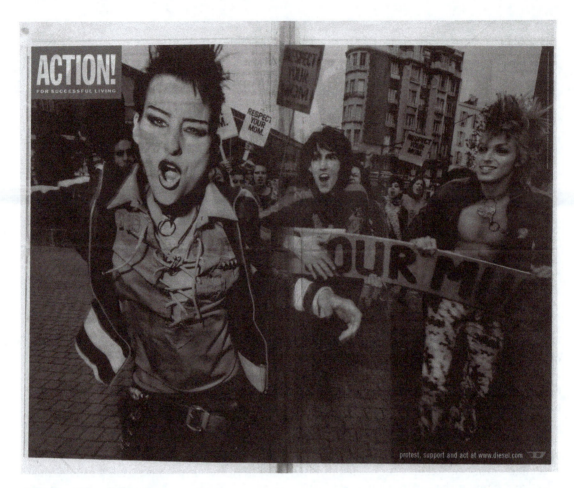

Figure 7.1 Diesel 'protest'

Source: Diesel

Conclusion

> Someday I want to be rich. Some people get
> so rich they lose all respect for humanity.
> That's how rich I want to be.
>
> (Rita Rudner, comedian)

We've seen in this chapter how capitalism is
both an economic and a cultural system. The
globalisation of capitalism has resulted in corporations having a growing influence, some would
argue dominating power, in taking and making

places across the world. Cultures of (corporate)
capitalism have thus had the effect of wielding
significant power over local communities and
nation states; they have generated new dependencies, challenged traditional democracies, and
commodified cultures and places. We have also
seen, however, how (corporate) capitalism has
bred new ways in which the traces of its culture
are being judged and resistance to them practiced. By transgressing the (b)orders of capitalist
culture (either through satire in the case of Rita
Rudner, through naming, non-violence, or full

frontal attack), its orthodoxies are revealed and raised into debate.

We can, and perhaps have to, make our own cases as to whether capitalist traces deserve to be dominant, maintained, or changed. How could alternatives be imagined? Can different notions of value offer a substitution to economic trading (e.g. 'time' dollars, see Boyle, 2000)? Is resistance of any sort futile to the 'only game in town'? By adopting a cultural geography approach to capitalism we can ask what traces do each of us make that (un)intentionally support either capitalism or anti-capitalism? What traces do you, and the wider public, want to contribute to either culture, and what trade-offs (in terms of violence, freedom, or democracy, for example) are acceptable in moving towards these ends?

Suggested reading

The following texts are useful to explore the issues raised by this chapter.

Books

Marx, K. and Engels, F. (1984) *Collected Works*. Lawrence and Wishart: London.

For a classic critique of capitalism look no further than Marx and Engels, in particular *The Communist Manifesto*.

Ritzer, G. (ed.) (2002) *McDonaldisation: The Reader*. Pine Forge Press: Thousand Oaks.

Ritzer's edited volume outlines what he sees as the 'McDonaldisation' of global cultures by transnational corporations.

Klein, N. (2001) *No Logo* ®. Flamingo: London.
Klein, N. (2002) *Fences and Windows: Dispatches from the Front Lines of the Globalisation Debate*. Flamingo: London.

Klein offers a more journalistic perspective on front line protests against corporate globalisation, but adopts an often explicitly geographical perspective in her arguments.

Kingsnorth, P. (2003) *One No, Many Yeses: A Journey to the Heart of the Global Resistance Movement*. Free Press: London.

Kingsnorth offers one of many insights into the range of protests against capitalist culture occurring across the planet.

Journals

Jackson, P. (2004) Local consumption cultures in a globalizing world. *Transactions of the Institute of British Geographers*, 29 (2), 165–178.
Smith, N. (2000) Guest editorials: global Seattle. *Environment and Planning D, Society and Space*, 18, 1–5.

Jackson argues how global capitalism can be indigenised into local contexts, whilst Smith et al. give their reactions to the global protests in their local places.

Websites

Information on the birth, development, and impact of corporate capitalism can be found at www.thecorporation.com/ whilst a range of websites offer up to the minute news and reactions to campaigns; these include SchNEWS.org and Infoshop.org.

Place and mobility

Introduction: cultural geography 'on the move' (after Cresswell, 2006)

> The world is constituted by mobility . . . From this perspective, movement is the foundation of geography.
>
> (Steinberg, 2013: 160)

In recent years, cultural geography has been on the move. The globalisation of ideas, services, money, and people has become omnipresent in our everyday lives, so much so that issues concerning mobility have come to forefront of geographical practice. As Sheller and Urry (2006: 207) argue,

> All the world seems to be on the move. Asylum seekers, international students, terrorists, members of diasporas, holidaymakers, business people, sports stars, refugees, backpackers, commuters, the early retired, young mobile professionals, prostitutes, armed forces – these and many others fill the world's airports, buses, ships, and trains.

As Appadurai neatly encapsulates, our cultural world is increasingly defined by the practices and consequences of mobility: 'everywhere [is now] shot through with the woof of human motion, as more persons and groups deal with the realities of having to move, or the fantasies of wanting to move' (1990: 297). Thus as Steinberg states (above), and Robins sums up: 'mobility has

become . . . the emerging world order' (Robins, 2000: 196). This chapter investigates the issues surrounding mobility, it outlines how cultural geographers have explored the importance of mobile people and commodities; it will refresh our ideas on the implications of mobility for the nature of places (see also Chapter 4); and finally it will offer suggestions about how mobility may alter the constitutive co-ingredience of people and place, changing our connections to the world from a preference for 'roots', to a desire for 'routes'.

Putting the meaning into movement: mobilising mobility through cultural geography

> Something is always coming from elsewhere, whether it be wind, water, immigrants, trade goods, or ideas.
>
> (Solnit, 2010: vii)

Over the course of the last decade, the study of mobility has emerged from the margins of academic study, with mobilities research now 'solidifying into an interdisciplinary field' (Hannam et al., 2006). This is not to say that issues concerning migration and movement had been overlooked in the past, rather it acknowledges that the nature of scholarly enquiry into their meaning, motives, and mechanics has recently undergone a substantial overhaul. As Sheller and Urry (2006: 208) put it, a 'mobility turn' is spreading into and transforming the social sciences, transcending the dichotomy between transport research and social research, putting social relations into travel and connecting different forms of transport with complex patterns of social experience conducted through communications at-a-distance.

Previous analytical approaches to movement, for example in transport, logistics, or migration studies, were often less about the nature of motion, and more about the 'push' and 'pull' factors prompting movement in the first place. If, as Solnit tells us, everything is moving in some way or another, scholars in migration and transport studies based their interest in the linear conceptualisation of this movement, seeking to understand its 'origin' and 'destination', 'how often' it happens, and 'how quickly' it occurs. Through doing so they failed to explore what this book has argued to be crucial in understanding our cultural world, namely the more-than-representational traces bound up with mobility. Sheller and Urry identified that the actual 'woof' of motion had become an often ignored 'black box' in the traditional approach (2006: 208). In Cresswell's terms, these studies 'have not been so good at telling us about the representations and meanings of mobility either at the individual level or at a societal level. Neither have they told us how mobility is actually embodied and practiced' (2010c: 19). In response to this, over the last ten years cultural geographers have begun to open the 'black box' of mobility. They have begun to conceptualise mobility differently; by investigating the nature of physical movement, its meanings and representations, as well as its non-material, affective dimension, cultural geographers have begun to 'take the actual fact of movement seriously' (ibid.).

Taking mobility seriously

As Adey (2010), drawing on Cresswell (2006), explains, the idea of 'mobility' is more than simply 'movement'. By using the term 'mobility' we immediately understand movement to have two related elements: firstly, mobility has cultural meaning, and secondly, mobility has consequences. Mobility may have cultural meaning through it representing ideas of freedom, exile, or reward, it may be yearned for, taken for granted, or suffered. What does mobility mean to you? Can you think of instances where mobility would be considered in these ways? The second element associated with the term mobility is that it has consequences. Adey explains this element of the practice as follows:

> Imagine you are sitting near a pond or a lake on a hot day. The water is perfectly still and you decide to put your feet in the water to cool them down. What effect will your feet have on the water? Every move you make will conjure ripples that move across the surface of the pond . . . We might take what has happened here as a useful little analogy of what happens every time we move or are mobile. Space is changed. The subsequent movement of the pond is displaced, charged, sploshed, frothed around due to our mobility. Of course, the space around us and through which we then move is disturbed, but it is also altered for others . . . our mobilities make waves.
>
> (Adey, 2010: 19)

These sploshes and froths are the consequences of mobility. These displacements and charges engender trace-chains that change the world around us. When we consider mobility we therefore cannot simply consider how mobility occurs, why, and in what direction, but we also have to raise questions concerning who is moving, what they are moving as a consequence, and what implications does their movement have for others? As Massey (1994: 149–150) states,

It is not simply a question . . . that some people move more than others, and that some have more control than others. It is that the mobility and control of some groups can actively weaken other people. Differential mobility can weaken the leverage of the already weak . . . Mobility, and control over mobility, both reflects and reinforces power.

(Massey, 1994: 149–150)

Understanding movement as mobility is thus a crucial step: it allows scholars to investigate the meanings and consequences of globalisation processes, and begin to understand how 'these diverse yet intersecting mobilities have many consequences for different peoples and places that are located in the fast and slow lanes across the globe' (Sheller and Urry, 2006: 208). Indeed, these globalisation processes involve important issues such as immigration, trade, pollution, and finance, and thus 'issues of movement, of too little movement or too much or of the wrong sort or at the wrong time, are central to [contemporary] lives, organisations and governments' (Hannam et al., 2006: 1).

Understanding mobility: moving along 'Constellation' Street

In order to investigate the importance of mobility to our everyday lives, Cresswell (2010c: 22–23) has suggested six questions that help us diagnose the material and non-material traces that constitute the practice of mobility, as well as the trace-chains this practice sets in motion. These questions include:

1. Why does a person or thing move?
2. How fast does a person or thing move?
3. In what rhythm does a person or thing move?
4. What route does it take?
5. How does it feel?
6. When and how does it stop?

Cresswell terms these various aspects of movement the 'constellations of mobility'. Diagnosing these constellations enables us to position mobility not as a black box, but rather as a set of trajectories that shoot through everyday places. On some occasions we are the objects that move, in others different commodities, non-humans, ideas, or people are mobile. In each case these constellations have (in)direct consequences on our geographies and cultures, as well as our co-ingredience with them. I reflect on these as I cycle to work along Constellation Street (see Figure 8.1).

I wait at the traffic lights to turn into Constellation Street. The 'green man' says the pedestrians can 'walk', and as the former blinks the latter skip across the street; then my light changes and I obey. There are no cycle lanes on Constellation Street, and the cars rev past. Up ahead a minibus is blocking my passage. It has its doors open to allow wheelchair access. I decide to stop. A car honks its displeasure. I see two grannies look up from the bus shelter opposite, they are basking in the sun till the Number 11 arrives; I see a mum

Figure 8.1 Constellations of mobility on the street

Source: Author

with a pushchair hurrying to join them: there are no drop kerbs on the sidewalk here; it isn't an easy feat to cross from one side of the street to the other. One of my students skates by, smiling. I notice his tan: 'Just back are you?' 'Yeah, still feel the jet lag!' he replies as he tic-tacs down the sidewalk. By the bus stop is a utilities van, parked up on a double yellow line: the 'ultra-fast, turbo-charged, 16th generation' broadband is being tinkered with in the junction box. The minibus ahead closes its doors and pulls away. I peddle on, imagining the ironic speed camera winking at me.

When I muse on Cresswell's constellations along Constellation Street I realise the range of ways in which mobilities (b)order our everyday life. I feel liberated by cycling on my chosen route to work, but this feeling does not seem to be shared by the mum with her pushchair, or indeed the car driver honking out his road rage. At times, my mobility may be impeded, or my safety threatened by other road users. We are all subject to the general (b)orders that regulate mobility in my culture (see Figure 8.2). What constellations of mobility enable your life to function? How reliant are you on the taken-for-granted mobility of people, commodities, information, and energy? What happens when one dimension of this mobility doesn't operate?

Mobility is now understood to make the cultural world go round. As we have seen in Chapter 4, this 'turn' to mobility has influenced the ways in which geographers conceptualise and study place. These changes are broadly outlined in Box 8.1. Following this, the chapter moves on to explore the new types of traces that cultural geographers have begun to study as consequence of the mobilities turn.

Mobile lives

Many of the ways in which mobility changes our understandings of the world are imaginative and stimulating; however, as we have seen, many aspects of the practice itself are banal or routine. As Vannini observes, 'much of mobility is a mundane accomplishment' (2011: 1032). We often take for granted the (b)orders which affect our mobility even though we thoroughly depend on commodities being shipped, freighted, pipelined, or cabled across the globe in order to service our lifestyles. We will see in Chapter 14 how taking mobility seriously has prompted innovation in the methods cultural geographers use to study place, but in this section we will explore the new types of traces and trajectories geographers have investigated as a consequence of our increasingly mobile lives.

As Binnie et al. tell us, everyday life is full of 'forms of mobility that are enduring, predictable, habitual, repetitive, and of brief duration and short distance' (2007: 166). However, despite the routine and mundane aspects of this everyday mobility, to accomplish these movements effectively requires skill and timing. Vannini (2011) has investigated these easy-to-overlook mobile achievements in his study on commuter ferries in British Colombia. In line with this book's contention that places are ongoing compositions of traces, Vannini argues that commuters 'craft' their journeys, with everyday mobility and travel becoming a form of art. In Vannini's words, 'passengers keen on making a ferry employ reflexive, adaptive, orientation skills, weaving artful journeys to the ferry terminal in order to make a specific sailing . . . Passengers' work is a creative performance' (2011: 1032). Consider how you successfully 'make' mobility in your own life; how is your own commute to work or study a complex piece of timing, motion, and logistics? How do you 'craft' your routes on a daily basis? We may overlook such everyday accomplishments, but as Robert, a Quadra Islander, outlines, we rely on these acts to realise our cultural geographies:

If you don't catch your ferry all hell breaks loose. I mean, sure, there are times and days

Figure 8.2 (B)orders of mobility

Source: Author

BOX 8.1 THE IMPLICATIONS OF MOBILITY FOR 'PLACE': THE ONGOING COMPOSITION OF 'PLACES OF MOVEMENT' (HETHERINGTON, 1997)

As we have seen throughout this book, places can be understood as an ongoing composition of traces. By considering the importance of mobility in the contemporary world – in focusing particularly on where traces have come from, how they are moving, and where they are going to – we have seen how 'traces' can be mobilised into 'trajectories' (Chapter 4). If we take mobility seriously, therefore, places can be understood as 'a constellation of on-going trajectories' (Massey, 2006: 92); traces move from one location, pause in our place, then continue on their journey elsewhere. From this perspective place is far from a static, stable, or fixed entity. Place is no longer reliable, consistent, or necessarily coherent, it is wholly provisional and

Picturing 'mobile' places
Source: Author

unstable. In this way, places are perhaps best thought of as train stations or airports, with mobile objects passing through at different speeds, on different journeys, and staying for different lengths of time. A place, at any moment, thus emerges from the web of mobile flows that meet at that particular connection. In this view, place becomes the 'point' in motion, a temporary pause or 'platform'. However, this 'platform' is permanently in the present, only temporarily 'fixed', and now something else.

Inspired by these mobile ways of thinking, Ingold (2005) compares places to line drawings. In the act of drawing a pencil across a page, places become the 'dots' (i.e. pauses) in a 'trail of action'. In this view, the *meeting* of lines at a place cannot be separated from the *movement* of the pencil, the 'dot' cannot be separated from the 'continuous line' (Ingold, 2005). In Ingold's words (2005: 47), even though:

> we are now more inclined to think of [place] as a complex of interconnected points [it is useful to think of place as a complex of] interwoven lines . . . The lines of the meshwork . . . are the trails along which life is lived. And it is in the entanglement of lines, not in the connecting of points, that the mesh is constituted.

Place therefore is the geographical and temporal 'dimension that cuts through . . . trajectories' (ibid.). Taking mobility seriously means that place has become a moment and a location, simultaneously shot through with lines of movement that constantly (re)combine to change its form and substance. With these points in mind, we can draw on Cresswell's 'constellations of mobility' to raise further questions concerning the 'mobilities of place'. For example, we can ask: how does a place 'move'? How fast does it change? In what rhythm? What route does it take? When and how does it stop? And, last but not least in terms of the interests of cultural geography, how does it feel for those experiencing these changes? (see Anderson and Erskine, 2014: 134.) Try to apply these questions to the place you live. How mobile is your place?

when a missed sailing is a beautiful opportunity to kick back, open a cold drink, and maybe light up a smoke. If you are departing – as opposed to returning – you might even be able to go back home, make yourself a sandwich, and try your luck later with the next sailing. But most days, life is not like that. You've got somewhere you gotta go. Missing a ferry is like bumping against one of those pyramid-shaped promotional displays of pop cans piled up on top of one another at the grocery store. You hit one can and the whole pile falls down.

(Vannini, 2011: 1033)

Practising mobility is therefore an art form. It requires pacing, timing, and logistical awareness – and has inconvenient consequences if we get it wrong. However, those of us who rely on public transport or a private car to get to work will know that crafting mobility is also sometimes about *im*mobility – we are sometimes static in traffic, or are rendered motionless and rudderless when waiting for that routinely delayed train. As Cresswell puts it (through adopting an appropriate nautical metaphor in light of our ferry example), 'moorings' are often as important as 'mobilities' (2010a). With respect to the questions raised above, these moorings relate to the frictions that impede mobility, they are the knock-on effects from the falling display of pop cans that mean we have to stop, with a further set of consequences for ourselves and others. As Bissell (2014: 191) outlines, passengers and commuters often only realise the omnipresence of mobility when these frictions occur and 'the normal functioning of mobility systems are interrupted'. Our expectation of mobility but our experience of immobility throws into relief our reliance on the smooth operation of transport systems, and their fallible nature. This is often the case at ferry ports, at bus shelters, but also at other place 'platforms', including air terminals.

Airport (im)mobility

As Adey (2007: 515) suggests, 'airport spaces contain important areas of not only mobility but also immobility'. In these places individuals queue, wait, and anticipate their moment to move on to the next stage of their journey. In these (im)mobile places, cultural geographers have begun to explore the different ways in which expectations of mobility, and the frictions that forestall it, have changed the relations between people and place. The dependence on mobility, particularly air travel, poses a number of risks to individuals and cultures. Threats from extreme weather, volcanic eruption, surface to air attack, and hijacking, have all increased in recent decades, and surveillance and security systems have advanced in an attempt to (b)order these threats. Those who wish to travel by commercial airline therefore have to forgo their individuality, as well as some of their civil liberties, in order to cross the globe at high speed. As Bissell (2010: 270) states, when we fly by air 'we temporarily submit ourselves to become part of a mobile collective'; through doing so we acquiesce to requests for identification, intimate searches, and body scanning which would be refused in many individuals' normal lives.

Picture yourself immobile in the airport, keen to begin your trans-oceanic flight. Here you change your identity: you are no longer a 'citizen' but a 'suspect'. As Adey (2008: 438) warns, you will be scrutinised by intrusive technologies: 'surveillance techniques of retina scanning biometric systems, algorithmic recognition and . . . passenger profiling software' go hand in plastic glove with 'drug-sniffing dogs, X-ray machines, metal detectors, mandatory

searches, restrictions on movement, security inspections, and intense screening' (Adey et al., 2007: 780). Whilst 'staring customs officers, sharp questioning, bioidentifiers, computerized facial recognition and other technological marvels are meant to produce an environment in which people's intentions are "revealed"' (Aaltola 2005: 263, cited in Adey et al., 2007: 780) it often results in stoic or frustrated resignation in the face of (un)necessary infringements of liberty. The (im)mobility of airport terminals forces us to consider how some prospective passengers surrender their sense of autonomy and citizenship to the collective power of the aeroscape. However, this surrender is not the same for all. As we stand waiting in line, Bissell reminds us of the 'trusted, regular, business-class travellers [who] are not subject to the same long and laborious surveillance strategies that other "unknown" and therefore "potentially risky" bodies may pose' (2009: 174). According to Graham and Marvin (2001), these are the 'kinetic elite', who – more often than not – fail to be 'encumbered by delays and the discomfort of crowds' (Bissell, 2009: 174). Whilst we are annoyed that this 'kinetic elite' get to enjoy ease of movement through the airport (and turn left at the cabin door rather that right), we thus imagine 'how it feels' to be an unencumbered traveller (and know how it feels to be a relatively encumbered one). In this way we can begin to acknowledge how experiences of mobility are highly unequal. In Bissell's terms, 'understanding this heterogeneity of passenger experience is central to achieving a more nuanced account of the contemporary mobile experience for those caught up in these "spaces of flow"' (ibid.).

Fast and slow lanes: (b)ordering mobile lives

Mobility in our everyday life is a highly unequal experience. From the lack of dropped kerbs on public streets which (b)order the routes of push- and wheelchairs, to militarised patrols governing access into countries (see below), mobility is (b)ordered by dominant cultural groups. I take for granted my ability to drive (not everyone's body facilitates such mobility, see Chapter 13, whilst not everyone may be able to afford the taxes, tyres, fuel, maintenance, insurance . . . that running a car involves). However, in some cultures, the opportunity to drive is a 'pleasure' that is (b)ordered (see Box 8.2).

We can see, therefore, how everyday mobility is (b)ordered in a range of ways, to a range of extremes. As Massey has told us above, this (b)ordering 'both reflects and reinforces power' (1994: 15). Through (b)ordering mobility, particular cultural groups have life opportunities enabled, or restricted: (b)ordering mobility therefore (b)orders lives. Imagine, for example, being forcibly extradited to another country, or being repatriated following asylum; imagine too being exiled in a 'space of exception' (see Agamben, 2005) – an area outside legal norms in which populations are subject to exclusion and summary violence (Gregory, 2004). These examples go some way to illustrate how some cultural groups travel in very different lanes on the mobility superhighway. Their mobility is (b)ordered, and their freedoms curtailed. Gogia (2006) has focused precisely on these heterogeneous experiences of mobility by contrasting the trajectories made by backpackers and seasonal workers travelling between Mexico and Canada. As Gogia claims, each group of individuals 'went to each other's country for completely different reasons, and they would be treated by the host country's citizens in vastly different ways' (2006: 359). Whilst backpackers from Canada had relative freedom in their mobility both in their host nation and in Latin America (notwithstanding airport delays and frustrations), Mexican workers' mobility was (b)ordered in far stricter ways. When in Canada, their farm 'patrones' not only (b)ordered the workers' labour hours and conditions, but also their eating, living, and recreational arrangements. As Basok (2002: 123) points out:

> the patrones discouraged their workers from having social activities outside of work. A worker who returned home too late or who got late visitors, particularly women, risked earning the disapproval of his patron, who believed that such activities might exhaust the worker and make it difficult for him to work well the next morning.

McGrath (2012: 1008) argues that such restrictions on 'workers' physical and social mobility' is part of a constellation which creates forms of slave labour. Denying the mobility of migrant workers (b)orders their lives.

BOX 8.2 STOP!

Most cultures (b)order driving in some way or other. Drivers have to move within certain speeds, in particular directions, and on specific roads. Individuals of particular ages or different abilities may not be allowed to drive. However, in some countries, your gender may be the cultural trait which (b)orders the use of a private car. In Saudi Arabia, for example, women are prohibited from driving. Although this activity is not often explicitly punished (women caught driving are often asked to 'pledge' their obedience to the law and abstain from driving in the future), in some cases draconian punishments are meted out. In 2011 a woman was sentenced to public flogging for flouting this (b)order (see Associated Press, 2011). Imagine you live in Saudi Arabia. How different would your life be if you were under the influence of this (b)order? What could, and what would, you do?

BOX 8.3 (B)ORDERING MOBILITY IN TOWN AND COUNTRY

Mobility for pleasure or profit? Skateboarding in the city

Ask yourself, why are people mobile in the city? One may argue that a significant proportion of mobility in the city is for work, for consumption, or for play. The third reason for mobility often comes into conflict with the first two. Skateboarding is one example of this. As Vivoni (2009: 133, 140, 133) states,

(B)ordering boarding
Source: Author/Kimberley Peters

> The modern city sports all sorts of physical terrain built mostly for the flow of capital and for the enjoyment of . . . prescribed [mobility]. Any sort of combination of smooth low-set ledges . . . that inspire both horizontal and vertical movements . . . are the preferred found spaces of street skateboarders . . . Landscaped public plazas [that most people want to move through] become sources of thrill-induced pleasure [for skaters] and [thus] sites for political and spatial contestation.'

Vivoni outlines how the geography of the city provides a unique terrain in which skateboarders can realise their mobility. However, their use of place directly challenges other forms of mobility, and the motivations behind them. As a consequence, skateboarding is often (b)ordered as 'out of place'. Signs, barriers on street furniture, and redesigned thoroughfares privilege mobility for profit, rather than mobility for pleasure. However, skateboarders resist such (b)orders, and seek to take and make their place by leaving their own cultural traces in the city: 'skatescuff' is their stuff, as Vivoni (2009: 141) states,

Skateboarders scout areas such as downtown financial districts, public squares, college campuses, and suburban strip malls in search of exhilarating terrain . . . Once a potential skatespot is identified, spatial appropriation is enacted through tactical modifications to the built environment. One copious tactic consists of the meticulous application of wax to hard edges in order to glide or grind with greater ease and increased speed. Palimpsests of board art, wax, and dirt are left behind as meaningful scuff denoting a skatespot in progress.

Where are mountain bikes out of place?

Mobility is (b)ordered in urban places in various ways, but our mobility is (b)ordered in nature too (see also Chapter 9). Consider what you think to be the most appropriate form of mobility when in the place of a forest, or on a lake, or up a mountain? On foot, by windsurf board, or on horseback? With these issues in mind, we can ask: what constitutes acceptable and unacceptable ways to leave one's mark when moving through

Non-urban mobilities
Source: Author/Cairngorms National Park

nature? Brown (2014) examines this (b)ordering of our mobility in the case of the Cairngorms National Park, Scotland. Ironically perhaps, given their nomenclature, many suggest that mountain bikes are an 'out of place' form of mobility here. In this place, it is argued that walking or hiking should be acceptable, but cycling should not. What do you think is 'wrong' with this form of mobility, in this place? Is it the tracks left by mountain bikes which erode paths and open areas? Is it the sight of lumo-lycra cyclists in the tranquil Scottish glens? Is it their pace which is out of place? What (b)orders do you think should be imposed to keep this place free of 'unnatural' mobility? (See Chapter 9.)

(B)ordering immigration: the wrong sort of mobility?

(B)ordering mobility has significant effects on people and places. The costs involved in (b)ordering effectively (or otherwise) can also be considerable. This is the case with regard to immigration. Summer in the Mediterranean is 'boat season'. This term does not refer to the elite mobility of super-yachts in the South of France (for this, see Spence, 2014), but rather illegal immigration into Europe from Africa. As Sherwood et al. report, during this time, 'hundreds of thousands of people prepare to make treacherous crossings on unseaworthy vessels from the north African coast to Greece and Italy' (2014). Illegal migration to the European Union is seen by these individuals as the first step to a safer life, despite the risks of capture, detainment, and deportation inherent to human trafficking, as well as the clear and ever-present danger of the sea itself. In 2011 alone, 1500 people drowned in their attempt to make this step across the sea (Shenker, 2013). To (b)order this mobility, and reduce the danger to vulnerable migrants, Miltiadis Varvitsiotis, Greece's merchant marine minister, said processing offices should be set up in Syria and Turkey which could 'examine who is eligible [to move] before people get on boats, put their lives in danger, and trespass our borders' (Sherwood et al., 2014). In the United States, illegal immigration is also a threat. In order to control the mobility of people across the Mexican border, the United States government has set up the 'Border Patrol'. The patrol has a $30 billion budget (more than the FBI, Secret Service, DEA, and Bureau of Alcohol, Tobacco, Firearms and Explosives combined) as well as 39,000 active troops which police the perimeter (see Helmore, 2014). This 'militarisation' of the (b)order is one way to deny the kinetic underclass their 'fantasies of wanting to move'.

BOX 8.4 RISKS OF MOBILITY: THE OFFSHORE CLASS AND TIME TRAVEL

As we have seen in Chapter 7, mobility continues to be a key fuel that drives globalisation. Mobility of capital, commodities, and people often generates jobs and wealth for many cultural groups, but this mobility also has negative consequences for the economy, for the environment (see below), and for the human body. In terms of the economy, there are some cultural groups who are able to use mobility to conceal their wealth from governments and secret it away to tax havens. Urry (2014) identifies this mobility as 'offshoring', and those who benefit from it as 'the offshore class'. According to Urry, this form of mobility has meant that 'powerful and pervasive offshore worlds have been generated, posing huge challenges both for governments

The plane's trail/razors open the sky/climate change is death/by a thousand cuts

Source: Author

and for citizens' (2014: blurb). Just like the case of immigration, these worlds of mobility trangress conventional (b)orders, and yet there is little 'friction' imposed to stop this sort of mobility; as Urry puts it, the offshore world is: 'literally and metaphorically over the horizon. Much is invisible, out of sight of states [and] law enforcers . . . [This form of mobility] provides a model of offshored neo-liberalism, with minimal . . . regulations' (2014: 13–14).

Another risk associated with mobility comes not from the ability of human minds to find loopholes through which kinetic elites can pass, but rather from our body's ability to cope with modern movement. Question: have you ever flown across time zones, arriving in a new place in the middle of the night, or even earlier in the day than when you left? How well have you coped with the language, weather, altitude, or pollution in this new place? Have you had problems fitting in with the familiar when you return home again? Like all forms of mobility, long-distance, high-speed, jet travel has consequences. These consequences occur for the planet and for our bodies. In the latter case, these consequences are often understood as jet 'lag' (see Anderson, 2013b). As Rozell identifies: 'all organisms, from one-celled creatures to humans, synchronize to the earth's 24-hour rotation cycle. This cycle dictates when they sleep, wake, secrete hormones and perform other bodily functions' (1995: 1261). When we travel at high speed into new time zones our cycles are disrupted, causing 'circadian dysrhythmia'. As Graeber (1988: 312) puts it:

> Normally, when we are living at home, our circadian rhythm resembles a finely tuned symphony, each rhythm rising and falling, not at the same time, but at its own prescribed time in harmony with one another. This internal synchronisation does not persist after transmeridian flight. Even though our myriad circadian rhythms are timed by only one or two clocks, they all do not resynchronise together.'

When experiencing circadian dysrhythmia, the body has to readjust its functions over a minimum of 2–3 days (see Graeber, 1988). A range of physical ailments are symptoms of this adjustment, as Hawkins (1987: 48) outlines:

> Sleep disturbance is experienced; bowel elimination and eating habits may be disrupted. Lassitude, anxiety, irritability and depression are often reported. Objectively, there is evidence of slowed reaction and decision-making times, defective memory for recent events, errors in computations and . . . a tendency to accept lower standards of operational performance.

Circadian dysrhythmia is, therefore, a significant physical consequence of high-speed mobility. The human body is 'out of sync' with its new geographical relations, a condition that may last for several days and fundamentally influence the way in which that body senses itself and its relation to the world. As Parker puts it, 'jet lag is penance, payback, protest – roughly equivalent in moral force, I'd say, to the hangover. It reminds us not just that we have bodies, but that we are bodies. Flout the time zones and you will feel unusual' (2010).

BOX 8.5 MOBILITY IN THE PALM OF OUR HAND: CELLPHONES

Alongside more explicit border controls, technology has a role in enabling and restricting our mobility. From GPS technology providing us with directions, to online timetables telling us about arrivals and departures, technology has become part and parcel of our mobile lives. Often this technology is at our fingertips in the form of our mobile (or cell) phones. Wi-fi linked phone technology enables us to get the necessary information to 'craft' our journeys from A to B, and gives us information where police patrols may be looking to curtail our speed of movement (see www.gosafe.org). However this technology also liberates the mobility of some cultural groups, as well as changing the way we consider our location and positioning.

For example, cultural geographers explore the ways in which cellphones release young people from parental (b)orders. Being universally contactable at any time changes the enclosures placed on young people and enables them to go to further and go differently across their cultural places – 'so long as they take their cellphone' (after Pain et al., 2005). Cellphones also enable us to make new forms of connection, rearrange our meetings, and improve the 'crafting' of our mobility. But this increased capacity to move can also change how we describe and think about our cultural and geographical positioning. As Laurier (2001: 485) recounts in relation to a railroad journey:

> You're sitting on a train, you're tired and irritable after a long journey. Someone else's mobile phone rings. You hear it get louder as they fish it out of their bag. 'Hi' they say. There's a pause. 'I'm on the train. About half an hour away from London.' 'Why', you sigh, 'why do they always do that?'

This story reveals how, in a world of mobility, simply stating where we are remains an engrained element of our lives, but is also an increasingly problematic statement to make. We may be 'on a train', but where is that train? On some tracks, somewhere between A and B? If we had the appropriate technology, we could give geographical co-ordinates, but we often simply approximate. How would we answer the question 'where are you' if we were on a plane? Are we in a particular country's airspace? In a different time zone? Or simply in the pressurised atmosphere of the aircraft cabin? So how do we understand our location and positioning in a world of mobile flows? Even being 'away' from 'home' makes less sense if we are always mobile, as our home may now be 'housed' in our technology, our websites, and our communication apps that are held in the palm of our hand. Yet Laurier's identification that individuals *want* to locate themselves, despite the difficulty in doing so, reminds us how vital the constitutive co-ingredience of people and place remains in this mobile era.

Mobile lives: flashpacking

We have seen how mobile technology can enable physical movement across (b)orders, and also how different types of travellers are (b)ordered in their mobility. Some geographers combine these elements through exploring how backpackers use cellphones and social media to change the constitutive co-ingredience of people and place. As Germann Moltz and Morris Paris (2013: 8) state,

> thanks to the digital immediacy of camera phones and the Internet, travellers can maintain a constant sense of co-presence with a dispersed social network. Snapshots and travel stories that would have been shared after the journey can now be shared in interactive

formats while the traveller is experiencing them. Travel mottos like 'I was here!' or 'Wish you were here!' seem obsolete in an age of 'I am here right now and you are (virtually) here with me!'

Germann Moltz and Morris Paris introduce the term 'flashpacking' to describe how locations near and far are connected through emerging forms of mobile sociality. These immediate and ongoing sociabilities are enabled by a new terrain of inter-activity as flashpackers are able to stay connected in the realm of the 'statusphere'. The statusphere – defined as 'the state of publishing, reading, responding to, and sharing micro-sized updates' (Solis 2009, cited in Germann Moltz and Morris Paris, 2013), provides flashpackers with a kind of 'virtual mooring' that keeps them connected to their original places whilst they are on the move.

The case of flashpacking highlights how the world is not only constituted by mobility, but also inhabited by mobile individuals. Another example of a mobile cultural group is the case of lifestyle travellers. Some of us may aspire to be lifestyle travellers, some of you may have already experienced a type of lifestyle travel ('gap years', for example, are a type of lifestyle travel). However, imagine for a moment 'gapping out' completely: this is what lifestyle travellers do. Whilst many students or tourists accept a return to 'normal' life once their travels are complete, the lifestyle traveller rejects this return, transforming their travel into their 'norm' and everyday way of life (see Cohen, 2010). The cases of lifestyle travellers and flashpackers are not only important as examples of cultural groups with mobile lives, but also because they change the ways we can consider the relations between people and place.

Constitutive co-ingredience: from roots to routes

We have seen in Chapter 4 how a 'constitutive co-ingredience' of people and place can be engendered through 'incoming' and 'outgo-ing' in a location. The material and nonmaterial traces in a place come to influence who we are, and our laying down traces in a location also comes to define the nature of the place itself. Tuan describes the emotional ties developed in people through this co-ingredience with the term 'topophilia' (1974). Topophilia refers to the affective bond that is registered in a person but created by their co-constitution with a par-ticular place. A positive relation between people and place (for example in a home setting) would be termed topophilia, whilst a negative feeling (perhaps in a strange or unwelcoming place) would be its counterpoint – topophobia. Due to the positive nature of topophilia we often seek to stop places changing and thus retain our sense of attachment with our cultural and geographical surroundings. As a consequence it is common for narratives of change to become pejoratively framed as threat. However, we know that the world is constantly changing – places are defined by movement, and by an ongoing composition of traces. Lifestyle travellers contribute to this change, not by their persistent incoming and outgoing in one location, but through their mobility within and between places. Their sense of constitutive co-ingredience is thus not gener-ated through a preference for roots, attachment, and topophilia, but rather for a desire for routes, mobility, and what I have described elsewhere as 'tropophilia' (see Anderson and Erskine, 2014). The term 'tropophilia', meaning the love of change, refers to the new ways in which the relations between people and place can be understood – not a love of rooted and static geographical relations, but a love of mobility, movement, and change in terms of the co-ingre-dient constitution of identity and geography.

Thus lifestyle travellers do not simply have mobile lives, but these lives challenge the sed-entary ways we consider person-place relations. As a consequence, lifestyle travellers are not sim-ply mobile subjects, but to use Braidiotti's term, they are also 'nomadic subjects' (1994). The

BOX 8.6 CO-INGREDIENCE AND LIFESTYLE TRAVEL

the point [of mobility] is not to find yourself, the point is to create yourself . . . I don't know if there was anything there pre-existing that I was supposed to find but I definitely created myself along the way.

(Marco, in Anderson and Erskine, 2014: 138)

Lifestyle travel is a form of mobility undertaken to transform the person moving. As Rojek states, whilst '[conventional] tourism is considered to actually confirm one's view of the world . . . travel is seen as a resource in the endeavour of self-realization' (1993: 175). Mobility is used to put the traveller in a new place, challenging them to create a new identity in this new set of co-ingredient relations. Due to the different paces and trajectories of place they encounter, individuals are required to adjust their selves to understand the languages, customs, rites, and practices which allow themselves to survive there. It is exactly this form of adjustment and change that lifestyle travellers thrive on, as 'Miriam' outlines:

there are so many people and so many countries that seem very similar but can be really different, and there's languages to learn and food to try and there's just an exhilaration of moving to a city and knowing that there's no-one in the whole country that you know that you could call if you had an emergency, that you don't have a job, you don't have a flat and you have $2000 in the bank and you have to somehow sort yourself out and you know . . . it's better than sky diving, the rush!

('Miriam', cited in Anderson and Erskine, 2014: 138)

For Miriam, the freedom to test herself in new co-ingredient relations is synonymous with fulfilling a personal challenge, to see how far she can push herself when isolated from familiar cultural traces. By submerging herself in an unknown place, she can explore the paths and networks that constantly become it, she can actively pursue those trajectories that challenge or resonate with her own in order to achieve goals of self-development and satisfaction. Through this testing and experimenting with aspects of herself, she is constantly engaged in getting to know herself when constituted differently in new person-place relations.

nomadic subject is defined not simply by acts of mobility, but also by the way they 'subvert[t] set conventions' (1994: 5). Not only do lifestyle travellers transgress sedentary norms by mooring in and to places as the exception rather than the rule, they also invoke new ways of describing 'home' and 'away', and subvert how we might consider a sense of place and belonging through mobility (see Box 8.6).

Conclusion

As we have seen in this chapter, cultural geographers have begun to open the 'black box' of mobility. We have identified how geographers have begun to conceptualise the different constellations of mobility, the varied sploshes and froths that (im)mobility causes, and the new traces and trajectories that can be forged as a result. Like cultures of capitalism (see Chapter 7), cultures of mobility cross-cut with many of the themes examined in this book. Mobility intertwines particularly with cultures of nature, and that is the destination to which this book now moves. Hold tight.

Suggested reading

The following texts are useful to further explore the issues raised by this chapter.

Adey, P. (2010) *Mobility*. Routledge, London and New York.

Anderson, J. (2013b) Mapping the consequences of mobility: reclaiming jet lag as the state of travel disorientation. *Mobilities*. DOI: 10.1080/17450101.2013.806392.

Anderson, J. and Erskine, K. (2014) Tropophilia: a study of people, place and lifestyle travel. *Mobilities*, 9 (1), 130–145.

Binnie, J., Edensor, T., Holloway, J., Millington, S., and Young, C. (2007) Mundane mobilities, banal travels. *Social and Cultural Geography*, 8 (2), 165–174.

Bissell, D. (2009) Conceptualising differently-mobile passengers: geographies of everyday encumbrance in the railway station. *Social and Cultural Geographies*, 10 (2), 173–195.

Cresswell, T. (2010a) Mobilities I: catching up. *Progress in Human Geography*, 35, 550–558.

Cresswell, T. (2010b) New cultural geography – an unfinished project? *Cultural Geographies*, 17, 169–174.

Cresswell, T. (2010c) Towards a politics of mobility. *Environment and Planning D. Society and Space*, 28, 17–31.

Cresswell, T. (2011) Mobilities II: Still. *Progress in Human Geography*. Available at: http: // phg/sagepub.com/content/early/2011/ 09/22/0309132511423349 (accessed 10 November 2014).

Germann Molz, J. and Morris Paris, C. (2013) The social affordances of flashpacking: exploring the mobility nexus of travel and communication. *Mobilities*. DOI: 10.1080/17450101.2013.848605.

Sheller, M. and Urry, J. (2006) The new mobilities paradigm. *Environment and Planning A*, 38, 207–226.

Place and nature

Introduction

As we have argued throughout this book, places are ongoing compositions of traces. These traces can be both material and mental in nature. They can take physical shape as things, events, and processes, and they can involve emotional, intellectual, and psychological connections. As we will see in this chapter, how different groups think and feel about places is as important as the physical form places take. This is because our thoughts and feelings about place often directly influence how we act in the cultural world. Nature is a key example of this process. In this chapter we explore how our thoughts and feelings about nature influence how we take and make its places.

Let's begin with an example: the place of Uluru, Northern Territory, Australia (Figure 9.1).

Uluru is a place. Materially it's a rock of natural sandstone in Northern Territory, Australia. To the white settlers of the nineteenth century it was named Ayers Rock by surveyor William Gosse, after the Chief Secretary of Central Australia, Henry Ayers. To the Aboriginal people, it is called Uluru. To many it's a place of tourism and commodification (Uluru attracts 400,000 visitors a year (Popic, 2005; Young, 2010)), to others it's a sacred place, similar to a synagogue or sanctuary. To some it offers a physical challenge, to others a place of worship, or it is simply a rock in the middle of nowhere. How should we make sense of this 'place of

Figure 9.1 The place of Uluru

Source: Yuko Kootnikoff www.flickr.com/photos/yewco/72071400/

nature'? How should we act in it? Should it be protected or exploited? Should it be worshipped, or climbed (see Box 9.1)?

Uluru then is a place of nature. But this natural place is (b)ordered by a range of often conflicting cultural meanings. Nature here represents tribal ideas of sacredness and dreamtime, as well as national ideals of adventure. Each set of (b)orders seeks to impose particular rules of behaviour on the rock, permitting certain actions but outlawing others. Where do these competing ideas come from? Can they be balanced? How would you adjudicate between them? We can see that, according to survey data, Europeans are more sensitive and sympathetic

BOX 9.1. SHOULD YOU CLIMB ULURU?

As Popic (2005) documents, 'Climbing the rock is now a point of contention between two different cultures.' The Anangu, the local Aboriginal people who in 1985 were recognised as the traditional owners of the area covered by Uluru Kata-Tjuta national park, do not want visitors to climb Uluru. Under traditional law, climbing is prohibited to everyone except senior men initiated into Anangu culture, but even elders are limited in where and when they can climb. '"That's a really important sacred thing that you are climbing," says one senior traditional owner. "You shouldn't climb. It's not the real thing about this place. The real thing is listening to everything . . . This is the proper way"' (cited in Popic, 2005).

The message of culturally sensitive tourism appears to be effective. Surveys conducted by geographer Richard Baker from the Australian National University showed the number of people who climbed or intended to climb Uluru had fallen from 43 per cent of park visitors in 2003 to 35.5 per cent in 2004. By 2012 numbers had fallen further to 17 per cent (Laughland, 2013). Interestingly, Baker found that Australians and the Japanese were the most keen to climb the ancient rock, while Europeans were the least likely.

As Popic (2005) records,

Would you climb Uluru?

Source: Millie Harris

for many Australians, the issue evokes heartfelt feelings of cultural ownership and personal freedom. Ken Duncan, an Australian landscape photographer who has produced some of the most iconic shots of Uluru over the past two decades, is fervently against any ban on the climb. 'Of course people should be able to climb it,' he says. 'You know what we call it now? Ulu-rules. Aboriginal people have no more claim on it than any other groups. We as Australians and as tourists are being locked out of this beautiful icon.' Although Parks Australia deny the suggestion, Duncan believes the powers-that-be will eventually close the climb. 'Where's the adventure left in this country? I'm embarrassed to take my daughter to a place that's so sanitised, so controlled. It's got to be spiritual to everyone, not just Aboriginal people. It's the heart of our nation.'

However, these claims for inclusiveness fail to take into account the wishes of traditional owners. Climbing Uluru transgresses the indigenous people's 'tjukurpa'. Tjukurpa refers to Anangu rites and folklore which tie the people into the land and its web of spiritual meaning. As Laughland comments, 'the desire to live harmoniously with the natural landscape is the key to understanding why climbing Uluru is deeply offensive' (2013). David Ross, director of the Central Land Council, argues how Aboriginal people have had their claims disregarded for generations, and points to the particular distress felt by traditional owners at the large number of climbers who have met their death or been injured on the Rock. 'It's like someone coming into your house and dying in your living room. It's not your fault but you still feel bad about it. That stuff just doesn't seem to be taken into account' (Ross, in Popic, 2005).

Climbing the rock has also created new environmental threats. Erosion caused by human traffic is causing visible deterioration of the rock surface, whilst discoloration and environmental harm is being caused by climbers relieving themselves during their ascent. As a protected site, Parks Australia are required to conserve the area from damage; should this environmental pollution be the reason to stop people climbing Uluru, or should indigenous cultural beliefs be enough? David Ross believes what is needed is a different kind of respect for this national icon, particularly from Australians, many of whom consider it their birthright to climb. 'You can't climb over the Acropolis any more,' he says. 'You can't climb the Pyramids, so how come you can still climb the rock?'

to the Aboriginal culture's construction of the rock, perhaps acknowledging the similarity of this natural entity to the cultural entities of the Acropolis and the Pyramids. (As Rosalind King, curator of the October Gallery, London, puts it, 'It's a sacred site. It would be like walking into a church and trampling all over the altar' (see Popic, 2005)). However, the Japanese tend to be more sympathetic to the settler community's construction of Uluru. Why is nature seen as a playground for human activity by these groups? Whose *Ulu-rules* should rule? Should Aboriginal people decide on how the rock is used, just because they were there first?

The case of Uluru illustrates that the place of nature is looked at very differently by different cultures. In this chapter we will investigate the different cultural (b)orderings of nature. We will see that dominant ideas of nature (from a Western perspective at least) border it outside and apart from human culture. We will see how this view is used to create a variety of 'common sense' actions in the natural world: some groups seek to 'protect' nature, some to 'control' it, and some to 'exploit' it. However, many cultures challenge this dominant idea of nature. Alternative (often non-Western) ideas of nature see the natural world not as apart from human culture, but rather as thoroughly imbued by it (and vice versa). These ideas of nature prompt us to freshly consider how we feel about nature, as well as the contributions that we make to the ongoing composition of 'natural' places. In other words, they affect our decision about whether we should climb Uluru.

The nature of nature

What is 'nature' to you? Perhaps what springs into our hearts and minds is a favourite environment, an image from our youth, or from a 'nature' documentary off the television. We might not be able to put into words a coherent answer to the question, 'what is nature?', but we know 'nature' when we see it (perhaps even when we feel it). From my own schooling, when we were asked to bring in 'natural' things into class, I remember someone brought in their photos of the barren, windswept Scottish Highlands (without any cyclists captured, see Chapter 8). To me, this was nature. But why? The only 'reasoning' I could come up with was that these photos had no human beings in them (the fact that I knew that there had to be a human there to take the photo I conveniently overlooked!). To me, then, nature was something where human culture was absent.

In this example lies not only the orthodox way in which the nature of nature has been framed in many industrialised countries, but also the problem with it. Nature has been traditionally framed as somehow *outside or beyond* culture. If culture is something humans construct, if it is something that humans *do*, then nature is something and somewhere they don't. Nature is something pristine and unsullied by human interference; it has yet to be tainted by cultural 'pollution'. So in this traditional framing there is a clear distinction between nature on one hand, and culture on the other. This binary dichotomy separates the developed from the natural, and the human from the non-human; it separates culture from nature. Maskit outlines this point of view; from this perspective:

> [Nature] must have a certain purity to it. It often must have a certain remoteness to it. It should show no (or few) visible signs of civilisation – roads, houses, power lines, and the like are all things that make a place less wild.
>
> (Maskit, 1998: 266)

The idea of nature being outside culture is the dominant way nature is framed in the West. It is the 'natural' way in which people respond to questions about nature, often assuming that there is no other way in which nature can be identified. The following comments echo this cultural perspective on the nature of nature:

B: What does [nature] mean to you?
A: Mostly uninhabited land. I would not expect to find much human activity there. I wouldn't expect to find structures other than maybe a makeshift camp that somebody put up at one point in time. Where the flora are totally in control of what's happening, and you're definitely a guest when you show up.

I feel it means that there's a place we can still go and get away from people and things, and everybody else hasn't been there and set up a McDonald's already.

(Bertolas, 1998: 267).

The notion that humans are apart from rather than a part of nature originated in the Judeo-Christian heritage that dominates Renaissance- and modern-scientific views. In the pre-modern world, nature was understood as the world of spirits and gods (as it remains in many aboriginal cultures today). Such divine and enchanted ways of looking at nature were displaced with the onset of Christianity (and, later, modern notions of science). Christianity regarded nature as created but not inhabited by God, and in God's absence humans were assigned the role of controlling nature. This role has been interpreted as one of stewardship by some (see Attfield, 1991), but also as a right to exploit by others (e.g. Francis Bacon, see Merchant, 1980).

As a consequence of this Judeo-Christian influence many argue that the separation of human culture from nature is profoundly ingrained in the West. As Castree and MacMillan argue, this separation has become 'so familiar and fundamental to be unquestionable' (2002: 208). From their point of view, this (b)ordering of nature has become doxa. Such (b)ordering brings with it not only geographical classification – nature is here, culture is there – but also cultural expectations about how the material world of nature should be treated. These expectations are outlined in Boxes 9.2, 9.3, and 9.4 that follow. Each of these practices competes to be accepted as the 'common sense' way to deal with the natural world.

BOX 9.2 NATURE AS OTHER TO HUMAN CULTURE: A PLACE TO BE PROTECTED

If nature is other from culture then many argue nature should be protected. Environmentalists and conservationists argue that the independence of nature is valuable, and as such nature should be valued in and of itself, regardless of its utility for humans. This 'ecocentric' (Devall and Sessions, 1985), or 'biocentric' (Goldsmith, 1996; Seed, 1988) approach often leads to the reinforcement of difference between nature and culture by 'ring fencing' nature protection into particular areas (such as national parks or nature reserves). Within these physical borders, human interference is limited. As these are areas reserved for nature, only certain human activities are deemed appropriate (such as mild recreation or education). From this perspective, nature is a place to briefly visit, where we should 'leave no trace' (see for example www.lnt.org). Nature is not a place in which we make our lives, just our recreation (after White, 1995).

Fell

Source: David J Glaves, Exeter, Whitendale Fell, Forest of Bowland, archived at www.flickr.com:80/photos/davejglaves/3012540106/

Colley Hill

Source: 604Plonker www.flickr.com/photos/11714573@N00/184380063/

BOX 9.3 NATURE AS OTHER TO HUMAN CULTURE: A PLACE TO BE TAMED

If nature is other from culture then some argue that it is not nature that should be protected from culture, but rather the other way around: it is humans and their culture that should be protected from nature. Here nature is framed as wild, uncontrolled, and a threat to humans; and as a consequence is a place that should be 'tamed'. This taming allows cultures to exert their control over nature to pacify its threat. Thus rather than human culture being excluded from places of nature, it is extended into them. Examples of this practice include land defences to control the worst excesses of the wild ocean, and the control of rivers to divert or limit flooding. The following extracts from McPhee's book *The Control of Nature* (1989) go some way to illustrate a number of ways in which nature is seen as a threat to human culture, and thus warrants particular measures to control it.

A fascinating report from three *battlefields in humanity's global war against nature.*
(*New York Times*, sleeve blurb, my emphasis)

Three epic struggles with nature form the subject of this book. Each takes place in a part of the world where people assert their right to live, despite nature's stubborn refusal to provide the conditions for safe human existence.

(ibid.)

Unaware of the people who depend for a living on the economy of the New Orleans and the Delta, the [Mississippi] river in one of its natural cycles is trying to shift its mouth more than a hundred miles to the west. The United States likes things as they are, and its Army Corps of Engineers has declared war on nature. A gigantic fortress, part dam, part valve, attempts to restrain the river's flow.

(ibid.)

Nature, in this place, had become an enemy of the state.

(ibid.: 95)

Our opponent could cause the US to . . . lose her standing as first among trading nations . . . we are fighting Mother Nature . . . it's a battle we have to fight day by day, year by year; the health of our economy depends on victory.

(Narrator of a US Army corps video, in ibid.: 98)

BOX 9.4 NATURE AS OTHER TO HUMAN CULTURE: A PLACE TO BE EXPLOITED

In framing nature as outside human culture, a third practice suggests that the common sense way to deal with nature is not to protect or tame it, but to exploit it. As a place yet to be influenced by human culture, nature offers a range of free resources that can be tapped by cultures to allow them to meet their own ends. Examples of this can be found in many places. Cultures exploit nature for its forestry products (e.g. timber

Three Gorges Dam, Yangtze River, China
TimS www.flickr.com/photos/tims/150280389/

Source: Etherflyer www.flickr.com/photos/etherflyer/ 99049363/

and land), its geological harvest (e.g. oil, gold, coal, heat), and increasingly its air space (e.g. wind energy). The following example illustrates how natural water courses can be dramatically exploited by human culture through the construction of dams to provide hydro-electric power.

The Three Gorges Dam across the Yangtze River in China is the world's largest hydro-electric project. The dam stands 185 m high and is 2.3 km long. The reservoir created behind the dam submerges 632 square km of land, and can be 660 km long. The dam provides 18,000 megawatts of electricity for China's growing economy. The power of nature is thus harnessed by human culture to forward the cultural need for power, industrialisation, and capitalist expansion.

The idea of nature being outside culture thus leads to a range of practices that compete to be accepted as the 'common sense' way to deal with the natural world. Ideas of protection, taming, and exploitation seem very different, but all resort to the idea of nature as somehow other than culture to forward their desired ends. As Castree and Braun (2001: 4–5) identify, all:

posit a foundational distinction between the social and the natural and assumes that the latter is, at some level, fixed and/or universal. Thus, where [environmentalists] urge us to 'save', 'live in harmony with', or even 'get back to' nature, [developers may] propose to 'manage', 'control', or 'dominate' nature as if [it] were a domain different to, and separate from, society.

Questions therefore arise as to how we should decide on the most appropriate way to deal with the material world of nature: should we protect, tame, or exploit nature, and why? These issues form the crux of cultural geography debates surrounding nature conservation, urban and industrial expansion, and sustainability. We will return to these important issues later in the chapter; at this point however we focus on the 'problem' with the orthodox idea about nature. As noted in the example of the Scottish Highlands photograph, due primarily to the absence of people, this vast expanse seemed like nature to me. However, I conveniently forgot that a human must have been in the landscape in order to take the snap. Similarly, in all of the examples outlined above, whether they involve protection or exploitation, humans are in some way taking and making the place of nature. Whichever practice is chosen, all these actions are affecting nature in a real sense. Even by simply recreating in nature, humans have the effect of re-creating it. Thus despite this dominant perspective (b)ordering nature as somehow *outside* culture, we can see that through practice nature becomes *thoroughly imbued by culture*. Culture and nature are irrevocably intertwined.

Making nature

The realisation that the material world of nature is thoroughly imbued by culture is something that the social sciences, including cultural geography, have explored in detail in recent years. Through practices of protection, taming, and exploitation, humankind has grown in its ability to fashion, mould, and manage the natural world, in ever more absolute ways. As Castree and Macmillan point out: 'Nature is increasingly being reconstituted materially, even down to the atomic level . . . nature is being physically "produced" . . .' (2002: 209).

Marxist geographers would consider this reconstitution as the destruction of 'first nature' – the fall of a pristine, wild, and unsullied natural world, and the gradual transformation of this world into a resource for the capitalist system. 'First nature' is thus transformed into 'second nature' – commodified as an input to the industrial system (e.g. coal as fuel, trees as timber, cows as meat, milk or leather producers), or into 'third nature', where humans manipulate the very genetic coding of non-human flora and fauna in order to gain more profitable or patentable commodities.

Such material reconstitution of nature has led many to suggest that nature in its 'traditional' form – as something outside of culture – no longer exists. McKibben, for example, heralds humans' transformation of the natural world as the 'end of nature' (1989). He states,

> We have changed the atmosphere and thus we are changing the weather. By changing the weather, we make every spot on earth man-made and artificial . . . We have deprived nature of its independence, and that is fatal to its meaning. *Nature's independence is its meaning*; without it there is nothing but us.
>
> (McKibben, cited in White, 1995: 182, my emphasis)

Some natural scientists would agree with McKibben. Crutzen (2006) for example has argued that humans have become so dominant in altering nature that we have moved into a new era, one he dubs the 'Anthropocene'. Unlike previous geological eras, such as the Pleistocene, it is now argued we live in a time when one species – us – plays the hegemonic role in determining the fate of the natural world. This framing echoes McKibben's conclusion about the end of nature, as Lorimer suggests: 'the recent diagnosis of the Anthropocene represents the public death of the modern understanding of Nature removed from society' (Lorimer, 2012: 593).

However, if we begin to think of humans influencing nature through our cultural actions, we can also begin to consider how nature affects human cultures too (for example through processes such as the weather, seasons, and droughts,

through the threat of predators, through the opportunities for food, shelter, recreation, etc.). With this reciprocal interaction in mind, can it really be said that humans dominate nature, or that nature is independent from human culture? Indeed, was it *ever* true that 'nature's independence is its meaning'? If we begin to seriously explore these questions then my common sense view of nature as outside culture is challenged. I'll have to look again at the naturalness of nature.

Challenging the nature of nature

As this book argues, from the perspective of cultural geography we need to interrogate common sense because it is never anything of the kind. Common sense understandings are not 'natural', even when they are about 'nature'; rather they are traces that are defined as such by a particular cultural group. As Wilson tells us, it is important to 'play close attention . . . when our physical surroundings are sold to us as "natural"' (1991: 12). According to Whatmore (in Cloke et al., 1999: 4–11), when we consider nature we should remember that:

1. The representation of nature is not a neutral process that simply provides a mirror image of a fixed reality.
2. Representations of the natural world should not be taken at face value, they have social meaning.
3. There are many ways of seeing the same thing – need to situate ways of seeing in social and historical contexts of representation.

My initial response to the question 'what is nature' – an apparently human-less vista of the Scottish Highlands – is thus not a 'natural' response at all. It was a response conditioned by the cultural context I have been brought up in. I was conforming to the orthodox way that the majority of my culture consider nature – as somehow outside human culture. However, if we look at other cultures we often see a very different approach to the idea of nature.

Competing cultural approaches to the place of nature

First Nations people in North America look upon the natural world not as something outside them and their culture, but as something connected both physically and psychologically to their way of life. As the following Cree Tribal Elder puts it:

RB: To some [wild nature] means an area where not many people live.
Cree Tribal Elder: Yes, that's true. That's why I don't like it.
RB: Is there a Cree word for wild [nature]?
CC: No, we don't – many times I've smiled about that word, and I just shake my head at it. It's a view from a different point of view. I say to myself, 'That's the white people – lost total spiritual contact with the world' . . . Because its home to me . . . it could never be 'wilderness'. Like wilderness means somewhere you were lost.

(Bertolas, 1998: 106)

As Frawley (1999: 272) recognises, the interconnectedness of nature and culture is also part and parcel of Aboriginal society:

For Aboriginal Australians, nature and culture are inextricably bound together in the Dreaming – the time when the world, including Aboriginal people and their law, was created. Belief systems associated with the Dreaming link specific places with Dreaming events, and give every person, living and dead, a place within a physically and spiritually united world. The landscape is not, therefore, a composite of external physical objects, but is made up of culturally defined features of mythical significance.

And in Haiti, nature and its species are central to cultural life:

When a child is born in the countryside, the umbilical cord may be saved and dried and planted in the earth, with a pit from a fruit tree placed on top of the cord. The tree that grows then belongs to the child. And when the tree gives fruit in five or six years, that fruit is considered to be the property of the child, who can barter or sell it. Trees in Haiti are thus thought to protect children and are sometimes referred to as the guardian angel of the child. However if the tree should die or grow in a deformed manner, that would be considered an evil omen for the child who owned the tree.

(Wolkstein, in McLuhan, 1996: 86)

By broadening our outlook to see how other cultures envisage nature, we can identify that, for them, nature isn't apart from culture at all. Not only is it not apart from culture now (due to a growing human interference marking the end of a 'first nature'), but also it never has been whilst *homo sapiens* have existed. If we look to other cultures in this way we can see beyond our own dominant framings of nature. We can to begin to think of the relationships between humans and nature in a different way.

Before and beyond nature-culture

As we have seen, there are many ways to look at the nature of nature. The common sense view of nature in the West may remain the 'normal' way many people see the non-human world, yet alternative ideas can be found in different cultures across the globe. These ideas focus on the ways nature and culture connect to one another, rather than on how they are separate and distinct. In these ideas, they focus on how nature and culture are mutually produced and entangled (see Box 9.5).

BOX 9.5 MORE-THAN-HUMAN GEOGRAPHIES: (B)ORDERING NATURE DIFFERENTLY

Rather than focusing on the relations between people and place (and often ignoring non-human agency and affect in this mix), cultural geographers are increasingly focusing their energies on 'the vital connections between the geo (earth) and the bio (life)' (Whatmore, 2006: 601). As Whatmore states, cultural geographers have begun to, 're-animate the missing "matter" of ["natural" places], focusing attention on bodily involvements in the world in which [places] are co-fabricated between more-than-human bodies and a lively earth' (2006: 603). This 'more-than-human' geography, as Lorimer (2010) summarises, has led to many revisiting ideas of intelligence, value, and even what we consider an organism to be. For example, perhaps we should no longer consider humans as discrete and distinct entities, but co-formed not simply with places, but also by viruses, bacteria, and disease. In Whatmore's terms, cultural geographers should 'interrogate "the human" as no less a subject of ongoing co-fabrication than any other socio-material assemblage' (2006: 603).

'Re-animating' the missing matter of nature has also involved geographers re-thinking the agency and intelligence of non-human species. In studying the elephant, Lorimer has been keen to draw our attention to the affective capacities of 'nature' in general. In his words, it has involved 'rethinking what forms of intelligence, truth and expertise count. Attention has turned away from cognition to issues of embodiment, performance, skill and affect, understood as relational and distributed forces and competencies' (2010: 238). Bekoff's work on birds, especially magpies, suggests that non-human species display emotion and could participate in what might be described as cultural practices; as he writes:

Elephants

Source: http://commons.wikimedia. org/wiki/File:Two_Elephants_in_ Addo_Elephant_National_Park.jpg

A few years ago my friend Rod and I were riding our bicycles around Boulder, Colorado, when we witnessed a fascinating encounter among five magpies. Magpies are Corvids, a very intelligent family of birds. One magpie had obviously been hit by a car and was lying dead on the side of the road. The four other magpies were standing around him. One approached the corpse, gently pecked at it, just as an elephant would nose the carcass of another elephant, and stepped back. Another magpie did the same thing. Next, one of the magpies flew off, brought back some grass, and laid it by the corpse. Another magpie did the same. Then, all four magpies stood vigil for a few seconds and one by one flew off. Were these birds thinking about what they were doing and what were they feeling? Were they showing magpie respect for their friend? Were they merely acting as if they cared? Were they just animal automatons? I feel comfortable answering these questions, in order: yes, yes, no, no.

(Bekoff, 2009: 1)

Magpies

Source: www.flickr.com/photos/ wildphotons/248008375/

Bringing non-humans to the forefront of cultural geography in this way involves 'affective, micropolitical experiments' that reconsider the (b)orders between nature and culture (after Lorimer, 2010: 239); furthermore it raises important questions concerning 'how humans and animals can and should live together' (ibid.).

An example of this new way of framing nature in the cultural world is outlined by White (1995: 184–185). He outlines a new way of framing nature by first critiquing the orthodox view, cited by McKibben (above), and then offering his own alternative:

When McKibben writes about his work, he comments that his office and the mountain he views from it are separate parts of his life. They are unconnected. In the office he is in control; outside he is not. Beyond his office window is nature, separate and independent. This is a clean division. Work and nature stand segregated and clearly distinguished.

But, unlike McKibben, I cannot see my labour as separate from the mountains, and I know that my labour is not truly disembodied. If I sat and typed here day after day, as clerical workers type, without frequent breaks to wander and look at the mountains, I would become achingly aware of my body. I might develop carpal tunnel syndrome. My body, the nature in me, would rebel. The lights on this screen need electricity, and this particular electricity comes from dams on the Skagit or Colombia. These dams kill fish; they alter the rivers that come from the Rockies, Cascades, and Olympics. The electricity they produce depends on the great seasonal cycles of the planet: on falling snow, melting waters, flowing rivers. In the end, these electrical impulses will take tangible form on paper from trees. Nature, altered and changed, is in this room.

For White, then, nature is not something separate from his everyday existence, but something intimately connected to it. It facilitates, limits, threatens, and improves his cultural life. In this view the place of nature is defined by its connections and ties to the places in which he lives. In a similar way to Massey's Kilburn High Road, or seeing the world in our own street (see Chapter 4), here the place of nature is both inside and outside his world, both intimate to it and abstract from it, both controllable and uncontrolled. In this perspective, therefore, a range of tracings can be identified between what once was deemed to be disparate culture and nature, and it is the particular composition of these tracings that come to define the place in question.

In this view, both nature as an idea and nature as a materiality are deemed inherently connected to the cultural world, and vice versa. Places are therefore ongoing compositions of traces that are not wholly 'cultural' in a limited sense (just of human construction), or wholly 'natural' (just of non-human construction). Instead we need to diagnose the different traces that are made in each place, the different human and non-human trace-makers, critically examining (as far as possible) their intentions, outcomes, and the meanings given to them by the actors involved.

Socionature

One geographer who has undertaken such an approach is Eric Swyngedouw (1999). In his study on how water can be (b)ordered, he coins the term 'socionature' in order to illustrate how just one container of water can be traced to include the intimate and the abstract, as well as the human and non-human.

For example, if I were to capture some water in a cup and excavate the networks that brought it there, 'I would pass with continuity from the local to the global, from the human to the nonhuman' (Latour, 1993, 121). These flows would narrate interrelated tales . . . of social groups and classes and the powerful socioecological processes that produce social spaces of privilege and exclusion, of participation and marginality; chemical, physical, and biological reactions and transformations, the global hydrological cycle, and global warming; capital, machinations, and the strategies and knowledges of dam builders, urban land developers, and engineers; the passage from river to urban reservoir, and the geopolitical struggles between regions and nations. In sum, water embodies multiple tales of socionature as hybrid.

(Swyngedouw, 1999: 445/446)

It is possible for us to take a place familiar to us and trace the cultural and the natural influences that compose its substance both in terms of its material form, but also the meanings associated with it. Adopting this perspective thus no longer focuses on ourselves and nature, but ourselves-in-nature, and indeed nature-in-ourselves (as White notes, above). In other words, it reminds us of the 'constitutive co-ingredience' of people and place (see Box 9.6 and 9.7).

BOX 9.6 TEMPLES AND TREES IN ANKHOR WAT, CAMBODIA

Ankhor Wat
Source: Author

In these photos, the ancient temples of Ankhor Wat, Cambodia, are literally entangled by the roots of tropical trees. These roots often dwarf the impressive temple structures, in some cases destroying the walls, in others supporting them. Here nature and culture are irrevocably intertwined.

BOX 9.7 A 'COMPROMISE' WITH NATURE: HUMANS AND NATURE AT LAKE MINNEWANKA, BANFF, CANADA

Lake Minnewanka
Source: Author

At Lake Minnewanka, near Banff, Canada (left), the sign on the foreshore (right) explains how human culture and nature have interacted in this place. The sign reads (in English and French) as follows:

> Lake Minnewanka – A compromise with nature. This lake has been artificially raised twice: 16' by the Devil's Canyon Dam in 1912 and an additional 95' in 1941. Beneath these quiet waters lie the remains of the old dam across Devil's Canyon, off the point across the bay the foundations of the old village at Minnewanka landing lie preserved beneath many fathoms of water. The natural scene has been altered but nature has done much to heal the scars wrought by man.

Here nature is deemed to be an aesthetic balm remedying the ugly interventions of humans. It is in this way that humans and nature interact in Minnewanka.

BOX 9.8 'HOW THE CANYON BECAME GRAND'

In the case of the Grand Canyon, Arizona, Stephen Pyne (1998) argues that the natural and cultural are co-ingredient in how this ongoing composition is understood.

> The canyon . . . most visitors actually see [is] a cultural canyon, the Grand Canyon as a place with meaning. This landscape has been shaped by ideas, words, images, and experiences. Instead of faults, rivers, and mass wasting, the processes at work involved geopolitical upheavals and the swell of empires, the flow of art, literature, science, and philosophy, the chisel of mind against matter. These determine

Grand Canyon
Source: Author

the shape of Canyon meaning . . . and that meaning depends less on the scene's physical geography than on the ideas through which it can be viewed and imagined. These ideas are not something added . . . like a coat of paint, or taken from it, like a snapshot, any more than the river was something added to a prefabricated gorge. They have actively shaped the Canyon's meaning, without which it could hardly exist as a cultural spectacle.

(Pyne, 1998: xii–xiii)

Viewing nature as 'socionature' thus requires a further refocusing on how we should think about the places of nature – not as general types, but as specific compositions of unique traces. Coupled to this, it also requires a further refocusing about how we should act in the places of nature: how do we contribute to these ongoing compositions, and should we protect or develop the unique connections found in this place? However, this perspective also gives us the opportunity to refocus our attention away from how we 'know' nature in a narrow representational sense (for example nature simply stands for the 'other') and begin to consider our understandings of nature *beyond* the representational (see Chapter 3). How does the socionature in specific places make us feel? How can we sense the ongoing composition of socionature?

'Place must be felt to make sense' (Davidson and Milligan, 2004: 524)

We ought to say of [nature] not only – this is what it looks like, but, this is how I feel it.
(Watson, 1983: 392)

As we saw in Chapter 3, engaging with the more-than-representational is important to understand the relational connections humans have with their world. From this perspective, the places of nature 'must be felt to make sense'. When we imagined nature in the question asked at the beginning of this chapter, part of this answer would have involved how we felt in the face of our ideas of nature. For me, the windswept expanses of the Scottish Highlands make me feel liberated, open to the elements and at once ener-gised, but perhaps slightly humbled and scared by the immensity of the landscape, the centuries over which it has evolved, and my insignificance in the face of it. Places, and their natural and cultural compositions, then involve a *relational sensibility*. They provoke emotions that define and connect nature and culture together (see Box 9.9).

BOX 9.9 THIS IS HOW I FEEL IT

There follows a number of extracts which highlight how some people articulate their relational sensibility to socionature.

Thinking back . . . I appreciate why I come to the mountains; not to conquer them but to immerse myself in their incomprehensible immensity . . . There is something about mountains that move the soul. They arouse a powerful sense of spiritual awareness and a notion of our own transient and fragile mortality and our insignificant place in the universe. They have about them an ethereal, evocative addiction that I find impossible to resist. They are an infuriating and fascinating contradiction. Climbing rarely makes sense but nearly always feels right.

(Simpson, 2003: 58)

I was standing on the bridge, not sitting, and it [the muskrat] saw me. It changed its course, veered towards the bank, and disappeared behind an indentation in the rushy shoreline. I felt a rush of such pure energy I thought I would not need to breathe for days.

(Dillard, 1975: 171)

Aboriginal cultures talk in different ways about their relational sensibilities to nature. With experience gathered by hunting and living in the natural world, their knowledge is built up through practical tasks rather than recreation (see Ingold, 1993). This knowledge is often articulated in stories, with places, spirits, and culture woven together into their fabric. In the following extract, Jack McPhee (cited in McLuhan, 1996: 113), an Aboriginal elder living in Australia, tells of his emotional and spiritual engagement with the natural world:

A place could be a cave, rock, a pool, anywhere where a big snake could be or where he comes now and then. I'm not talking about a real snake in the sense of something you can see, I'm talking about a very old spiritual thing . . . Our sacred sites are more to do with the spirits, and they can't be dated because they've always been there. There used to be a sacred site on the way back from the Comet Mine . . . it has been there for hundreds and hundreds of years . . . you could always count on getting a drink of water there, even in drought time. Unfortunately, white people didn't understand how special this place was. Someone went and dug a hole there, probably a prospector, hoping the water would build up, but of course it didn't, it just died away. You see, in doing that he killed Gadagadara, a snake with a strange head shaped like a horse's, who had placed his spirit there to live and keep the water for the people.

As we noted above, in Whatmore's terms, cultural geographers have 'interrogate[d] "the human" as no less a subject of ongoing cofabrication than any other socio-material assemblage' (2006: 603). Here Whatmore is questioning whether our unit of analysis is too narrow if we simply look at 'the human', especially if any 'human' is always 'more-than' this, co-formed with other viruses (etc.) in practice. This is also the case if we think about human pursuits in nature, for example, mountain biking, rock climbing, or even surfing on the sea. For instance, are surfers isolated and simply human? Or are they defined in part by their involvement with 'nature'? One could argue they are discrete human organisms, riding waves and conquering nature. But many surfers don't define themselves in this way. Some surfers suggest that they are a different entity when they surf, co-joined not only with their surfboard and wetsuit, but also by the wave, swell, and surfbreak (see Anderson, 2012). From the perspective of the surfer, and for the length of their wave-ride, there is no 'surfer' and 'wave' in isolation, but only a 'surfed-wave' convergence.

Lessons from the place of the surfed wave. Places change over time: they ebb and flow, they flourish and decline. Places are not stable but constantly changing, processes of convergence may seem to endure but are always becoming something else. There is movement, re-emergence and divergence. There are mergers and emergings. We interpenetrate, coalesce and form; yet do so only temporarily. This is place as transient convergence.

(Original photo: Gecko, 2009; Anderson, 2010)

Lessons from the place of the surfed wave
Source: Author

Engaging with our relational sensibilities to socio-nature therefore highlights another key tracing in the constitutive co-ingredience of people and place. These traces can 'move' us, prompting re-evaluations of relationships, and also perhaps how we should act in the place of 'nature'.

How should we act in the place of nature?

[H]ow can we defend some uses (and non-uses) of nature over others? How can we protect the environment if everything is up for grabs?

(Cronon, 1995: foreword)

If places are ongoing compositions of the natural and the cultural, prompting a range of relational sensibilities and emotional responses, we return inevitably to the key question we raised at the beginning of this chapter: should we climb Uluru? In other words, how should we contribute to the entangled traces in the 'place of nature'? What cultural ideas of nature should we support, what traces should we leave, and how do these affect the nature of that place?

To some extent the orthodox view of nature as outside and distinct from culture helps us answer this question. By creating an external reality outside culture we have a guide to which our actions in nature can be compared (in a similar way to how Ratzel and Kropotkin did in the nineteenth century, see Chapter 2). From this comparison, we can make an argument as to what practices are appropriate in nature (i.e. protection, tam-ing, or exploitation). However, if we accept that culture and nature have always been subject to ongoing mutual interference we can no longer 'step outside culture' to understand nature as a separate entity. This leaves us with no 'right' answer as to how we should act in nature. How we treat natural traces is up for grabs. What should we do? Are natural traces more important than cultural ones? If so, what actions does this legitimate? Are cultural traces more important than natural ones? In what circumstances, where? (see Box 9.10).

BOX 9.10 KILLING NATURE OR KILLING CULTURE?

Whaling has a long tradition in Japan. The oldest Japanese book, *Kojiki*, written in the eighth century, mentions whaling activity, and harpoons have been used since the seventeenth century to kill these mammals. However, such cultural tradition was banned in 1986 under international legislation. This legislation (b)orders the sea into particular zones, for example the Southern Ocean Whale Sanctuary (established in 1994), in which all commercial whaling is banned. Japan resisted the imposition of this sanctuary, and has been given special permission to hunt whales for scientific purposes within these waters. In 2008, the Japanese fleet killed just under 1000 whales (McCurry, 2008). The Sea Shepherd Society, and its leader Paul Watson, attempts to uphold the spirit of the sanctuary. They are clear about how they should act in defence of nature; as Watson states, 'I would choose life over death, and the life of the whale mean[s] more to me than a cultural right to kill' (1994: 68).

Clashes between these cultural (b)orders, what Mitchell (2000) terms 'culture wars', are often dramatic. In 2008, 'stink bombs' containing butyric acid were thrown onto whaling ships by the Sea Shepherd Society (McCurry, 2008). Watson himself was shot by Japanese coastguards whilst attempting to stop the hunt (only his Kevlar vest stopped the bullet hitting his heart). Should cultures be allowed exemption from legislation in line with their traditions and customs, even if this means non-human animals are threatened with death, or even extinction? Is direct action appropriate to stop the killing of nature?

On land, the radical environmental group Earth First! argue that due to our interdependence with nature, the use of violence is legitimate when defending these traces. Wild places should be defended as we are part of nature – they are our historical and spiritual home. As Abbey outlines:

> If a stranger batters your door down with an axe, threatens your family and yourself with deadly weapons, and proceeds to loot your home of whatever he wants, he is committing what is universally recognised – by law and morality – as a crime. In such a situation the householder has both the right and the obligation to defend himself, his family, and his property by whatever means are necessary . . . For many of us, perhaps most of us, [nature] is as much our home, or a lot more so, [than] our ever expanding industrial culture. And if [nature] is our home, and if it is threatened with invasion, pillage and destruction – as it certainly is – then we have the right to defend that home, as we would our private rooms, by whatever means are necessary.
>
> (in Foreman and Haywood, 1987: 7–8)

In your view, is direct action, violence to property (or even people) justified in defending natural traces from cultural ones?

How we choose to act in nature, or whether we decide to climb Uluru, thus says something not only about ourselves, our own culture, but also about how we value the cultures of others. It says something about how we value nature as an idea, as a relational sensibility, and as a material thing. It answers questions concerning the place of nature, and our place in it.

Conclusion

As we have seen in this chapter, how different groups think and feel about places is as important as the physical form places take. Our thoughts and feelings about nature can directly influence the actions that we take and how we contribute to the natural world. There are many different ways in which nature can be (b)ordered. As we have seen, the orthodox (b)ordering of 'nature' in the West is as something outside and apart from human 'culture'. This idea leads some cultural groups to protect nature, some to control it, and some to exploit it. However, in other cultures nature is seen as thoroughly imbued by culture, and vice versa. In these cultures humans are a part of nature not apart from it. This view prompts us to consider anew how we should act towards the imbroglios of natural and cultural traces in the world.

Places then are composed by the ongoing interdependence of natural and cultural traces. These traces involve the emotional and the cognitive, and include cultural beliefs and practices. In some ways perhaps cultures enhance nature; in others nature will threaten culture. Our actions in places thus will contribute to these ongoing compositions, and how we think nature should be may influence these actions most of all.

The material nature we inhabit and the ideal nature we carry in our heads exist always in complex relationship with each other, and we will misunderstand both ourselves and the

world if we fail to explore that relationship in all its rich and contradictory complexity . . . The struggle to live rightly in the world is finally not just about right actions, but about the ideas that lie behind those actions . . . to protect the nature that is all around us, we must think long and hard about the nature we carry inside our heads.

(Cronon, 1995: Foreword)

Suggested reading

The following texts are useful to explore the issues raised by this chapter.

Books

Soper, K. (1995) *What is Nature? Culture, Politics and the Non-human*. Blackwell: Oxford.

Castree, N. (2005) *Nature*. Routledge: Abingdon.

Soper and Castree both provide ideal insights into nature as geographers study it. They introduce different cultural understandings of nature and how ideas and ideals of nature contest the minds and hearts of cultures.

Castree, N. and Braun, B. (eds) (2001) *Social Nature: Theory, Practice, Politics*. Blackwell: Oxford

Cronon, W. (1995) *Uncommon Ground: Toward Reinventing Nature*. W.W. Norton and Co.: New York.

Castree and Braun outline how nature can be considered social, introducing ideas of hybridity into the nature debate, whilst Cronon, particularly in the Introduction to his volume, raises important questions about how we should consider the place of nature in the situation where cultures materially influence its form and composition.

Devall, B. and Sessions, G. (1985) *Deep Ecology: Living as Nature Mattered*. Gibbs M. Smith: Layton.

McKibben, B. (1989) *The End of Nature*. Bloomsbury: London.

Devall and Sessions offer a radical view on nature as outside and independent from culture, whilst McKibben suggests that this nature is at an end due to the influence of culture in once-natural places.

Szersynski, B., Heim, W., and Waterton, C. (eds) (2003) *Nature Performed: Environment, Culture and Performance*. Blackwell: Oxford.

Szersynski et al. demonstrate the importance of emotions and practices to how we think about and produce the culture-natures around us. Chapters focus on activism, performing environmental identities, and the different architectures and infrastructures that connect the natural and the cultural together.

The place of race and ethnicity

Introduction

As we have seen, cultural geography is all about taking and making place. Every action we engage in contributes to the cultural world, leaving a trace that affects the identity of who we are and where we are. We have seen in the previous chapter how our ideas and ideals about place – how we think places should be and what traces are appropriate in them – affect our actions in place. Places come to be ordered and bordered in line with our ideas as regulations are constructed to classify places for particular people and uses.

In this chapter we explore questions concerning the form and substance of the cultural traces which (b)order our places. Cultural values, ways of behaving, even ways of looking come together to order and border places, making some people and behaviours acceptable in one place, but unacceptable in another. Traditions, appearances, and actions can make us feel comfortable or give us a sense of home, but also make us feel as if we are a fish out of water where cultural traces are strange and novel. Traces in places can therefore act to unify and bind us into a group, into an 'us' with our own place, but they can also disconnect and divide us, separating 'us' and 'our place', from 'them' and 'theirs'. This process of 'othering', of culturally and geographically distinguishing between cultural groups on the basis of action, value, and appearance, is the subject of this chapter. What happens when one of 'them' is in 'our' place? How do we feel and respond when 'they' transgress our (b)orders? What are the bases for acceptable existence in a place: who must you be? What must you look like?

In this chapter we explore how 'othering' occurs on the basis of ethnicity and race. We will investigate the history of this process, exploring how cultural ordering on the basis of biological difference was instigated, and how geographical bordering has subsequently occurred. We will see how three main (b)ordering mechanisms are used to maintain a geographical and cultural purity between ethno-racial groups, namely: *genocide*, *tight control*, and *partnership*. We will turn to ideas of the '*stranger*' to understand how and why these mechanisms remain within cultures, as well as looking at how a '*mixophobic*' attitude creates the nature of the places we live in. The chapter ends by raising questions concerning the absolute nature of ethno-racial divisions and how alternative conceptions of identity may recognise similarities between ethnicities and races, rather than difference. Should ethno-racial groups have a right to an alternative culture in your place? Should ethnic difference be tolerated, or regulated? Does diversity threaten or enhance culture? Who gets to decide, and why?

What is race and ethnicity?

Race can be understood as a cultural and geographical identity based on biological and physical characteristics. Different races have genetic variations that result in different skin colours, a predominance of certain eye colours, and other biological characteristics (e.g. hair types). Ethnicity is a broader concept that emerges when a group share a common ancestry, origin, and tradition. Ethnicity connects to geographical territory, world view, custom, ritual, and language, and ethnic difference can be identified when a cultural group has linguistic, anthropological, and genetic variation when compared to others. Ethno-racial differences are therefore widely visible; they can be seen through skin colour, but also through customs, clothing, rituals, and practices. We all have ethno-racial characteristics, so we can ask ourselves what it feels like to be part of our particular ethno-racial group. Is it an important element in *your* identity? We can consider what the dominant ethnic and racial group in our own neighbourhood might be, or in our own country. We can consider if there are any ethno-racial minorities, and what sort of places they seek to create. In our increasingly populous and globalising world, it is common for people from different ethnicities and races to try to take the same places. How is this process (b)ordered, and what affect does it have for how we think about, and act in, these places?

We have seen in Chapter 2 that, during the history of colonialism and cultural geography, the process of 'othering' was widespread. Using theories of environmental determinism, racial and ethnic differences between groups were used to distinguish between cultures, their place, and acceptability. To European scholars and colonists of the time, difference in skin colour meant difference in intelligence, civilisation, and ability. These conclusions had no evidence to support them; they were simply culturally constructed responses to 'markers of difference' (K. Anderson, 2012: 4). Unfortunately, despite the prejudicial and discriminatory aspects of these responses, these cultural constructions still have resonance today – indeed as Koboyashi tells us, race remains 'a way of life' and part of our

'taken-for-granted reality' (2013: 57). As a consequence, it is worthwhile pausing to reflect on why such apparently arbitrary biological markers, and their culturally constructed interpretations, have gained such dominance in categorising human people. The (now discredited) theories of environmental determinism had a key role to play here; as Kobayashi (2013: 59) states, cartographers in the nineteenth century,

mapped the world as a grid of political power, early human geographers speculated on whether climate was the dominant factor explaining the putative superiority of the white European man over the black African. In so doing, they legitimized and fed the notion of race that would by the end of the . . . century become a thoroughly naturalized and normatized part of modern Western life.

Prejudicial notions of European superiority were thus employed by geographers, explorers, capitalists, and politicians to legitimate unequal relations between humans from different races. Even though this prejudice had no evidence base – it was simply constructed from the ignorance and fear of the time (see K. Anderson, 2008), that is not to say it was without meaning or influence; as Kobayashi confirms, 'in retrospect, [this prejudice was] entirely complicit in strengthening a racialized – and racist – society' (2013: 59). Racial and ethnic difference therefore was, and still remains, crucial to (b)ordering culture and place (see Box 10.1).

For the geographer . . . it makes as much sense to speak of *spatialization* as it does of racialization.

(Kobayashi, 2013: 60)

If classifying people in terms of skin colour is a key way in which some cultures order their world, this ordering has clear geographical consequences. Kellerman (1993) notes three key ways in which dominant ethno-racial groups

BOX 10.1 THE ARBITRARINESS OF OUR RADICALISED WORLD

People generate prejudice; prejudice governs place.

(Watson, 1983: 388)

The categorizations don't work. They're not based on any biological anything. But the fact that it is [culturally] constructed . . . doesn't necessarily make it less real. It makes it less independently verifiable. But if people treat it as real, then it becomes real.

(Oliver, Director of an anti-racist think tank in London, cited in Paul, 2014: 710)

Over recent decades, anti-racism expert Jane Elliot has conducted a series of workshops designed to show both the arbitrariness of categorising peoples by their skin colour, as well as how easily humans take on culturally constructed difference as a 'taken for granted reality', especially if they benefit from such arrangements. Elliot's workshops begin with volunteers being labelled 'brown' or 'non-brown', not in terms of skin colour, but rather in terms of eye colour. In this simple move Eliot questions the assumption that skin colour is the 'normal' or 'natural' means through which to establish difference between groups, and that difference could be identified in terms of other characteristics (e.g. eye colour, hair colour, length of finger, size of feet, etc.). By using a binary 'either-or' classification, individuals with blue, grey, or green eyes, or those with eyes which are different in colour, are grouped together; their individuality and difference is silenced into one classification that is defined not by what it is, but rather what it is not. From this basis, the 'non-brown eyed' group is marked with a collar to denote their inferiority; they are restricted to a holding room, and constantly monitored by security guards. The 'brown eyed' group is given favourable treatment in the exercise, be that in terms of quality of food, where they can move, or their position in the decision-making hierarchy. Cultural superiority is thus attributed to those with brown eyes, simply on the basis of this arbitrary biological difference.

Through the process of this experiment, discrimination against the non-brown eyed group often becomes normalised, the brown eyed girls (and boys) quickly and easily assume that their superiority is not an arbitrary act, but rather a divine and natural right. How would you act in this cultural experiment? How different is Elliot's experiment from what has occurred to many humans on the basis of their skin colour? Through this experiment it is possible for both groups to see how processes and practices that are deemed to be fair and equitable in real life in many cases favour particular ethnicities and races over others. To paraphrase Watson: prejudice (b)orders place.

(b)order minorities: through *genocide*; *tight control*; and *partnership*. This chapter focuses on these three (b)ordering mechanisms, all of which seek to 'appropriately' accommodate ethnic and racial difference. Each seeks to (b)order the identity of groups in a specific way, attempting to generate a 'common sense' view of these groups and the geographical and cultural places they should occupy within society.

Genocide

The material (b)ordering of place is often a brutal business. Although one might consider it highly unlikely to attempt to (b)order cultures through the mass murder of people on the grounds of ethno-racial difference, in both past and contemporary societies there are instances where genocide is practiced. Historically,

European colonisation of Africa and South America involved the direct and indirect genocide of millions of indigenous peoples. In the sixteenth century, for example, 24 million people in Central America were killed through disease, impoverishment, and subjugation at the hands of the Spanish, French, and British (see Johnston, 1986). In the twentieth century, Nazi Germany practiced genocide against Jews, most notably during World War Two. Cases of genocide have also occurred during the Balkan wars, in clashes between China and Tibet, and in East Timor and Indonesia. In these cases, the difference represented by the cultural 'other' has been deemed a significant threat to the existence of the dominant culture. The presence of a different ethnic and racial group, and the cultural, linguistic, and religious alternatives they offer, have been seen as a risk to the hegemony of the practices within that place. There follows a case to illustrate this process. It focuses on Rwanda in the 1990s, specifically on the civil conflict between the rival ethnic tribes of the Tutsi and the Hutu.

> All Tutsi will perish. They will disappear from the earth slowly, slowly, slowly. We will kill them like rats. When you kill rats you don't spare the babies.
>
> (Radio broadcast, cited in McIntyre, 2001).

In Rwanda in 1994, the Hutu held power through military government. In this position they employed a range of tactics to marginalise the Tutsis, the minority ethnic group. One innovative, but wholly effective, mechanism they used to (b)order their place was through the control of a national radio station: Radio Télévision Libre des Mille Collines. This was a Western style radio station, broadcasting popular music and comedy, and was amazingly popular. As Frances D'Souza, a human rights expert notes, this radio station was used by the Hutus to transmit anti-Tutsi propaganda:

What the government needed to do was to try to incite the people about the 'danger' that was facing them, as an excuse to slaughter their perceived enemies. This very trendy radio station began to convey messages of absolute horror: 'Cannibals!' 'They eat men these Tutsi! They will tear out your stomach, your liver, your heart! They will eat your children!' I think that this fear must translate itself eventually into a release that is very very violent.

(in McIntyre, 2001)

Through radio broadcasts, alongside soldiers on the ground, a doxa of opinion was spread that stated the survival of the Hutu ethnic culture could only be realised through the elimination of the Tutsi. As Denis Bagaruka, a Hutu, remembers: 'We heard the radio telling us to "cut down the tall trees". These trees were the Tutsis. We were listening to the radio, and because of this and what the soldiers were urging, we started to kill our neighbours' (ibid.).

Such explanations may seem terrible, but they were effective. Over 800,000 Tutsi and moderate Hutu were killed during the Rwandan genocide. Such shocking treatment of difference was realised through a combination of three factors: the absence of a free and open media (how would you respond if the radio stations in your country told you that a certain ethnic or racial group represented a clear and present danger?), a latent distrust and fear about 'other' ethnicities, and a dominant government seeking to exploit this situation in order to (b)order place. In the same way as 'dirt' can be used to signify inappropriate and inferior groups and behaviours (see Chapter 5), dehumanising rhetoric was used to compound the 'otherness' of the Tutsi. No longer were Tutsi valued as human beings, albeit those with a different ethnic world view, they became different and alien, no better than 'rats' or 'trees'. Through comparing Tutsi to non-human species, species that the agricultural workers of the country knew required control and removal in their everyday lives (Rwanda's economy is predominantly based on coffee plantations), the Hutu attempted to normalise and naturalise both the threat posed by the very presence of Tutsi in Rwanda, and the necessary response. Such dehumanisation thus made the practice of genocide – or ethnic cleansing – literally removing the 'dirt' from their cultural place, somehow an appropriate and 'natural' solution to the existence of a rival ethno-racial group.

Tight control

The second mechanism dominant cultures use to (b)order places of ethnicity is what Kellerman terms 'tight control'. Tight control involves the creation of laws and policies to regulate the movement and activities of particular ethno-racial groups in specific places. In other words it involves the use of legal statute and enforcement to take and make place for 'normal' society on one hand, and segregating 'deviant' ethnicities into 'other' places. This process, as we will

see in the upcoming examples, is often used as a step towards the annihilation of that ethno-racial minority – not in the same way as that used in the genocide example – but rather through control then assimilation. By enclosing minorities into specific areas mainstream groups remove their ability to take and make place in line with their cultural beliefs. As a consequence these cultures are effectively neutered, often lie dormant, and sometimes die. In this scenario the remaining groups have little alternative but to be assimilated into mainstream culture and adopt orthodox beliefs and practices.

A key example of 'tight control' was practised on Native American Tribes in what became the United States, and Aborigines in Australia.[1] Box 10.2 outlines the former example, drawing on Razac's (2002) account of the role that a specific (b)ordering tool – the application of barbed wire – had in structuring place for the settler population and against the indigenous tribes.

In this example we can see that in order to create the 'new nation' of America it was felt necessary

---

BOX 10.2 TIGHT CONTROL: THE (B)ORDERING EFFECTS OF BARBED WIRE ON AMERICAN INDIANS

[Barbed wire] has always been efficient enough to perform its designated tasks: to define space and to establish territorial boundaries.

(Razac, 2002: ix)

Barbed wire has been used in a range of cases, from the First World War trenches, to the concentration camps of the Second, as well in defining places in contemporary society e.g. military compounds, prisons, or even private backyards. It is effective in keeping people in, and keeping people out. This segregation capacity renders barbed wire a specific (b)ordering tool to realise at once the geographic and cultural control of an area by one cultural group. In the case of North American Indians, the use of barbed wire by the white settler societies was crucial in (b)ordering place to the detriment of the indigenous populations. As Razac argues, in this case barbed wire's 'direct political impact has been crucial to . . . the physical elimination and then the ethnocide of the North American Indians' (ibid.: 4).

By the end of the century, barbed wire was used to parcel off . . . Indian lands. It chopped space into little bits and broke up the communal structure of Indian society. Barbed wire made the Indians' geographical and social environment hostile to them, so that it became a foreign territory where the tribal way of life was unimaginable and where nomadic wandering and hunting were impossible.

(Razac, 2002: 22–23)

The installation of barbed wire was thus the first step in changing the (b)orders of the Indians' culture and geography. It had the effect of cutting Indians off from their hunting grounds, neutering their warfare tactics (based on surprise attacks on horseback), and impeding their nomadic way of life. Tight control was then further instigated as the 1887 Dawes Act authorised the zoning and segregation of land to the now settled tribes. Barbed wire was thus used to keep Indians within a designated place: the reservation. Despite the physical limits of these reservations, some aspects of tribal life remained. In Grant's 2003 book *Ghost Riders*, a useful account is given of two senators returning from Oklahoma's reservations shocked by the culture that persisted there:

'There is no selfishness, which is at the bottom of civilization. The tribes were still sharing all their food and possessions with their kinfolk. They were more impressed by displays of generosity than the accumulation of private wealth'. 'We need to awaken in [the Indian] wants. In his dull savagery he must be touched by the wings of the divine angel of discontent . . . Discontent with the tepee and the starving rations of the Indian camp in winter is needed to get the Indian out of the blanket and into trousers – and trousers with a pocket in them, and with a pocket that aches to be filled with dollars! . . . this is the first great step in the education of the race.'

(Grant, 2003: 195)

It was realised therefore, that the existence of the tribe, albeit in (b)ordered reservations, still offered a functioning alternative to the burgeoning capitalist and nationalist system being introduced by the white settlers. Instead of cultural assimilation, a pluralism existed: the Indian culture may have been limited but it remained intact alongside that of the settlers. To the white population, this pluralism meant the threat from Indian culture remained. The reservations required further transformation:

The next task was transforming the [Indians] into American citizens. As long as a tribe's sense of community stayed strong, the tribe could resist the Indian's new and forced status as a private individual. For this reason, the tribe had to be obliterated. The Indian could be an individual integrated into American society only when his tribe no longer existed. The project of the reformers, humane people concerned about the Indian's fate, was intended to change them into farmers who fully believed in the concept of private property . . . After implementing measures which threatened to physically eliminate the Indian, the United States turned its Indian policy to cultural destruction; in other words, ethnocide.

(Razac, 2002: 18)

to wipe out 'other' races that did not conform in culture to the values and practices of the white settlers. The thorough challenge of nomadism, sharing, and co-operation of Native Indians, compared to the sedentary, self-interested (see Adam Smith, 1979), and competitive white culture was seen as threatening the success of the new nation. As Jackson and Penrose (1993) argue, such tight control is undertaken to blunt the resistance of the existing group to the imposition of new cultural (b)orders – and thus allow the new group to impose their values and accrue benefits from so doing. Through this process the possibility arises to establish a new 'sense of place', a new collective identity without internal threat, and establish a collective strength as a consequence (see Chapter 6).

This 'tight control' is not only an historical but also a contemporary phenomenon. There follows one such example, focusing on how gypsies and travellers are treated in the UK in the twenty-first century.

Tight control: the (b)ordering of gypsy culture in contemporary Britain

The case of gypsy culture presents an archetypal example of how mainstream society engages with a minority group. Gypsies represent a completely transgressive challenge to the cultural orthodoxy of Western, settled, white society. Neither gypsy nor mainstream culture makes concerted attempts to accommodate the difference that the 'other' culture presents. The subsequent tight control of gypsy places by the mainstream makes explicit both the geographical and the cultural nature of (b)ordering mechanisms.

> The gypsies are a race . . . like any ethnic people they want their own community among their own people, they have their own heart, thoughts, their own culture.
>
> (David Jones, Welsh Gypsy, cited in Waterhouse, 1993)

As Jones outlines, gypsies have their own culture. Like all cultures, they take and make place in line with their own values and principles. As Sibley (1999) identifies, gypsies integrate rather than separate their work, residence, and leisure spaces. Instead of commuting to work, or having different locations for their home and recreation, gypsies combine these places into the temporary sites they occupy. (Like 'old-fashioned' houses

in mainstream society, on these sites gypsies separate their toilet areas from their living areas, deeming the close proximity of defecation and living space to be 'dirty'.) Thus the 'threat' posed by gypsy culture to the mainstream is not that of 'anarchy', they do not threaten structureless places in an absolute sense; they do, however, embody and create an alternative cultural order to the mainstream.

Perhaps the key defining order of gypsy culture is its nomadic nature. Gypsies travel about, traditionally in horse-drawn Romany caravans, but more recently in modern vehicles. This nomadism is perhaps the most important difference between gypsy and mainstream culture, and offers a particular threat to mainstream orthodoxy. It leads to the absence of 'territorial fixity' in their lives – they are not anchored in place like 'normal' society. This freedom of movement gives gypsy culture the capacity to elude conventional (b)ordering practices. How do you (b)order something that has freedom of movement?

Mainstream culture in the UK has sought to (b)order and tightly control gypsy culture in three key ways. It has '*stopped movement*', '*stopped sites*', and '*enforced settling*'. Particular legislations such as the Public Order Act 1986 have attempted to '*stop movement*'. These laws have given police the authority to deny access and use of roads to certain cultural groups. This legislation has been used to restrict the mobility of gypsy and traveller communities, as Chris Fox, when Assistant Chief Constable Warwickshire Police, stated,

> Control [of traveller and gypsy groups at one camp site] was lost because large numbers of people arrived from different directions. We're working with local forces to prevent these gatherings. We can actually stop anybody going somewhere if there's a liability to cause a breach of the peace.
>
> (Waterhouse, 1993)

Thus through the use of this legislation, gypsy nomadism has been curtailed in the UK – the dominant ethic group has been successful in '*stopping movement*'. Secondly, gypsy and traveller culture has been tightly controlled through '*stopping sites*'. The spontaneous creation of gypsy camps has been restricted through the introduction of specific legislations such as the Criminal Justice and Public Order Act (1994). This not only removes local councils' statutory obligation to provide gypsy sites, but also restricts one of the only options left to travellers: unauthorised camping on undesignated areas. If gypsies do not desist from camping on any site 'as soon as is reasonably practicable', 'under Section 62 vehicles may be seized and removed' (McKay, 1996: 162). Refusal to leave can result in three months' imprisonment.

Thirdly, gypsy and traveller culture has been controlled by '*enforced settling*'. With places effectively (b)ordered through movement and site restriction, gypsy and traveller communities have been left with little option but to buy their own land to maintain their culture (in a similar way to the reservations in the US). However, now in settled locations, existing planning laws work to restrict the type of culture that can be practised there. Where travellers do not conform to planning law, sites have been compulsorily purchased, often at sub-market prices, and then leased back to travellers under stringent conditions. In these cases, sites are often surrounded with barbed wire and security lighting, tightly controlling this culture in a very real sense (see for example, Undercurrents, 1996; Monbiot, 1996).

The threefold process of *tight control* thus has the effect of geographically and culturally restricting gypsy culture. Anne Bagehot, once a gypsy liaison officer with Save the Children and later a worker for the Gypsy Council for Education, describes the consequences of this (b)ordering process: 'Speaking as an individual', she says, 'my personal view is the phrase "ethnic cleansing" is not over the top' (McKay, 1996: 162). *Tight control* is thus used in both historical and contemporary society against ethnic minorities

that are deemed a threat to the mainstream. This control does not lead to ethnic cleansing in the sense of genocide, but seeks this goal through restriction then assimilation. Without the ability to take and make place in their traditional way, the platform from which ethnic minorities can exercise their culture is removed. If groups can no longer physically take and make place, successful (b)orders can no longer be established, and cultures can no longer exist. Ethnic minorities are forced to adapt to the cultural (b)orders of the mainstream.

Summary: stranger danger?

The two (b)ordering mechanisms we have examined thus far – *genocide* and *tight control* – have been used by the dominant culture to remove minority ethnic groups from the culture and geography of their places. In these cases, an alternative ethnicity within their place is deemed inappropriate, it does not fit with the ideal of place as constructed by the mainstream. To them, places must have a degree of ethnic purity. As a consequence, this inappropriate group – this dirt, this stain – must be cleansed from their place. In these cases, the ethnic other is a 'stranger'.

The 'stranger' is an idea suggested by Simmel (1950) and refers to those who are out of place within a specific cultural and geographical (b)ordering. As Bancroft (1999) outlines, the key aspect of the stranger is this out of place presence: 'Perhaps most important[ly], the Stranger is disturbing through proximity . . . the proximate other is disturbing, threatening to overthrow that psychological unity [of place].'

In cases where genocide or tight control is practised, ethnic others remain strangers to the dominant group. Ethnic difference is seen as transgressive, its presence threatening the hegemonic stability in that place. The threat of the stranger thus disturbs the meanings and ownership of place, to paraphrase Cresswell (1996: 60), the presence of the stranger means that:

1. The meaning of the place will change.
2. If the meaning of the place changes, the place itself will change.
3. The new meaning will be their meaning (the meaning of the stranger).
4. The place in question will become their place (the place of the stranger).

In the empirical examples outlined above, neither the dominant nor the minority ethnic or racial culture seeks a common ground with each other. The idea of absolute, pure places and identities holds sway – both groups seek to maintain the coherence of the culture and geography of their places and are unwilling or unable to adapt or compromise their sense of place to include the 'other'. As a consequence, the strangeness of these cultures is likely to remain a danger to mutual co-habitation in a place. As neither seek to understand each other's traces, nor make efforts to tolerate or allow for their difference, the chances of a partnership are slim (see Box 10.3).

Partnership

According to Kellerman, the third method used to control racial and ethnic groups is that of 'partnership'. In cases of partnership each culture does not necessarily wish to remain 'strange' to the other, nor seek the complete purity of its culture and geography. In these cases, an element of common ground is sought and different ethnicities seek to co-exist in one place.

If 'partnership' is practised by the dominant culture, often through governments and state institutions, integration of diverse ethnicities is not deemed 'novel' or 'bad', rather it is orthodox and normal. In this scenario an ideal society is created when ethnic minorities are granted equality, with each culture taking the opportunities the 'other' offers (e.g. through employment, economic development, and cultural stimuli). Key examples of this approach occur in many of the cosmopolitan cities of the West. London, New York, Toronto, and Paris all have a broad range of ethnic and racial cultures within their places. These places herald tolerance and multiculturalism, as Ken Livingstone, when Mayor of London, celebrated:

BOX 10.3 BETWEEN TIGHT CONTROL AND PARTNERSHIP: TALKIN' 'BOUT THE GHETTO

The practice of (b)ordering places on ethno-racial lines often produces 'ghettoes'. As Jaffe (2012) explains, the concept of the ghetto originated in sixteenth-century Venice, where an island within the city called 'Ghetto Nuovo' was designated as a secured area where Jews were confined. Inventively, Sennett (1994: 228) describes the ghetto as an 'urban condom', a form of (b)ordering which allows the dominant ethnicity to 'interact "safely" and profitably with a group they considered dangerous and contagious' (see Jaffe, 2012: 675). In this sense, the ethno-racial minority is both 'controlled' and 'partnered' (see above). In this light, Wacquant suggests that ghettoes serve as a 'vehicle to extract [one ethno-racial group's] labour while keeping [these] bodies at a safe distance, to the material and symbolic benefit of [the dominant race]' (in Jaffe, 2012: 676). Ghettoes are therefore one form of ethno-racial (b)ordering, but they also have a symbolic and psychological affect. As Wirth (1927: 71) argues, the ghetto is 'not merely a physical fact, but also a state of mind'. Ghettoes can institutionalise a sense of geographical confinement, and ethno-racial difference. As a consequence, ghettoes become the 'original no-go areas: outsiders and even the police are scared to go in, while insiders are unable to get out' (Jaffe, 2012: 676).

'This city [London] . . . typifies what I believe is the future of the human race and a future where we grow together and share and we learn from each other' (GLA, 2005) . . . 'There was a saying that city air makes you free . . . and the people who have come to London . . . all races, creeds and colours . . . have come for that. This is a city [where] you can be yourself as long as you don't harm anyone else. You can live your life as you choose to do rather than as somebody else tells you to do. It is a city in which you can achieve your potential . . . London is the whole world in one city.
(cited in Massey, 2007: 1–4)

The claim that the world's cultures can be found in London is difficult to dispute; three hundred languages, fifty non-indigenous communities with populations of 10,000 or more, and virtually every race, nation, and culture can claim at least a handful of Londoners (see Benedictus, 2005). Despite this, how well integrated are these different cultures? Do cultures mix in London, do they remain plural, or do they assimilate into one broad culture? (see Box 10.4).

Attempts at partnership have resulted in a diversity of ethnic cultures in London. However, as Box 10.4 illustrates, it is clear that different ethnicities take and make place in ways that result in a degree of segregation between them. This segregation is due in part to both historical and economic issues:

The large concentration in south London on the Black Caribbean map is said to have originated in 1948, when many people from the Empire Windrush, the first boat bringing large numbers of migrants from the Caribbean, were housed temporarily in old air-raid shelters in Clapham. The Caribbean areas in west London and north-east London are centred on the other two cheap housing districts of the time, in Notting Hill and Hackney.
(Lewis, Senior Demographer at the Greater London Authority, 2005)

Although historical and economic factors can explain a degree of segregation, 'self-segregation' is also argued to be a cause of the residential choices illustrated in Box 10.4. According to Lewis, 'self-segregation' can partially explain the 'Black African' map's similarity to that of the 'Caribbean': 'If you were a Nigerian student coming to London in the 1960s you might think you stood a better chance in south London, where there are already lots of black people' (Lewis, 2005). Self-segregation thus refers to the process where ethnic and racial groups decide of their own free will to congregate together, and not disperse to create a more 'multi-cultural' neighbourhood. Although it may be unsurprising that ethnicities and races want to be amongst groups they feel share their own beliefs and take place in a similar way, such a process results in the absence of cultural difference. In such cases, despite a formal wish for partnership on behalf of the state and governmental institutions, a sense of the 'other' as a 'stranger' may remain.

Do we live in a post-racial world?

Earlier in this chapter, two rhetorical questions were asked: What does it feel like to be part of an ethno-racial group? And is race and ethnicity an important element in *your* identity? I would like you to consider your responses to these questions now, and at the same time, consider whether in recent years your answers may have changed. In the twenty-first century, many argue we have entered into a time when issues of ethnicity are less important than they once were; it is suggested that the influence of the prejudice of the nineteenth century has diminished as races begin mixing, and those from once minority ethnicities increasingly take up positions of power. In short, it is argued we have entered a 'post-racial era'. As Howard and Flennaugh (2011: 105) put it,

some [say we are in a] 'post-racial' . . . a new and progressive time in . . . life, law, and

BOX 10.4 DIVERSITY AND SEGREGATION?

Source: www.guardian.co.uk/graphic/0,5812, 1395103,00.html

politics, where race is no longer relevant; an era where [the] history of racial strife and discrimination could be considered issues of the past, with little to no contemporary meaning as to how an increasingly racially diverse society experiences life in the twenty-first century . . . How could have Obama's election occurred if racial discrimination still exists?'

How would you respond to Howard and Flennaugh's final question? Do we live in a 'post-racial' era? What is your experience? Paul (2014) suggests that, regardless of any 'post-racial' era, the legacy of racism still has very real consequences today. In the UK, for example, the colour of your skin still has a real influence on how you will be surveilled and dealt with by law enforcement. If you are 'White', you have a 1.620 per cent chance of being stopped and searched by police (2009–2010 figures, see Dodd, 2013), if you are 'Black', this figure jumps to 10.8 per cent. In terms of drug arrests, 44 per cent of 'Whites' in London were charged with drug offences, whilst 78 per cent of 'Black' people where charged for similar offences (see Walker, 2013). In the US, similar discrepancies can be identified. According to the Bureau of Labor Statistics (2009), the average 'Black' worker earns just over $600 per week, about 80 per cent of what the average 'White' worker earns (Howard and Flennaugh, 2011). The US Department of Justice (2009) reports that 'Black' men are jailed at 6.6 times the rate of 'White' men, with almost 1 in 20 'Black' men spending some time in prison at some point in their life. There are, therefore, mixed signals regarding whether our cultures offer equal partnership to all ethnicities and races (see Boxes 10.5, 10.6 and 10.7).

We can see in the above examples that even within cultures where partnership and equality are favoured, informal processes of fear, prejudice, and loyalty towards specific cultural ideals lead to practices of informal (b)ordering such as self-segregation and racism. In these cases, even if ethnicities share the place of the 'other', they rarely interact with 'them'. A Community Cohesion Review report, commissioned by the UK Home Secretary, sums the situation up as follows: 'many [ethnic] communities operate on the basis of parallel lives. These lives often do not seem to touch at any point, let alone overlap and promote meaningful interchanges' (2001: 9).

Ethnicities thus may share places, but often lead 'parallel lives'. They live in diversity, but also in 'solitude' (following how the French-speaking and English-speaking cultures in Canada have often been described). To some extent each

---

BOX 10.5 CULTURAL RACISM: ABORIGINES IN AUSTRALIA

Despite land rights gains in Australia, Mellor (2003) identifies key attitudes that maintain ethno-racial (b)orders within Australian society.

'About 2 months ago we went to Alice Springs and on the train we had this one person who was giving us the history of Aboriginal culture, all the way from Adelaide . . . Yeah, and he was tellin' the people on the train the ways of the Blacks, that they were dirty and stink.' (K33)

'I've been called a "nigger," "abo," and things like that, things that are said to other Aboriginal people, and I think [today], you know, we shouldn't be dealing with that any more.' (K05)

'I reported it to the coppers, and um, and I had one female copper stare at me in the face, and she told me that I loved it, and that being Black, and I quote, 'being Black, I asked for it,' and 'I was that drunk, how the hell do I know that I was raped?'

'I remember I went up to the police station up here once to [register] car and after about 45 minutes of just standing there, any way, I said, "excuse me lady, I've been here bout 30 minutes, 40 minutes." And she says, "the only reason I didn't see you was because you're not standing under the light." (K23)

How would you feel being subject to such treatment? How would you feel if your cultural and geographical position was defined in this way? How would you take place in response? Is such racist behaviour 'fair game' to the 'strangers within'? Should attempts be made to have more respect for difference?

BOX 10.6 REDNECKS AND REDSKINS

In 2014, the US Patent and Trademark Office ruled that the National Football League team the Washington Redskins could not license their name as a trademark. According to their ruling, the term 'Redskins' 'disparages' a group of people on ethno-racial lines, and 'brings them into contempt or disrepute' (see Gambino, 2014). As campaigner Amanda Blackhorse stated: 'The team's name is racist and derogatory . . . If people wouldn't dare call a Native American a "redskin" because they know it is offensive, how can an NFL football team have this name?' (see ibid.). This controversy highlights the importance of the 'power to name' (see Chapter 5), and who has the capacity to label an individual, group, or practice. Language that is acceptable in some contexts and cultures becomes offensive and discriminatory in others – often depending on who is saying it and what histories are invoked through their utterance (see also Box 10.7). For example, in 2013 American Football player Richie Incognito called fellow Miami Dolphins player, Jonathan Martin, a 'half-nigger piece of shit'. As Bandini (2013) reports, although an altogether unpleasant form of words, it was the ethno-racial reference in this insult that 'crossed the line'. Incognito is 'white', Martin is 'mixed-race', and in this insult echoes centuries-old prejudice based on the arbitrary category of skin colour.

It is possible, of course, for ethno-racial discrimination to be cross cut with class (b)orders. For example, if it is inappropriate to call a Native American a 'redskin', is it okay to call a 'white' person a 'redneck'? In work investigating ethno-racial difference at the demolition derby, Falls (2013: 429) notes how the rural working class is often 'publically referred to by terms like "white trash" or "redneck"'. Falls states how the acceptability of such language in the public arena, 'underscores the construction of the working-class white as embodying a complex mixture of racial privilege and class disadvantage' (ibid.). In this light, does the post-racial era actually mean prejudice against all ethnic differences?

BOX 10.7 CULTURAL RACISM: BLACKS IN AMERICA

Despite advances of black politicians in senate and presidential elections, prejudice towards non-white ethnic groups remains strong in some states. As Mangold (2007) reports, although on the face of it all ethnicities are equal, 'stealth racism' is still apparent in the Deep South.

> Three rope nooses hanging from a tree in the courtyard of a school in a small Southern town in Louisiana have sparked fears of a new kind of 'stealth' racism spreading through America.
>
> It all began at Jena High School last summer (Jena has a mixed community, 85 per cent white, 12 per cent black). A black student, Kenneth Purvis, asked the school's principal whether he was permitted to sit under the shade of the school courtyard tree, a place traditionally reserved for white students only. He was told he could sit where he liked. The following morning, when the students arrived at school, they found three nooses dangling from the tree. Most whites in Jena dismissed it as a tasteless prank, but the minority black community identified the gesture as something far more vicious. 'It meant the KKK, it meant "niggers we're going to kill you, we're gonna hang you 'til you die"', said Caseptla Bailey, one of the black community's leaders. Old racial fault lines in Jena began to fracture the town. It was made worse when – despite the school head recommending the noose-hangers be expelled – the board overruled him and the three white student perpetrators merely received a slap on the wrist.

As racial tension grew last autumn and winter, there were race-related fights between teenagers in town. On 4 December, racial tension boiled over once more when a white student, Justin Barker, was attacked by a small group of black students. He fell to the ground and hit his head on the concrete, suffering bruising and concussion. He was treated at the local hospital and released, and that same evening felt able to put in an appearance at a school function. District Attorney Reed Walters, to the astonishment of the black community, has upgraded the charges of Mr Barker's alleged attackers to conspiracy to commit second degree murder and attempted second degree murder. If convicted they could be 50 before they leave prison.

The above example illustrates how places in Jena are (b)ordered on grounds of ethnicity (from the courtyard tree, the school, to the broader town and perhaps even the judiciary). Has cultural assimilation in Jena worked? Are all ethnic groups equal? Whose culture dominates in this example? How is resistance practised?

maintains their own cultural practices, language, religions, and customs. They have little desire to fully integrate into one culture, as Steiner notes in a different context, each group are together but apart, 'standing with their backs to each other' (2001: 145, cited in Kymlicka, 2003).

Mixophobic or mixophilic?

Do we fear difference and want to preserve existing identities and prevent potential threats to us? Or do we celebrate difference?
(McGhee 2003: 394)

Through the three (b)ordering processes we have explored in this chapter (i.e. *genocide*, *tight control*, and *partnership*), dominant ethno-racial groups seek to exercise their power through realising geographically their cultural idea(l)s. Through genocide and tight control, groups look to keep ethnicities apart, emphasising division rather than diversity. These characteristics correspond to what Bauman calls '*mixophobia*' (1995: 221). In these places ethno-racial minorities are not treated equally, but fear, prejudice, and a loyalty to the 'in group' results in cultural marginalisation, or even cleansing, of the 'out' group. In these places, ethnicities demonstrate

loyalty to 'us' and antagonism to 'them'. These idea(l)s about cultural difference create separate places for each group (see Box 10.8).

In attempting to adopt more 'partnership' engagements, some cultures are demonstrating a more '*mixophilic*' approach (Bauman, 1995). Here, ethnicities are more outward looking, looking to loosen their (b)orders and have more flexible and open idea(l)s about their places and concomitant identities. Here the desire for what Young (2000: 221) refers to as a 'differentiated solidarity' is in evidence; difference is accommodated on both sides, with tolerance and mutual respect being sought through interaction and practice (see Kwan, 2004). In a globalising world, increasingly affected by capitalist culture, media communication, and population mobility, how durable and fixed are absolute notions of ethno-racial identity? Should we, as McGhee asks, seek to defend traditional ethnic positions, or look to forge new connections and celebrate difference? As we have seen in previous chapters, ideas of hybridity and mutual interference of once-separate identity positions are increasingly acknowledged. What implications do these ideas have for how we think about ethnicity, race, and place?

Phillips suggests that in the modern world, notions of fixed and absolute ethnic identity are 'over-simplified constructions' (2006: 30). As

BOX 10.8 CULTURAL RACISM IN WARTIME

Following the attack by Japanese forces on Pearl Harbor on 7 December 1941, the United States government (b)ordered 120,000 Japanese-Americans into concentration camps, one of which was the Manzanar War Relocation Center, California.

Although many were legitimate US citizens, and were 'completely American in speech, dress, and manner', those with Japanese ethnicity were deemed to be the 'stranger within' (NPS, 2007). In 1943 the War Relocation Authority Director Dillon S. Meyer, stated it was the self-segregation of ethnicities in California that bred distrust and ignorance of the 'other': 'It would be good for the United States generally and I think it would be good from the standpoint of Japanese-Americans themselves, to be scattered over a much wider area and not to be bunched up in groups as they were along the coast' (NPS, 2007).

Following a legal case, Japanese-Americans were allowed to return to their homes on 2 January 1945. Speaking during his Presidency in 1988, Ronald Reagan said the following about this wartime relocation policy:

> 120,000 persons of Japanese ancestry living in the United States were forcibly removed from their homes and places in make-shift internment camps. This action was taken without trial, without jury. It was based solely on race . . . Yes, the nation was at war, struggling for its survival, and it's not for us today to pass judgement upon those who may have made mistakes while engaged in that great struggle. Yet we must recognise that the internment of Japanese Americans was just that: a mistake . . . It is not in spite of, but because of our polyglot background that we have had all the strength in the world. That is the American way.
> (NPS, 2007)

Manzanar is now a heritage site. Illustrated right are some comments from visitors to that site.

Manzanar
Source: Author

the example of the Notting Hill Carnival outlines (Box 10.9), ethnic identities are often no
longer pure or absolute in their composition,
rather there is a diversity in their constitution and a permeability in the (b)orders which
construct their places and identifications. Ethno-
racial identity, and identity in general, is thus
argued to be more flexible, diverse, and less
coherent than commonly suggested. Following

Massey (2007: 4), it is possible to suggest that
the place of the Notting Hill Carnival is open
rather than bounded, hospitable rather than
exclusive and excluding. Through moves being
made towards partnership on all sides, Notting
Hill Carnival has become an ethnic and racial
'meeting place' (Massey, 2005). Does such an
argument mean that 'pure' places no longer, or
have never, existed? If (multi)cultures need to

BOX 10.9 MIXOPHOBIA AND MIXOPHILIA AT THE NOTTING HILL CARNIVAL

As we have seen above, Notting Hill in West London served as a destination for many black Caribbeans
in the late 1940s and 1950s. In 1959, the first Notting Hill Carnival occurred, taking and making place to
initially celebrate African and Caribbean culture, then predominantly Caribbean music and culture in the
1970s, and now arguably presents the face of a multi-cultural London. As outlined in the extract below,
how this event has taken and made place over the course of its history charts how the attitudes of Britain, its
population and its institutions, have changed from being mixophobic in nature, to increasingly mixophilic.
As Melville (2002: 1) states:

> The Notting Hill Carnival is the largest street festival in Europe, attracting two million people. The Carnival
> parade is great fun – a succession of floats, lorry mounted sound systems, and platoons of peacock garbed
> masquerade 'camps'. These are the elements familiar from the founding carnival traditions of the Atlantic
> Rim: New Orleans, Mardi Gras, and the Carnivals of Port of Spain, Trinidad, and Rio, Brazil.

As the advert below suggests, 'everyone is welcome' at the Notting Hill Carnival. The young woman can
be considered to exemplify the range of cultures mixing there: white in this case, but not exclusively so, this
white person's identity is fused with Caribbean roots typified by her dreadlocked hair and colours of t-shirt
and jewellery.

> In its promotional imagery, as in popular imagination, the Notting Hill Carnival is confirmation of the
> health and vitality of postcolonial, multicultural Britain. Images of young people, black and white, in
> shared celebration, of policemen grinding with passing ragga girls, suggest a trans-racial coalition. Such
> an interculture surely exists in London, [yet the] carnival is no celebration of a pre-existing harmony, but
> an attempt to found a multicultural community, sometimes in the face of extreme adversity.
>   The Carnivals of 1975 and 1976, in particular, were marked by confrontation between black youth
> and the police. Tensions had always been there. With some justification, black Londoners regarded the
> British police as forces of arbitrary oppression. Throughout the 1970s . . . there was scarcely a single
> black person in London who had not been stopped and searched by police, many numerous times. This,
> combined with . . . police containment tactics, made confrontation inevitable . . . Some years, barriers
> appeared on major roads; attempts were made to close the carnival before darkness fell; and the tactics
> of closing off exits in order to channel the crowd in certain directions always led to frustration and anxiety.

But the history of carnival also marks out Britain's multicultural evolution – an unplanned process Stuart Hall has called 'multicultural drift'. . . The genuine multiculture that has emerged around the music of Afro-America, the Afro-Caribbean and in the past decades Afro-England is nowhere plainer in evidence. [The carnival is] where Brits of all complexions cast off the[ir] social inhibitions; where the fundamental hybridity, the ultimate heady concoction of London's multiculture, is finally, clearly understood. Notting Hill Carnival draws on the music of the Black Atlantic and the forms of the Afro- Caribbean in particular. But it is forged in the local circumstances of London . . . We would do well to remember, too, that it has been born in struggle and conflict, and that the multiculture it represents needs to be renewed in every generation. It needs defending, not only from those who still view difference with suspicion and mixture as betrayal, but also from the corporate raiders who are keen to harness the vitality of multiculture for their own ends.

(Melville, 2002: 1–4)

Notting Hill Carnival
Source: Author

be 'renewed in every generation', how will your generation contribute to the ongoing composition of ethno-racial traces in places?

Conclusion

In this chapter we have explored questions concerning the form and substance of the cultural traces which (b)order our places. We have seen that, in many places ethno-racial difference offers a systematic challenge to the (b)ordering practices of dominant cultures. As a consequence, cultures have responded through adopting three main (b)ordering mechanisms to maintain a geographical and cultural purity of place, namely: genocide, tight control, and partnership. Each

seeks to (b)order both the identity of ethnoracial groups in specific ways, attempting to generate a 'common sense' view of these groups and the geographical and cultural places they should occupy within society.

In many cases, the difference represented by the cultural 'other' has been deemed a threat to the dominant culture. The presence of a different ethno-racial group, and the cultural, linguistic, and religious alternatives they offer, have been seen as a risk to the strength of the 'natural' practices within that place. As a consequence, dominant cultures have practised a 'mixophobic' attitude to the 'other', viewing them as a 'stranger', a place invader who threatens their ownership of culture and geography. However, there are alternative ways in which ethno-racial difference is accommodated. In some cases equality between ethnicities is championed and difference celebrated. Examples such as the Notting Hill Carnival demonstrate that it is possible for races to mix, producing traces that at once defend durable values but also invent new solidarities within cultures. Such events offer the opportunity for 'the enactment of alternative, utopian social arrangements' (Shields, 1991: 91), and through their success offer potential insights into how future ethnic idea(l)s, and the new places that go along with them, could be made.

Note

1   A further example of tight control in the form of new settling society in a nomadic cultural area is the case of white settlers in Aboriginal Australia. A brief empirical example is outlined below from Pilkington's (2002) account of her family's experience of this (b)ordering.

> All those who arrived with Captain Stirling, and others who settled before 1830, had the right to choose an area of land wherever they fancied . . . The Nyungar people, and indeed the entire Aboriginal population, grew to realize what the arrival of the European settlers meant for them: it was the destruction of their traditional society and the dispossession of their lands. Bidgup and Meedo complained to Yellagonga after several attempts at unsuccessful hunting trips. 'We can't go down along our hunting trails,' Bigdup told him, 'They are blocked by fences.' 'And when we climbed over the fence, one of those men pointed one of those things – guns – at us and threatened to shoot us if we went in there again,' said an irritated Meedo, 'There are huts and farms all over the place. Soon they will drive us from our lands.' Cut off from their natural food source, the Nyungar people expected these white settlers to share some of their food with them . . . When the brothers were caught spearing a sheep they were the first of many Nyungar

men to be brought in to be sentenced under English law. They received several years' imprisonment and were transported to the Rottnest Island Penal Colony. The white settlers were a protected species; they were safe with their laws and had police and soldiers to enforce these rules . . . It became apparent then, that the Aboriginal social structure was not only crumbling, but it was being totally destroyed.
> (Pilkington, 2002: 13–15)

In 2012, contemporary Australian politicians began the process which would remove clauses from their constitution which sanctioned state racism against Aborigines. Section 25 of the constitution recognises that states can disqualify groups, such as Aborigines, from voting, whilst Section 51 says federal government can make laws based on a person's race (Rourke, 2012). Tight control of different races is not simply an historical phenomenon.

Suggested reading

The following texts are useful to explore the issues raised by this chapter.

Books

Sibley, D. (1995b) *Geographies of Exclusion*. Routledge: London.

Here Sibley argues that Western culture is based on exclusion. He suggests that powerful cultural groups place 'others' (including marginal ethnicities) as outsiders, 'purifying' their own locations, breeding fear, and excluding marginal voices.

Jackson, P. and Penrose, J. (eds) (1993) *Constructions of Race, Place and Nation*. UCL: London.

Jackson and Penrose offer sustained analyses of how 'race' and 'nation' vary from place to place, how these spatially construct cultural life, and territorially express racist and nationalist ideologies. Particularly useful chapters include those on Aboriginal Australians and Muslim culture.

Bauman, Z. (2002) Foreword. Individually, together. In Beck, U. and Beck-Gernsheim, E. (eds), *Individualization*. Sage: London, xiii–xx.

Bauman, Z. (1995) *Life in Fragments*. Blackwell: Oxford.

Bauman's work illustrates his argument that cultural life is increasingly fragmented, liquid, and negotiated. He gives key insights into the role that individualisation may play in the construction and exclusionary nature of our cultural geographies.

Journals

Phillips, D. (2006) Parallel lives? Challenging discourses of British Muslim self-segregation. *Environment and Planning D, Society and Space*, 24, 25–40.

Paul, J. (2014) Post-racial futures: imagining post-racialist anti-racism(s). *Ethnic and Racial Studies*, 37 (4), 702–718.

In these papers, Phillips challenges ideas of ethnic mixing and self-segregation in the UK, whilst Paul interrogates the reality of a post-racial present.

Place and youth

Introduction

As we have seen, cultural geography is all about taking and making place. Every action we engage in contributes to the cultural world, leaving a trace that affects the identity of who we are and where we are. Places are (b)ordered in line with our ideas, with cultural and geographical regulations constructed to classify places for particular people and uses. We have seen in previous chapters how places become (b)ordered in line with orthodox ideas about culture and nature, about race and ethnicity, and about nationalism and religion. Our own individual senses of place are created within and increasingly across these (b)orders; our own places sometimes fit snugly amongst the (b)orders constructed by the mainstream, at other times we have to create new places in order to be ourselves. This chapter focuses on one cultural group within Western society that traditionally has no sense of place within the mainstream. As a consequence, this group takes and makes place on the margins, redefining themselves in ways that gives us new insights into the cultural geographies around us.

This chapter explores the cultural geographies of youth. Youth culture is an example of a group that defies conventional (b)ordering mechanisms. They are 'in-between' and often across (b)orders. Youths are no longer children, but they are not yet adult. This '*liminal*' positioning has consequences for the places that they can take and make within the cultural world. In this chapter we will look at some of the ways in which

cultural geography has looked at youth culture. How youths can be angels or devils, goodies or hoodies. We explore how youths have attempted to make place within and beyond conventional (b)ordering mechanisms through practices such as graffiti, free running, and urban exploration. From these explorations we consider how youth culture can influence how we think about, and act in, place.

(B)ordering young people: act your age

As we have seen in previous chapters, binary classifications between entities or things is one way in which cultures (b)order places (for example, the world is classified into culture *or* nature, and the economy *or* culture). Another such example is the distinction between age groups; one way we can classify ages is into 'children' *or* 'adults'. On the face of it such a (b)ordering mechanism is useful as it designates between 'grown ups' and 'kids', between those who are morally and legally responsible for their actions, and those who may need some sort of protection or education as to the cultural 'ways' of the world. However useful this dichotomy may be, it's either/or format provides a problem for some age groups. 'Older people', middle aged people, as well as youths and teenagers do not obviously fit into this conventional categorisation. The example of young people illustrates this. In some senses youths are still children – some may be in full-time education or remain legally 'minor'; yet in others they are not: they are too old for playgrounds, they can stay out late(r), and do not need constant parental supervision. At the same time, however, youths can also be classed as not yet 'adult'; in some cases they can't vote, drink, or drive, yet they can still experiment with their own music, drugs, languages, and practices. Youths therefore live in a 'grey area' between the distinctions of 'child' and 'adult', when they are told to act their age, are they being told to grow up, or realise that they are not grown up yet? Youths are therefore in an 'in-between' position within modern culture. As Skelton recognises, they are: 'at once children (in full time compulsory education), teenagers (socially defined as difficult, moody, rebellious and trouble-making) and young people (celebrated as the future, full of energy and life)' (Skelton, 2000: 82).

Youths' ambiguity in relation to this child/adult (b)ordering mechanism thus leads to problems for their sense of place within mainstream culture. Whilst youths may be told to 'act their age' they may also be told to 'know your place'. In this ambiguous position between child and adult, what is the place of youth within our culture and geography?

The geographies of youth culture

The place of 'children' in mainstream culture is clear. As Valentine (1996a) has shown, the idea of children as 'little angels' has led to places being especially (b)ordered for them. The concern for children in open public space, from 'stranger danger', abduction, or paedophilia has led to children being appropriately (b)ordered with a carer or parent, in the home, in education, or in a designated play space. The child is therefore adequately controlled within mainstream places – restricted for their own protection, but also kept away from adult places so these can function as smoothly as possible (see Figure 11.1).

Adult places are also clearly defined within the mainstream. Places of the pub, the club, work, or gambling are not arenas in which you would expect to see children; these are places to which you only gain access when you reach a certain age. Indeed, in general terms, Valentine has suggested that all 'public' places have become naturalised as adult in nature, the orthodox use of these places is restricted for those outside this categorisation (Valentine, 1996b, see also Hil and Bessant, 1999). In this situation, with children's places and adults' places appropriately separated and (b)ordered, where is the place for and of youth cultures? As Wyness (2006: 24) argues, youths are 'not part of the social world that counts'; due to youths' in-between age a concomitant cultural and geographical 'grey area' is also produced. Youths are 'out of place' in the mainstream.

Youths as in-between: young people as 'liminal' beings

Youths therefore occupy a grey area within the mainstream; in terms of their age and the places they can go they do not quite fit into the orthodox (b)ordering mechanisms of our cultural geography. Youths are therefore in-between conventional categories; they can be understood as 'liminal' beings. Liminality, according to van Gennep (1960), refers to a position on the threshold or margin, at which activities and conditions are most unstable and uncertain. For Turner, liminal beings are constituted through their existence on 'a margin or *limen*, when the past is momentarily negated, suspended, or abrogated, and the future has not yet begun, an instant of pure potentiality when everything, as it were, trembles in the balance' (Turner, 1982: 44).

Youths can be seen as existing on this margin. Their past as children is 'negated, suspended, or abrogated' yet their potential as adults is not quite yet realised. Youths are therefore 'threshold

Figure 11.1 (B)ordering age

In the second image we can see how this play area in Changi Airport, Singapore is (b)ordered just for children. If children have this place provided for them, and adults have bars and 'over 18' venues, where do youths 'hang out' in the specific place of the airport, or in public places more generally?

Source: Author

people' (Turner, 1969), they are located between cultural (b)orders, 'literally occupying a threshold or borderland space, straddling, subverting, and disrupting boundary lines of separation and contact' (Jones, 2007: 17). As Turner states, they are 'neither here nor there; they are betwixt and between the positions assigned and arranged by law, custom, convention and ceremonial' (Turner, 1969: 95). In Bhabha's terms, youth's status is liminality made flesh, they are 'neither one nor the other, but *something else besides, in-between*' (Bhabha, 1990: 224, emphasis in original).

Youths thus have a liminal status, and although this means they have no place made for them by the mainstream it does not mean they have no place at all (see Figure 11.2). Liminal status, as Turner tells us, is a place of 'pure potentiality', it is place that can be oppressive, but is also enabling (see Foucault, 1980). As we have seen in Chapter 5, wherever there is domination there is also the opportunity for resistance, thus the place of the (b)order is 'filled not only with authoritarian perils but also with possibilities for community, resistance and emancipatory change' (Soja, 1996: 87). In the words of Bhabha, it offers the potential to 'give rise to something different, something new and unrecognisable, a new area of negotiation of meaning and representation' (Bhabha, 1990: 211). Later in this

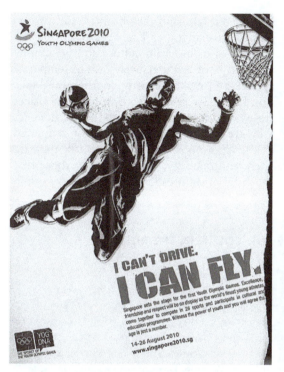

Figure 11.2 I can't drive, but I can fly . . .

As this image from the Youth Olympic Games suggests, by virtue of their age, youths are limited in the range of cultural activities they can practise, but this does not necessarily limit their capacities to influence the cultural world

Source: Author

Figure 11.3 Le street c'est chic

Source: Author

chapter we will explore practices through which youth cultures contribute 'something different, something new' to the ongoing compositions of places. We will broadly introduce these by beginning in the location where youths generally take and make their place: they take their culture to 'the street'.

Take it to the streets

Most of the police would [ask:] 'Why are you hanging around the streets?' And the kids say, 'This is where we live.'

(Sinclair, 2012: 85)

Without a place (b)ordered for them by adult culture, youths have to take and make their own places wherever they can. Matthews et al. (2000) use the term '*the street*' as way of encompassing the places in which youths claim to articulate and express their culture (for youth culture, the streets are 'chic', as graffiti in Barcelona suggests, see Figure 11.3).

We use the term 'the street' as a metaphor for all public outdoor places where children are found, such as roads, cul-de-sacs, alleyways, walkways, shopping areas, car parks, vacant plots and derelict sites . . . To young people the street constitutes an important cultural setting, a lived space where they can affirm their own identity and celebrate their feelings of belonging. In essence, these places are 'won out' from the fabric of adult society, but are always in constant threat of being reclaimed.

(Matthews et al., 2000: 281)

In our own lives I'm sure we can remember how we have taken and made 'the street'. Through my everyday practices I have turned waste ground into my playground, car parks into skate parks, industrial estates into velodromes, and underpasses into hangouts. These are places on the margins of mainstream culture and geography and thus have been relatively easy to temporarily take and make for (my) youth culture (see Box 11.1). To an extent, these marginal places are where (b)orders are more loosely defined, they are not necessarily policed or regulated in a strict, authoritarian way. As such, they are open to trangressive uses, especially those that are trivial or unthreatening. Even so, as a youth

---

BOX 11.1 CREATING THE KETEN: A PLACE FOR YOUTH IN THE RURAL NETHERLANDS

For those young people who grow up in rural rather than urban areas, taking and making the street proves problematic simply due to the different conurbations and street spaces that exist in the countryside. In the Netherlands, rural youths tackle this issue through the phenomenon of 'keten'. Keten are static caravans normally located on family farms, and are used by youths in close friendship networks. Keten comes from the Dutch word 'keet', which refers to both a building (a den, a shed, a caravan, or part of a barn), and the activity of hanging out and having fun. As Haartsen and Strijker observe, keten provide rural youth, 'with a place of their own, where they can do whatever they want without having to consider the rules of parents and other educators ... [they] can be seen as one of the many types of informal places where young people try to create their own subculture' (2010: 163).

Keten are therefore rural equivalents of 'the street'. Can you think of other clandestine or liminal places that youth cultures take and make in urban and rural locations? (See for example, Valentine et al., 2008.)

Keten
Source: Beljaars

I enjoyed changing the nature of these places. It was in some small way a challenge, a risk. Feeling marginal to the mainstream made me seem almost invisible, I could play with, flirt, and subtly transgress these places without ever feeling I was wholly challenging them. My friends and I could explore the limits of the mainstream almost beneath its gaze, creating places where we were in charge for a short period of time – making a *'temporary autonomous zone'*, to use the words of Hakim Bey (1991). Our practices thus did not resist the marginal position given to us by the mainstream, but rather revelled in it. If we became too much of a threat – breaking into the school yard or private property – we knew we could play on our liminal status and pretend we

were just kids larking about, and as 'little angels' we'd probably get away with it.

In true liminal style, therefore, I liked to hang around on the edges of both culture and geography. However, many youths directly test their own and the mainstream's (b)orders by attempting to take and make adults' places. Matthews et al. (2000) identify the 'mall' as a key example of this. Through this process, Matthews et al. argue that the mall becomes a 'cultural boundary zone', a place where youths contest and transgress the conventional cultural and geographical (b)orders of the mainstream, whilst adults seek to reaffirm and reconstitute their (b)orders by removing youths from this place (as Box 11.2 describes).

BOX 11.2 THE MALL AS A CULTURAL BOUNDARY ZONE (AFTER MATTHEWS ET AL., 2000)

Matthews et al. outline how, 'young people's occupancy of some settings, such as the mall, can be interpreted as . . . a manifestation of a cultural politic that challenges their marginality'. From this perspective, they argue that, 'the mall becomes a symbolic site of contest and resistance, a place infused through and through with the cultural trappings of [liminality]. When there, young people are asserting the right of the "hybrid", no longer child but not quite adult, to be active agents in the [taking and making] of place' (2000: 282).

The mall is predominantly an adult place. Often you need cars to get there. You need money or credit to buy things. These resource requirements thus have the effect of informally (b)ordering the place of the mall as a place for grown ups. However, youths are also out of place in other ways. The mall is a place for small groups, for couples, for families. It is a place for walking and shopping, not skating or cycling. These activities and modes of behaviour are strictly controlled. Through security patrols, and closed circuit television, youths can be moved on or thrown out of the mall for transgressing these codes of behaviour. Jones (2007), in her work on youths' cultural geographies, asked young people about their experiences that took the place of the mall:

We went . . . into the shopping centre and we went to Starbucks, and we sat there but again none of us bought anything because we were all skint so we got chucked out. All we wanted to do was sit down but they wouldn't let us . . . they were like 'someone is going to have to buy a drink', so Phil bought a drink and they were like 'the rest of you are still being kicked out'. Phil stayed because he had the drink but it didn't count for all of us.

(Helen and Molly, age 15).

M: People don't like it . . . usually older people just walking around town doing their shopping; they just don't like big gangs of us walking around.

T: Even if you're not doing anything bad. If you're just walking around,

M: It makes them feel vulnerable.

S: They think you're causing trouble. So you feel that other people are like judging you and thinking bad of you even if you're just doing nothing,

M: Because they stereotype us teenagers as trouble makers.

(Mark, Tom, and Samir, age 16)

People think oh because we're young people, people think we could do something wrong. Cos like when you go shopping in a group, they always think you're gonna take something. Or when you hang out maybe near a shop, people think you're gonna do something wrong to the shop or break their windows or whatever . . . sometimes people just look at you really weirdly like, what are you doing?

(Matilda, age 14, all cited in Jones, 2007: 119–120)

By taking place in the mall, youths experience the difficulty of their liminal position. The place of the mall is not a youth's place: they are assumed to be transgressive because of their cultural difference to the mainstream, and adults on the whole successfully defend their (b)orders; through monitoring and regulation they reclaim their place. Youth's occupation of the mall is thus under adult control. Even if they do behave, they may be out.

Why pick on us? It's not fair we're not doing any harm . . . Look at them there. They're standing and talking [pointing to two adult couples] why don't they get picked on? We have a right to be here like anyone else.

(Girl aged 14, Newlands, from Matthews et al., 2000: 291)

Why should we go? We're human too.

(Girl aged 16, Grosvenor Centre, in ibid.)

In the UK, legislation has been recently introduced that further marginalises youth culture. 'Injunctions to Prevent Nuisance and Annoyance' (or IPNAs) have been agreed by the government which have, according to the Children's Commissioner Dr Maggie Atkinson, the effect of 'promot[ing] intolerance of youth' and 'punish[ing]' youth culture (see Helm, 2013). Under this legislation young people will receive an injunction if a police officer judges that 'on the balance of probabilities' (i.e. a weaker balance of proof than that required for adult offences), a young person 'has engaged or threatens to engage in conduct capable of causing nuisance or annoyance to any person' (UK Parliament, 2013). As Lord Macdonald, formerly the Director of Public Prosecutions,

summarises: 'it is difficult to imagine a broader concept than causing "nuisance" or "annoyance"' (Macdonald, 2013). Once enacted, an IPNA restricts where that young person is able to go and what they are able to do. If a youth flouts the injunction, they face a three-month prison sentence. As Monbiot suggests, this law 'will be used to stamp out plurality and difference, to douse the exuberance of youth, to pursue children for the crime of being young and together in a public place' (Monbiot, 2014).

Summary

A (b)orderland, 'is a vague and undetermined place created by the emotional residue of an unnatural boundary . . . The prohibited and forbidden are its inhabitants'.

(Anzaldua, 1987: 25)

Youths therefore live on the cultural (b)order. Due principally to their liminal age they are classified as different, they are not quite children, but not yet adult. In this situation, they reside in a 'vague and undetermined' position, they are the 'prohibited and forbidden'. The dominant culture (in this case adult culture) has attempted to remove from youths the ability to articulate their cultural idea(l)s and practise their activities. In mainstream places their geography has been 'delimited' and their 'invisibility enforced' (after Breitbart, 1998: 306). As a consequence, youths have to get by in the margins of mainstream culture.

So far in this chapter we have focused on youth culture from the outside, maybe even from an adult perspective. From this point of view, perhaps, youth culture all looks the same. However, if we look more closely at youth cultures we can see that there is not just one 'youth culture' but many. Youth culture is not homogeneous in nature, rather it is heterogeneous. As we have seen in Chapter 6, processes of individualisation mean we all find our own ways of weaving together our senses of place and identity positions, and these create a range of cultural groups that is as diverse for youths as any other culture.

Neo-tribal places

Just cos some people do it doesn't mean we're all the same. Just cos we're the same age doesn't mean we're all like that.

(Sophie, age 16, in Jones, 2007: 120)

Every generation of youth culture, in every place, produces its own groups. From 'Teds', 'Mods', and 'Rockers' (see Cohen, 1979; Shields, 1991), to 'Chavs', 'Emo-kids', 'Goths', or 'Jocks', youths have constructed for themselves their own identity positions and senses of place. In my school there were 'punks' and 'casuals'. They each had their own culture, their own look, and their own place. Punks had mohicans, some skinheads, all had leather jackets with stainless steel studs. Many had tattoos, and all loved the Sex Pistols. Casuals wore trainers, jeans, they had neat, gelled coiffures, they wore Sergio Tachini or Pringle jumpers. The punks congregated at one end of the corridor, the casuals at the other. My tutor room was in the middle: 1st year, new blazer, no-man's-land. Who owned the corridor was the subject of lunchtime punch-ups. These were the 'gangs' of my school.

Youth groups coalesce in gangs in order to express their interests, tastes, and preferences, and demarcate themselves from others. As with all cultures, these gangs are often tied into geographical places – be they particular parks, streets, or areas of the mall, which they take and make to their own ends. As one young gang member in London states:

We all want to feel like we belong, especially when we're young, and there's no greater sense of belonging than being in a gang. We all want to live a life of meaning too – but if you're not being offered the chance of really achieving something, of following your dreams, of making your mark, then you've got to find your own meaning, create something that matters out of what you've got. So you end up with people fighting, even dying, over their own made-up rules, their postcodes, their turf.

(Lucas, 2012)

As Lucas outlines, if young people are not offered opportunities, ideas, and meanings by dominant cultures (in this case adult cultures), they will try to carve out their own (b)orders; gangs become one instrument through which they do this. Maffesoli (1996) refers to these gangs of young people as 'tribes', or 'neo-tribes' (Shields, 1992a, 1992b, 1996). As experienced at my school, these groupings have their own fashions, music, and linguistic jargon; they are maintained through shared beliefs, rituals, and consumption practices (Shields 1992b). Each neo-tribe defends its orders and borders, and often disidentifies with other neo-tribes that do not share their idea(l)s. Yet despite the importance of these neo-tribes, and the strength with which their (b)orders are felt and defended, these gangs need not be permanent groupings; rather they are often temporary collectives created or joined for a short period of time. True to the instability and creativity of the liminal position, these tribes are often transitory both in terms of membership and existence. They can perhaps be thought of as experiments with attachments, the 'trying on' of different cultures to see if they suit the identity of the young person. If they do, they are likely to be carried forward into adult life.

BOX 11.3 SCALLIES TAKING AND MAKING PLACE

As we have seen, a 'neo-tribe' not only has different cultural practices, but through them takes and makes different places. Jones (2007: 136) found that different dimensions of 'the street' were taken by different groups, in different ways. As a consequence, she states, 'the spaces in and around [any] town centre . . . comprise a diverse topology of appropriated spaces (Lefebvre 1991) or temporary autonomous zones (Bey 1991) that have been claimed . . . by neo-tribes'. This can be demonstrated through the following example of 'scallies'. According to Jones' research, scallies were revealed as a significant grouping that occupied the central spaces of the town centre. As she states,

> Scallies are instantly recognisable through their distinctive 'group uniform', gang formations and practices of aggressive rituals. As a local variation of the national 'chav' phenomenon, scallies have a distinct clothing style and similar to chavs favour designer brand clothing, usually tracksuits and sports jackets, baseball caps and heavy gold jewellery.
>
> (Jones, 2008: 139)

This was evident in conversations when teenagers were asked to describe the scally neo-tribe:

> People who wear their collars up, and tracksuit bottoms and their socks pulled up over their trousers. It's quite amusing because their socks are pulled right up to their knees. They wear Burberry hats and they wear chains and that. (Hayley, who describes herself as 'normal')

> You know people who wear trackies in their socks and that. And caps like half way off their head . . . they all walk round with their caps like resting on their heads, thinking they're dead hard. (Ben, 'normal').

Alongside these fashions and bodily practices, Jones identified key places that were taken and made by this 'neo-tribe'. She stated,

> Scallies have a dominant presence in the town centre particularly on a Saturday afternoon, where they congregate in large groups outside McDonalds' restaurant. For scallies, McDonalds acts as a socially central meeting point, as Susan describes: 'When they [scallies] go into town, there's like a million of them. It's like they're on this pilgrimage to McDonalds where they all hang out' (Susan, goth).

As one scally put it:

> Being a scally, it's just like your mark, it's the way you are. Like if you're a scally, one part of it relates to where you hang around. It's basically your area. No one will mess with your area. Me and him hang around, don't we? About 20 of us just standing there it's dead good.' (Jamie, scally)
>
> (Ibid.)

Moral panics

Scallies are therefore a good example of a neo-tribe, a youth group with their own cultural and geographical practices. Such tribes can be seen as a threat to the mainstream. Through the traces they contribute to mainstream places – congregating in large groups and often demonstrating aggressive rituals – it can be argued that scallies assert a 'geography of fear' (Pain, 1997, 2001). This threat can be felt by the range of others that use that place – either other youth 'neo-tribes', or the dominant adult group. In such situations, youths are no longer the 'little angels' of the child/adult binary, but become 'little devils' (see Jenks, 1996; Valentine, 1996a). Such characterisation has led to '*moral panics*' (Cohen, 1979), where youth tribes are seen to be posing such a serious challenge to the physical, psychological, and moral (b)orders of the mainstream that they require increased legislation and control (as we have seen above). One example of a moral panic has been the backlash against 'hoodie culture'.

BOX 11.4 HOOD 'RATS' AND MALL 'RATS': YOUTHS IN THE HOOD, UP TO NO GOOD?

The moral panic against hoodie culture is clear in the UK. As McLean states, the overwhelming perception is that, 'the hoodie is the uniform of the troublemaker' (McLean, 2005); David Cameron, when Conservative Party Leader, put it this way: 'the hoodie represents all that's wrong about youth culture in Britain today' (2006). Such prejudice against an item of clothing and the youth culture that wears it has led to many shopping malls banning hoodies from their premises, and often the young people (or mall 'rats') who wear them. Any clothing that obscures the face (hooded tops, baseball caps) will not be allowed. Rachel Harrington, vice-chair of the British Youth Council, says such decisions demonstrate a growing demonisation of the culture and geography of young people.

> It's yet another example of a trend – tarring all young people with the same brush and overreacting to any behaviour by young people. You can understand shopping centres' desire to please their customers, but it doesn't seem to me to be the best response. It's very easy to create the stereotype of the young thug as emblematic of society's problems, rather than seek out the root of the problems.
>
> (Cited in McLean, 2005)

Sharmarke Hersi, a black youth, was stopped by police officers for wearing a hoodie. He states, 'It's like some kind of moral panic. I was on the train not long ago and a lady was holding her bag tight because of my dress code. You sometimes see people crossing the road to avoid you or putting their phone away' (in Casciani, 2005).

However, attempts to demonise and outlaw the hoodie often have the consequence of making it a more popular trace amongst youths. As McRobbie identifies,

> [The hooded top] is one in a long line of garments chosen by young people, usually boys, and inscribed with meanings suggesting that they are 'up to no good' . . . Moral panics of this type have only ever made the item, and its cultural environment, all the more attractive to those who prefer to dis-identify with establishment figures and assorted 'moral guardians' and who enjoy the outlaw status of 'folk devil'.
>
> (McRobbie, Professor of Communications at Goldsmiths College, in McLean, 2005)

Hood

Source: http://static. sky.com/images/ pictures/1665087.jpg

Hood demo

Source: http://londonist. com/attachments/ London_Ricky/hoodie_ sign.jpg

The demonisation of the hoodie by mainstream culture, making it a 'novel', unorthodox and unwelcome cultural trace, has not had the intended affect on youths in the UK. Rather than making the hoodie unpopular, dominating powers have unintentionally reinforced the cultural kudos of the hoodie. Rather than acquiesing to the (b)orders preferred by adultist cultures, youths have reasserted their claims to this cultural trace, either revelling in or resisting the outlaw status it gives their presence in places.

Summary

Youth cultures, like all others, take and make places. These places can be on the margins or in the mainstream. Youths loiter in the margins to experiment with their liminal status, flirting with adult practices and trying out new cultural expressions. In the mainstream, these cultural traces are often not understood, deemed as a threat, and vilified. Moral panics ensue, seeking to demonise this 'other' group, further marginalising their sub-culture and making any transition or assimilation into the mainstream more problematic. The places of youth culture can thus, as Soja has identified, be, 'filled with authoritarian perils' (1996: 87), but at the same time offer sites of creativity, invention, and potential. When taking and making the street, youth culture can offer us new ways of experiencing and thinking about the world around us. In its diversity and plurality, youth cultures offer bubbles of refreshment to the cocktail that is place. This chapter goes on to explore some of these bubbles: free running (or parkour), graffiti, and urban exploration. All can be seen to offer new approaches to thinking about and acting in the places around us, and illustrate further the complex interaction between minority and dominant cultures in the ongoing composition of place.

Liberating the city: free running

I tried skateboarding, I tried BMXing, but when freerunning came along, it sort of took over my life. I think the relationship between all of these sports is maybe the fact that i[s] there's no set rules, whether you've got a skateboard or whether you've got a BMX or whether you've just got a good pair of shoes, you can just go and move you know? There's no one telling you 'you can't do that', 'that's not in the rule book', 'that's not in the official guidelines' . . . you're free with these sports, and that's what attracts a lot of people to them and what keeps them doing them.

(Paul 'Blue' Joseph, 2011)

Free running suggests we have a choice, do we allow our cities to trap us, or liberate us?

(Tomlinson, in Christie, 2005)

Free running or parkour (the 'art of movement') is a youth culture that has its origins in Paris, France.[1] Free running seeks to use the built objects of a place as a kind of gymnasium, through running, climbing, and acrobatics the aim is to move through these objects as stylistically and efficiently as possible (see Figure 11.4). In France free runners are known as *traceurs*, through their movements they leave *traces* which influence how we come to see, experience, and think about the places around us.

You just have to look, you just have to think, like children. This is the vision of parkour.

(Sebastien Foucan, free runner, in Christie, 2005)

Free running has thus taken the urge to jump between kerbstones or vault over bollards to a more organised and expert level. The orthodox practices through which we make the street (walking, cycling, or avoiding) are thus transgressed through free running. 'Traceurs' do not follow these conventions, they use the built form of the street as something to experiment with, playing with limits and physical (b)orders, and through so doing, influencing the way the street is understood and used.

Central to this new sport is a philosophy that changes the way free runners view their towns and cities. Obstacles become a kind of . . . springboard to launch off, leap from, or land on.

(Tomlinson, in Christie, 2005).

Cities have always been places of escape. When they bed down, when they cease to be that, when tramlines become too settled, everything is too stereotyped, these things start to die. What these free runners are saying, in the most original and unexpected way you can think of, 'You [grown ups], you're in tramlines. Come and escape with us. Think afresh.'

(Will Hutton, in Christie, 2005)

Through the difference that free running constitutes to our uses and approaches to the street, how does it pose a threat to orthodox uses of place? As traceurs climb over bollards, bus shelters, or churches, do the conventional meanings and uses of these traces change? Are they subverted, or even corrupted? Should the street be used in this way? In a more experiential sense, can free running liberate the city from its 'tramlines' and bring different emotions to the daily commute? Can it take and make the city into something different? For its practitioners, free running offers a

certain freedom and expression, something very real and very raw . . . When you take people out of the ordinary you create something magical.

(Bear Grylls, climber, in Christie, 2005)

After the jump you're in a state of . . . you feel a surge. I don't know how to explain it. It's the happiness of creating something,

Figure 11.4 Traces and traceurs: free-running places

Source: Statue sujathafan www.flickr.com/photos/sujatha_fan/2671004494/

'Above traffic' Daniel Ilabaca, www.daniel-ilabaca.co.uk, Jon Lucas I Photography www.flickr.com/photos/jonlucas/204150906/

'Stairs' Mute www.flickr.com/photos/mutephotoblog/103108395/

'Building' Jean-Christophe: Parkour GMX

achieving something and doing it well. There is a sense of accomplishment that you've succeeded in everything you planned and you think 'I've done it', and you can see it in my face. In your mind you're walking tall.

(Sebastian Foucan, free runner,
in Christie, 2005).

Free running is therefore an example of a youth culture that offers potentially positive ways for both youths, and indeed other cultures, to think and engage afresh with the street. It contributes traces that are not necessarily durable in a material sense, but may leave a permanent trace in your memory if you see an acrobatic leap from a parking meter, or in your heart if you perform one yourself. In this sense, therefore, it emphasises the combined nature of the substantial, symbolic, and psychological composition of traces that come to define our places. Did you jump over bollards as a kid? Do you practice free running on your journey to university? If so, why? Why has the street become a place of walking or driving? Should free running be encouraged, or banned?

The writing's on the wall: graffiti traces

Graffiti has been used to start revolutions, stop wars and generally is the voice of people who aren't listened to. Graffiti is one of the few tools you have if you have almost nothing. And even if you don't come up with a picture to cure world poverty you can make someone smile while they're having a piss.

(Banksy, 2001: 3)

Graffiti is 'unauthorized writing or drawing on a public surface' (Srivathsan et al., 2005: 422), and as such it explicitly seeks to leave material and visible traces in places. These traces give a 'voice' to groups – they get to offer their opinion or state their case through their writings or drawings. In one sense therefore the message of graffiti can easily be read, graffiti's message is in the words or images stencilled or sprayed on the wall (Figure 11.5 and Box 11.5).

However, as McLuhan (1967) recognises, cultures, 'have always been shaped more by

Figure 11.5 Graffiti messages

According to this graffiti in Cardiff, we should use the electoral system to keep out right-wing fascist parties. On this train in Rome, Italy, graffiti writers want us to know that 'life is fast'. (The difficulty of writing graffiti on a train is referred to later in this chapter.)

Source: Author

BOX 11.5 GRAFFITI TRACES

Graffiti can be simple but colourful tags, but can also involve images. As we will discuss below, whatever the nature of these traces, all say something about the writer, and the location in which it occurs. How do you read these traces?

Shush

Source: Author

Resistance

Source: Author

Banksy: Bristol Crown Court

Source: Banksy/Lazarides

Drunk tag

Source: Author

Designated graffiti area

*Source: Joel Saunders
www.flickr.com/photos/joel_
saunders/2885993244/in/
pool-banksy*

Metro tag

Source: Author

Flower 1

Source: Author

Flower 2

Source: Author

Peace?

Source: Author

the nature of the media by which men [sic] communicate than by the content of the communication' (cited in Watson, 1994: 19). McLuhan is suggesting that the medium is as much a part of the message as what is said or drawn; and this is particularly true for the case for graffiti. As Banksy (a famous UK graffiti 'artist') states above, graffiti gives a voice to people who aren't normally heard in the mainstream. The medium chosen to articulate this group's message is not the ballot box or the mainstream media, these places are seen to disenfranchise rather than represent them. Resorting to taking and making place through graffiti therefore says more than simply the stencilled words written: it also says, 'I am not part of your culture, but I still have a right to my say. This place is mine too and you can't silence me.' The message offered by the medium of graffiti is thus often deemed a threat to the mainstream as it symbolises the lack of absolute control any dominant group can have over a minority.

Thus in contrast to the non-material traces left by 'traceurs', graffiti offers a visible and durable claim to the ownership of place. As Srivathsan et al. state, 'the [street] is ordered in the map but the contestations within are made visible . . . through graffiti' (2005: 427). Claiming their place within the street is a key reason youth cultures practice graffiti. As Nandrea (1999: 113) outlines,

> Graffiti takes many forms and serves many functions, but perhaps the most prevalent and certainly the best publicized function of urban scrawl is to stake out gang territory, to lay claim to an alley, a corner, a roof, or an entire area symbolically fenced off by gang signatures. In an amazing and wholly audacious gesture, these urban gangs redraw all the lines of the city, simply overwriting proper legal and political boundaries.

From this perspective graffiti challenges the orthodox (b)orders of the street and re-places them with new traces that redefine ownership and rules. Graffiti therefore symbolically and literally takes place for particular youth gangs. This reading of graffiti is illustrated in Box 11.6. Box 11.6 is an extract from the film *Falling Down* (1993). In this scene, the main protagonist (William 'D-Fens' Foster, played by Michael Douglas) sits on a rock traced with graffiti tags, in what he considers to be waste ground. Two unnamed gang members confront Douglas about his right to be in their 'backyard'.

BOX 11.6 THIS IS MY PLACE: GRAFFITI AS A SIGN OF CULTURAL AND GEOGRAPHICAL OWNERSHIP

Gang 1: What you doing, mister?
Foster: Nothing.
Gang 1: You're trespassing on private property.
Foster: Trespassing?
Gang 2: Loitering too.
Gang 1: That's right.
Foster: I didn't see any signs.
Gang 1: What you call that?
Foster: Graffiti?
Gang 1: No, man.

Gang 2: That's not fucking graffiti!

Gang 1: That's a sign.

Gang 2: He can't read it, man.

Gang 1: I'll read it for you. It says this is private property. No fucking trespassing. This means fucking you!

Foster: It says all that?

Gang 2: Yeah!

Foster: If you wrote it in English, I'd fucking understand it!

Gang 2: Thinks he's being funny.

Gang 1: I'm not laughing.

Gang 2: I'm not either.

Foster: Hold it, fellas. We're getting off on the wrong foot. This is a gangland thing, isn't it? We're having a territorial dispute? I've wandered into your pissing ground or whatever the damn thing is . . . and you're offended by my presence. I understand that. I mean, I wouldn't want you people in my back yard, either. This is your home . . . and your home is your home. I respect that. So if you would just back up a step or two . . . I'll take my problems elsewhere. Fair enough?

Gang 1: What do you think?

Gang 2: He should pay a toll . . .

Gang 1: Give us your briefcase, man.

Foster: I'm not giving you my goddamn briefcase.

Gang 1: Motherfucker, give us your motherfucking briefcase!

Foster: I was willing to mind my own business. I was willing to respect your territory and treat you like a man. You couldn't let a man . . . sit here for five minutes to rest on your precious piece-of-shit hill? Want my briefcase? I'll get it for you, all right? You want my briefcase? Here's my briefcase! [he attacks the men with a baseball bat]. Clear the path, you motherfucker! Clear the path! I'm going home!

(From: www.script-o-rama.com/movie_scripts/f/falling-down-script-transcript-douglas.html)

In this extract there are many examples of resistance and domination (and for an extended discussion of these in the film more broadly, see Cresswell (2000)). In this chapter, however, it is worthwhile noting the symbolic, territorial, and proprietary function of graffiti. Graffiti traced onto the rock by the gang members claims this place as theirs. This taking and making of place is clear to all those who can read this (b)ordering trace, and invokes trouble for those who transgress it.

Graffiti tags therefore can be read as signs that construct new (b)orders and ownership for a place. As such, graffiti defiantly issues a challenge to the question 'whose place is this?' By explicitly taking place, and failing to acknowledge or value existing uses or ownership, graffiti seeks to 'draw' other claimants out into the open, making them question their own 'sense of place', and testing the power of their (b)order controls. In many places, the challenge presented by graffiti to the dominant culture gains a swift and strong response. In Melbourne for example, where a culture of graffiti was widespread, the authorities clamped down on all forms of wall writing in 2006. Police Minister Tim Holding stated the dominant cultural line when he said, 'Graffiti's not art, it's vandalism and it's something we all deplore' (cited in Anon, 2006). Young (Head of Criminology at Melbourne University) emphasised the battle that was occurring over who owned the streets, 'The clean-up is an imposition of a supposedly mainstream, or dominant, cultural view. It is a denial of the diversity of cultural styles that actually exist within a city space' (ibid.). The ongoing battle for place, and the questions that it raises about our own and others' sense of place was a process I experienced on the South Bank in London (Box 11.7).

BOX 11.7 THE UNDERCROFT, QUEEN'S WALK, LONDON

Across the Thames from the Houses of Parliament in London, UK, a few hundred metres from the London Eye, is the Southbank Centre. This Centre is a set of buildings home to classical music, ballet, and 'high' culture. At river level, on the Queens Walk promenade, under the mezzanine entrance to the Centre, is the Undercroft.

South Bank
Source: Author

Since the 1970s, youth culture has dominated the Undercroft. Skate and graffiti culture has taken and made this place, right under the noses of political and high culture (see also Chapter 8). This site has come to be known as *the* place to participate in and watch UK youth culture as it happens. However, in 2013, the owners of the Southbank Centre announced plans to develop the site. The Undercroft was to be removed, and the area sold for retail units. The youths who had carved out their own place in the Undercroft responded to the plan as follows:

It's like they['re] mov[ing] the lions from Trafalgar Square.
(Ben Marsh, 25, BMX rider and bartender, cited in Fox and Hazellton, 2013)

I've been skating here since I was 15 . . . Take [it] away, and they'll be taking away a chunk out of us.
(Biko Dadzie Issah, 32, tattooist, in ibid.)

I would not say this is a skate park, I would say this is a temple. It's a mythic place, not just for UK or London, but for the whole world. Surf pros travel to Hawaii to surf; this is Hawaii in Europe for skateboarders.
(Pedro Emanual, 29, graphic designer and musician, in ibid.)

The Undercroft was a place that youth culture took and made from the mainstream, a place which defined skateboard culture in the capital, and across the UK. As such, it had become a 'mecca' for this culture, a place of almost religious significance; its history, stories, and traces, a priceless landmark for both participants, and spectators. The idea that this culture could be *dis*placed to service both capitalist and high culture (in terms of the opera, dance, etc.), that is catered for in the Southbank Centre, signified to many the continual marginalisation and silencing of youth culture in public places.

Just because this is a counter culture, it doesn't make it any less legitimate . . . It would be like knocking down a 100-year-old clock tower in the middle of a village and replacing it.
(Davies, in Barford, 2013)

Sensitive to the importance of the Undercroft, the Southbank Centre planned to create a new site for skaters 120 metres along Queens Walk (see Wainwright, 2013; Festival Wing, 2013). In many senses, their attempt to fund, consult, and sensitively design a new site evidences their wish to ensure that youth cultures retain their presence in the city. However, can you simply displace an important cultural site and maintain its value? Can you design in history, constitutive co-ingredience, and belonging?

Commodified by the mainstream: graffiti, art, and recuperation

As we have seen in previous chapters, dominant cultures adopt a number of tactics to (b)order minority cultures. In the case of graffiti, its threat can be painted over and scrubbed out, and its practice outlawed. However, graffiti has also been controlled by being assimilated into the mainstream. This process, akin to the 'partnership' mechanism seen in Chapter 8, involves the commodification of this cultural trace into something that can be marketed and sold (see Chapter 7).

graffiti today is more than an illegal sub-culture – it has become big business and . . . corporations are cashing in. Red Bull, Adidas, Puma, 55DSL and Lee Jeans have all incorporated graffiti into their marketing campaigns, so it is hard to walk down a major shopping street without seeing spray-can art decorating shops. (Asthana, 2004)

Graffiti then has moved from 'the street' to the 'high street', and indeed even to high culture through galleries and auction houses. The popularity, rebellion, and marginality encompassed by the medium and message of graffiti is being used by capitalists as a new way to sell products. Graffiti and youth culture has thus been turned into a commodity. Galleries stock graffiti images on canvas (see Figure 11.6), graffiti artists themselves have their stencils and spray paintings hung in galleries, and their work (or the buildings it

is sprayed on) now sells for top dollar. As a consequence some graffiti artists have become rich and famous through commodification, rather than outlaws or criminals.

The commodification process thus raises questions about the meaning of graffiti. On the street we have seen how graffiti traces can be threatening, but in the high street they are transformed into money-earners and, in the gallery, works of art. This process thus reminds us that it is not simply the words and images of graffiti that are important, or the medium itself, but also the *context* in which graffiti is found. In Cresswell's words what tells us most is 'the crucial where of graffiti' (1996). In 'the street' graffiti is 'temporary',

 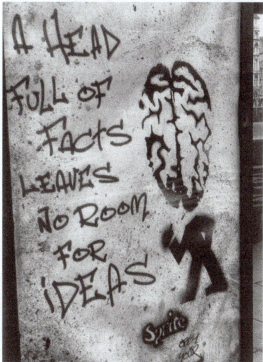

Figure 11.6 A head full of facts . . .

In these images graffiti is sold as art (in a shop in Cardiff) and is used to sell a fizzy drink (on a bus stop bill board in Dublin, Ireland). Has graffiti gone from a novel to a normal, even a natural, trace through such usage?

Source: Author

Figure 11.7 The graffiti artist graffiti-ed

This piece of graffiti satirises and condemns the stencil artist Banksy due to the perception that he has moved to the
mainstream, either by choice, or via recuperation.

Source: Author

'wild', 'unexpected', and a 'noncommodity'. In
the 'high street' or 'gallery' it becomes 'permanent', 'tame', 'expected', and a 'commodity'
(1996: 50). In many ways, Cresswell is arguing
that the assimilation of graffiti into the mainstream has the effect of 'recuperating' it (see
Chapter 7). The threat and transgression offered
by the practice on 'the street' is removed when it
is *placed* in the mainstream. Is this because graffiti
itself has changed? Perhaps its words and images
no longer seek to challenge us? Or those who
are speaking have 'sold out' and thus are themselves less threatening (see Figure 11.7)? Or has
graffiti's presence in different places changed the
way we see this trace, and, as a consequence, we

BOX 11.8 GRAFFITI GIVEN, STOLEN, THEN SOLD

In 2004, Banksy graffiti-ed a mural on the
Prince Albert pub in Trafalgar Street, Brighton.
It depicted two policemen kissing. In 2011, as
the value of Banksy's work was growing, the
bricks upon which the mural was sprayed were
removed, and a replacement image was put in its
place. In 2014, the original sprayed bricks were
sold in a Miami auction house for $575,000
(£345,000) (The Argus, 2014).

In this example, what is actually being sold at
auction? The artistic skill of the artefact? The cultural kudos of the artist? The transgressive nature
of the act of graffiti? The message of the image?
The place and history of the original site? In your
view, is something lost when a cultural product is
'translocated' from its original place? Is the meaning of this piece of graffiti changed through its
mobility? Has its value been lost, or enhanced?

'Kissing coppers'
Source: Duncan Hull (Flickr)

have become more tolerant and accepting of this
minority voice (see Box 11.8)?

The material effects of graffiti are significant.
They take and make place for youth cultures, and
often prompt trace-chains within the cultural
world as dominant cultural groups retaliate or

commodify these claims on *their* place. However, in fixing our attention on the material traces of graffiti and their cultural and geographical significance, we may ignore the non-material practices that contribute to their creation. As in the case of free running, it is often these non-material practices that motivate graffiti. It is the experience of performance that is the desired goal, rather than the product ultimately created.

The performance of graffiti

The rush you get from doing graffiti and the respect you get from your peer group is certainly very addictive . . . Like the Streets song says, 'Geezers need excitement', and if you're young then graffiti is an easy way to get it.
(Fiasko, cited in Addley, 2007)

The performance of graffiti is for many 'artists' the point of its practice. Accessing your 'canvas' can be a challenge, indeed the more (b)ordered the place, the more exciting the project (see Ninjalicious, 2004, and Box 11.9). This, coupled to the buzz of creating the tag, image or message itself, can be affirming.

'Why do we do it?' It's about respect. Obviously there's the adrenaline from painting in a tube yard, they are really hard to get into these days . . . It's probably similar to breaking into a house or something, it's such a big mission. Finding it, breaking in, painting, getting away scot free . . .
(Twisted, cited in Asthana, 2004)

Halsey and Young acknowledge the importance of the performative element of graffiti. They

---

BOX 11.9 DISCOVERING THE CITY: URBAN EXPLORATION

Free running and graffiti involve using the city differently, invoking new traces on the urbanscape, and often transgressing or resisting cultural (b)orders. Another activity which is often associated with youth culture and explicitly trespasses in un(der)used areas is the phenomena of urban exploration. As Garrett explains, 'Urban exploration is a practice of researching, rediscovering and physically exploring temporary, obsolete, abandoned, derelict and infrastructural areas within built environments without permission to do so' (Garrett, 2014: 1).

Urban exploration gained initial popularity in the early 2000s through the practices of Ninjalicious in Canada. As Ninjalicious (2004: 3) states, urban exploration draws on the enthusiasms and daring of youth cultures: it

inspires people to create their own adventures, like when they were kids, instead of buying the pre-packaged adventures too many of us settle for . . . No wonder people feel unfulfilled and uninvolved [in the city] as they are corralled through the maze of velvet ropes on their way out through a gift shop . . . [Urban Exploration] nurtures a sense of wonder in the everyday spaces we inhabit that few local history books could ever hope to recreate . . . explorers can feel a vivid, exhilarating awareness of the urban environment that can be almost overwhelming in its intensity during and following an enjoyable expedition. It's a rush.

Urban exploration therefore raises questions about transgressing cultural (b)orders in relation to trespass and private property. Should urban spaces (like an increasing number of rural spaces) be subject to debates on free access (as long as property is respected and damage avoided)? Should urban exploration be outlawed or legalised? If it were legalised, would some aspect of its performative 'rush' be lost?

state, 'graffiti writing [is] an affective process that does things to writers' bodies (and the bodies of onlookers) as much as to the bodies of metal, concrete and plastic, which typically compose the surfaces of urban worlds' (2006: 276). In this way, the difference of graffiti culture to the mainstream is accentuated. In graffiti culture, the painting produced is not necessarily the key to the practice. The point is the act of painting itself. Ingold recognises that such an emphasis is similar to the priority given to artistic performance in non-Western societies.

> In many non-Western societies . . . what is essential is the act of painting itself, of which the products may be relatively short-lived – barely perceived before being erased or covered up . . . The emphasis, here, is on painting as performance. Far from being the preparation of objects for future contemplation, it is an act of contemplation in itself . . . in [Western] painting, gestures leave their traces in solid substance, the resulting forms may last much longer, albeit never indefinitely.
>
> (Ingold, 1993: 161–162)

In this sense graffiti does not only stand as a statement against dominant culture, but also represents a different way of 'framing' the act of art- and place-making. The performance is the act that gives the graffiti artist credibility within the culture, not necessarily the finished product. It is the identity created by the performance that will be durable within the culture, even once the graffiti has been removed:

> People are always going to remember him because he was Ozone. People will always have a lot of respect for him. That's the truth in graffiti. And that's the reason I do it. I know when I die I won't be forgotten.
>
> (Addley, 2007)

Summary

What does graffiti tell us about place (after Nandrea, 1999)? It reminds us that, in the first instance, where things happen is crucial. As emphasised in Chapter 1, 'so much depends on context' (Cook, in Clifford and Valentine, 2003: 127). The places that graffiti takes and makes are at once altered by its presence, but at the same time, the meaning of the trace itself is influenced too. Thus graffiti can be seen at once as 'temporary', and 'permanent'; 'wild', and 'tamed', 'noncommodity' and 'commodity' etc.; but in the style of the liminal youth who practise it, perhaps it is 'neither one nor the other, but *something else besides, in-between*' (Bhabha, 1990: 224, emphasis in original). Graffiti is an act that makes us confront our own sense of place, our own (b)orders. Graffiti turns lines into question marks, forcing us to consider whose world we are living in, what our (b)orders should be, and how we should act as a consequence. With this in mind, how should we read the piece of graffiti in Figure 11.8, found on a wall in Cardiff?

Conclusion

We have focused in this chapter on a group who traditionally have had no sense of place within the mainstream. Youth culture exists 'in-between' (b)orders, between the conventional categories of 'child' and 'adult'. Through the individualisation process (as outlined in Chapter 6) youths experiment within their liminal place, trying out different identities by forming 'neo-tribes' and staking their claims in 'the street'. As one graffiti 'artist' states, the street is 'where like all of us, sort of, the younger people, or people in graffiti totally think it's somewhere where we can represent' (in Halsey and Young, 2006: 285). The material places youth culture takes and makes are thus often marginal in nature, yet they are also explicitly public – as a consequence their representations are often visible and widely read as a direct threat to the mainstream.

Figure 11.8 A 'completely pointless and mindless piece of vandalism'?

Is this graffiti an act of vandalism? Is it art? Is it political? What culture does it stand for, and what does it challenge? Does it change how you consider the act of graffiti?

Source: Author

However transgressive youth cultures may be we have seen that they are not explicitly resistant in nature. Their intention is not to edit or revolutionise the mainstream, but simply to take and make place for their own culture. Youth cultures may be categorised then as different, but also *indifferent* to the mainstream. Youth cultures do not necessarily care about how their traces are seen by the mainstream, they are more concerned about how they are seen and read by their own culture. Perhaps in this disregard of mainstream values lies their real threat.

However, when we look to the margins, we can see innovation and ideas which offer insights into how we think about and act in places. Through practices such free running and graffiti, youth cultures leave traces in places that alter our experience of orthodoxy. They raise questions concerning how we acquiesce to and accept the 'naturalness' of the places around us, and how these (b)orders can be tightened or made more flexible in order to remove or accommodate diversity. Youth cultures, like other minority cultures, can help us think differently about the world around us. Through their traces they change both the culture and geography of place.

Note

1 Interestingly, Paris has had major problems with its youth cultures. As Stewart outlines, many of its suburbs have suffered from 'a toxic combination of poor education in poor areas, and in some cases sheer prejudice, creat[ing] a vicious circle which means that those from ethnic minorities, heavily concentrated in large estates on the outskirts of cities, can become trapped outside the job market' (Stewart, 2005). Youths, particularly ethnic youths, have thus become the 'prohibited and forbidden' in French society. When Interior Minister, the ex-French-President Nicolas Sarkozy threatened to 'clean' youth cultures 'off the streets' and said: 'the louts will disappear – we will clean this estate with a Kärcher' (cited in Willsher, 2005). Kärcher is a make of high-pressure hose used to clean buildings. Youths are dirt to be washed out of mainstream French places.

Suggested readings

The following texts are useful to explore the issues raised by this chapter.

Books

Skelton, T. and Valentine, G. (eds) (1997) *Cool Places: Geographies of Youth Cultures*. Routledge: London and New York.

Skelton and Valentine give insights into the cool places of youth culture ranging from experiences of clubbing, backpacking, and gang culture, and related issues of safety, danger, and moral panic.

Holloway, S. and Valentine, G. (eds) (2000) *Children's Geographies and the New Social Studies of Childhood in Children's Geographies: Playing, Living, Learning*. Routledge: London.

Holloway and Valentine offer a sociological take on the competency of youths as cultural actors and outline how this contributes to our understanding of their geographies, senses of belonging, and issues of liminality. This latter issue can be explored in more depth in Bhabha and Shields.

Bhabha, H. (1990) *The Location of Culture*. Routledge: London.
Shields, R. (1991) *Places on the Margin: Alternative Geographies of Modernity*. Routledge: London and New York.

Journals

Nandrea, L. (1999) Reflections. 'Graffiti taught me everything I know about space': urban fronts and borders. *Antipode*, 31 (1), 110–116.
Halsey, M. and Young, A. (2006) 'Our desires are ungovernable': writing graffiti in urban space. *Theoretical Criminology*, 10 (3), 275–306.

Nandrea, and Halsey and Young tell us 'everything they were taught about place' by traces of graffiti, whilst Cresswell's 1996 text *In Place/Out of Place* has an excellent chapter about the writing on the wall.

Place and language

Introduction

In this chapter we explore language as a cultural practice. Speaking of which we are going to focus predominantly on linguistic forms of communication, i.e. the words and vocabularies, accents and dialects that we use to identity ourselves and our relations to the world. There are, of course, other forms of language; as we have seen in Chapter 3, dance, design, or art can be considered a form of language – these forms *may* be non-linguistic in nature, but they are still languages of communication. We have also seen throughout the book how places, buildings, performances, and signs all function as means of communication (and we will return to investigate the role of visual languages in particular in Chapter 14). Here, however, we focus on spoken language; it is predominantly through spoken language that we imagine, represent, and make (sense of) our cultural world. This form of language is therefore a clear form of 'representation'. Spoken language stands in for objects, feelings, practices, and events, like a lawyer stands in for a defendant. However, language is more than 'simply' representation, it is a multisensory experience: we can consider the type of language that cultural groups speak, what words they choose to communicate their ideas, and what worlds they bring into existence through these vocabularies. We can also consider what these languages sound like, we can respond to their dialect, tone, and accent. Language can

therefore be a product of culture – language can be written to identify, transform, and give the last word on how we think about the world; but language is also a practice, it is a medium to communicate culture, forming our identity, and locating where we are, and where we're from. Language is therefore vital to understanding our cultural world, as Kanngieser (2012: 336) puts it: language is 'inseparable from . . . political, cultural and economic landscapes'.

Historically, language has been studied by geographers in a way reminiscent of 'traditional cultural geography' (see Chapter 2). In other words, languages have been conceptualised as if they were cultural artefacts or material traces, with geographers cartographically annotating the distribution of language use, carving them into locations, and setting them in stone (see Desforges and Jones, 2001). However, in recent decades the languages of the word have been examined using the languages of the world, analysing how languages produce meanings, and how words themselves are chosen, popularised, or silenced, in order to stimulate or discard particular voices and ways of seeing the world (see Chapter 3). In the words of Kanngieser (2012: 338), geographers have begun to explore how 'space, time and social constructions are productive of, and produced by, languages and their usage'. In this chapter, we will focus on how language is a form of cultural practice, and how the words we use help to create the worlds we live in. We will see how language forms our identity, our relations to the world, and the broader collectives we feel a part of. We will investigate how languages 'sound' to us, and how languages (b)order different cultural groups. We will see how we can be polyvocal with our words, and talk back through hybrid voices which resist and articulate our own cultural geographies. But enough intro already, let's talk turkey.

Language as culture

Languages shape our cultural world. To paraphrase Valentine et al., 'language can change the way that spaces are [b]ordered' (2008: 375). Languages therefore *place* us, they give us ways of communicating our cultural world view, and we make sense of the world through our languages. We name traces in our world that we think are important in order to identify them and communicate them to others. Some 'things' can be named quite easily, with various degrees of simplicity or sophistication, other 'things' (and more often practices or affects) are more difficult to put into words. What happens when we cannot adequately identify a 'thing' or 'affect'? Is it more difficult to make someone understand what we mean? Do we resort to a different type of vocabulary to get our feelings across (e.g. using music or bodily movement[1] for example?) With these issues in mind, it is worthwhile considering the different ways in which cultures name practices and things: not all cultures have equivalent words that can be easily translated with minimal loss of meaning; whilst some cultures construct words for things and affects that other cultures do not. Does this mean those cultures value these feelings and practices in different ways to us? For example, we have seen in Chapter 4 that Welsh culture has a word for 'that sense of belonging or homesickness that Welsh people feel in their hearts when a long way from the Valleys of the south or the mountains of the north' (Elmes, 2005: 98). As Elmes states, 'it *may* be that this special sense of emotional attachment to a place,

of longing and belonging, is much the same the world over' (ibid., my emphasis), but it is not by accident that the Welsh language has named this affect. To them, 'hiraeth' is a valuable experience, one that identifies, validates, and affirms this relational connection between people and place. The English language doesn't have an equivalent word for hiraeth, the experience is not directly translatable. I have just typed 'hiraeth' into an online translation engine; it tells me that three English words are required to convey this word's singular meaning: nostalgia, homesickness, and yearning. Why hasn't English got an equivalent word for this affect? Does this mean that the English don't experience this feeling, or have as strong a sense of place as the Welsh? Or does it mean that they simply don't value this feeling in the same way? Despite its diverse origins, is English simply too 'thin' a language (or as the Welsh would put it *yr iaith fain* (after Williams, 2002: 173)) to really encapsulate the full range of senses that humans experience?

> I can still remember how puzzled I was to find that the French word *aimer* spans a range of feeling that English speakers designate by both 'love' and 'like'. To me, loving and liking are separate emotions. Not so, it appears, in Paris.
> (Abley, 2003: 48)

A range of cultural languages are similar to Welsh in identifying human experiences that English has failed to document and name. As Lavers articulates:

> Boro, a language spoken in north eastern India and parts of Bhutan, Nepal and Bangladesh, is graced by the verbs *gobray*, 'to fall into a well unknowingly' (is there any other way?), and *egthu*, 'to create a pinching sensation in the armpit'. Ever since I can remember, I've needed the Boro verb *asusu*, 'to feel unknown and uneasy in a new place'. A language with a one-word verb *onsra* for 'to love for the last time' simply cannot be allowed to die.
> (Lavers, 2004)

BOX 12.1 MAKING WORDS, MAKING WORLDS

'our [world] shall be a little Academe, still and contemplative in living art'

(Shakespeare, *Love's Labour's Lost*, Act 1 Scene 1)

English is not an essential, pure language. It has many origins. As something that humans 'do', language is the quintessential cultural practice, it is dynamic and in a state of ongoing composition. English is formed by Latin, Greek, Norse, and Anglo-Saxon, whilst its modern form has many influences. As Nunn (2014) argues, the language has adopted German words for experiences and objects that simply aren't defined by English as it stands; for example: *schadenfreude*, *zeitgeist*, and *doppelganger*. In history, the playwright William Shakespeare is said to have scribed many 'new' words that are still in common use today. Shakespeare coined 'academe' into popular currency, alongside 'madcap', 'moonbeam' and for the purists amongst you, 'pedant'. Modern technology has also led to a reinvention of English, including new words such as 'hashtag' and new meanings for existing words such as 'meme', 'spam', and the ubiquitous 'troll'. Technology has also led to the inventive use of punctuation to express a range of emotions :)

We will see later in this chapter how in contemporary culture many words are hybridised or globalised by the mobility of people, ideas, and languages (see also Chapter 8). Youth cultures in London, for example, talk a version of English called MLE or Multi-cultural London English. MLE includes words such as 'low batties' and 'crepes' to define fashion apparel. Do you know what these words refer to? (Low batties are trousers worn low, very low, on the waist, whilst crepes are trainers). Can you guess the origins of these words? We can see, therefore, that the cultural traces of the English language are a 'living art'; they are constantly re-worked and influenced by a number of sources. Are you one of these sources? Have you ever invented your own words to describe experiences or people? If so, as you'll know, sometimes, only made up words will do.

I've never fallen down a well (by accident or on purpose), so with luck I'll never need to articulate the experience and stumble over the lack of 'gobray' in my 'mother' tongue; however, having the word 'asusu' on its tip would broaden and deepen my ability to communicate what arriving in a new destination feels like (see Chapter 8). Such words then are not simply decorative embellishments for poets and playwrights; they are tools with which cultures communicate what is important to them. The things and practices which are not named (and therefore are left uncommunicated and unvalidated) are effectively silenced: the message is sent by omission that these experiences are not important because we do not speak of them. Thus words create worlds: if we don't have a word for it, how can we articulate it? If we cannot articulate it, how can we mobilise around it? If we can-

not mobilise around it, how can we change the world?

Words therefore change our relations to the world around us, how we understand that world, and how we categorise it. It is usual in my culture, for example, to categorise time in terms of days, months, and years, and to separate those years into four seasons (the English, for example, have 'Autumn', whilst Americans prefer the 'Fall'). There is nothing natural about these categorisations, but they serve to structure my cultural life. However, on Bathurst Island, Northern Territories, Australia, time is not always categorised in the same way, as Abley (2003: 28) explains:

English speaking Australians in the region divide the year into two or three seasons: the dry, the wet, and (sometimes) the insufferably humid build up before the wet. [The] Tiwi

BOX 12.2 DIFFERENT WORDS MAKE DIFFERENT WORLDS

According to the BBC (2010), approximately 6000 languages are spoken around the world, with each one contributing to the wealth of ways we can pinpoint and skewer important elements of the human experience. For example, Russian gives us the word 'toska', meaning 'a more complex ennui where the soul aches with a longing and vague restlessness' (see Nunn, 2014), whilst I wonder how I have done without the Indonesian word 'jayus' up to now, which refers to a 'joke so poorly told that one cannot help but laugh' (ibid.). With reference to the Wintu language, Rebecca Solnit outlines how words not only help us to define our world, but also help us position ourselves within it, literally and metaphorically:

Malcolm, apropos of nothing at all, brought up the Wintu in north-central California, who don't use the words 'left' and 'right' to describe their own bodies but use the cardinal directions. I was enraptured by this description of a language and behind it a cultural imagination in which the self only exists in reference to the rest of the world, no you without mountains, without sun, without sky. As Dorothy Lee wrote, 'When the Wintu goes up the river, the hills are to the west, the river to the east; and a mosquito bites him on the west arm. When he returns, the hills are still to the west, but, when he scratches his mosquito bite, he scratches his east arm.' In that language, the self is never lost the way so many contemporary people who get lost in the wild are lost, without knowing the directions, without tracking their relationship not just to the trail but to the horizon and the light and the stars, but such a speaker would be lost without a world to connect to, lost in the modern limbos of subways and department stores. In Wintu, it's the world that's stable, yourself that's contingent, there's nothing apart from its surroundings.

(Solnit, 2006: 17)

[language] recognises not only those three but also thirteen lesser seasons, flexible and overlapping in length, for which English has no words.

Words therefore categorise the complexity of the world around us, structuring that world to allow us to function in it, but also influencing how we sense our place within it. Consider the following statement, for example:

in Huron, the verb we might translate as 'to calm down' means 'to make the mind like a field prepared for planting'.

(Abley, 2003: 157)

In the Huron language, anger and frustration is tackled by using a term that compares our mind and emotion to uncultivated (and perhaps infertile) ground. This ground requires tending in order for it to be productive. At once this word ties its users into a culture that is indelibly connected to the soil and to place, and to a history that has (transient) agriculture as a key component of it. In encouraging calm in this way, it helps the Huron to become more attuned to their moods and their ability to (b)order themselves. With this in (a calmed, prepared for planting) mind, it is not going too far to suggest that languages have a strong influence over our philosophies of life. As Christie suggests,

There are amazing things about Aboriginal languages . . . their concepts of time and agency go right against our ideology of linear time – past, present and future. I reckon they'd completely revolutionise Western philosophy, if only we knew more about them.

(Michael Christie, Northern Territory University in Darwin, cited in Abley, 2003: 277)

As Christie suggests, languages are intimately tied to philosophies of knowledge. We have seen this to date in Chapter 8 – when considering places as 'mobile', and throughout this book when framing places as ongoing compositions – on these occasions we are using language to shift our understanding of the world away from fixed and discrete categories, and towards more fluid and shifting processes. As stated in Chapter 4, it is now more common to consider place as a 'verb' rather than a 'noun', and this shift in language is key. As Abley states, 'it runs deep, this disparity, or disjuncture, between the Indo-European emphasis on nouns, and the American Indian [for example] reliance on verbs' (2003: 51), and as Bhattacharya confirms, 'if we truly recognised that we occupy a universe of constantly transforming things, people and relationships, then we would have no choice but to discard our heavy reliance on nouns to capture and describe it' (cited in ibid.: 125).

Language and identity

> People are who they are because of (among other things) the way they talk.
> (Cameron, 1998: 272)

Words therefore display our taken for granted assumptions about the nature of the world, and what our culture thinks is important. By implication, they also go some way to inform who we are and where we are: as a cultural act, language (b)orders our identity. As Valentine et al. (2008: 375) state, language plays a key role in how 'people make sense of their identities and affiliations', and as Carter and Aulette argue, 'language is a fundamental part of our character, an expression, and a mirror of what and who we are' (2009: 214). As stated above, language is a multisensory experience, thus we recognise similarity and difference between groups and ourselves based on the language people speak, the worlds they create through their language,

as well as how these languages sound to us. As Bradley (2002) suggests, our attitudes to language connect to our attitudes about the culture itself, we may (b)order a person's character and background based on their language (see also Carter and Aulette, 2008). Consider your own response to particular dialects or accents: are there areas of your country whose dialects are music to your ears? Are some accents less harmonious, and how does this influence how you (b)order the person speaking it? Perhaps these elements of language affect how we listen and what we hear when communicating with others? Kanngieser (2012) argues this to be the case, she considers that how we 'voice' our language is crucial in (b)ordering our and other cultures; in her words, 'the voice creates worlds . . . voices and their linguistic articulations are produced by, and productive of, relations, geographies and subjectivities. These are tied to projections and positions of class, race, education, culture, social value, sexuality and so forth' (2012: 337).

Language as polyvocal

However, consider how you use your own language. Do you speak differently with different people? Does the language, words, or accent you adopt depend on where you are? In my life my mum always seemed to switch to 'received pronunciation' when she answered the phone – her 'phone voice' was far 'posher' than her normal tone, whilst I always seem to drop my 'aitches' when I'm relaxed with friends. Perhaps you speak differently at home compared to your voice at university? Do you use a different accent when you're angry? These flexible uses of accents and words point to the polyvocal nature of language – in other words, we often speak in many tongues.

In a sense, therefore, geographical context influences how we use language. It is possible to speak one way at work, use another tone at home, and yet another vocabulary with friends. These

BOX 12.3 PLACING AN ACCENT ON 'ACCENT'

Baratta considers 'accentism' – or the pressure on people with regional accents to switch to a more stand-ard English – to be the 'last taboo' (see Moss, 2014). Baratta likens accentism to racism (see Chapter 10), arguing that, 'people make snap judgements based on [nothing more than] accents' (ibid.). The following example of Deborah illustrates how her working-class Cardiff accent alienates her from fellow Welsh people living in the north of the country:

> I did 'ave a negative response when I went into a shop and the lady just blanked me; so I asked 'er again what I wanted and she still blanked me . . . And we come out from the shop and we joined the people from the coach an' they were sayin' the same, and we said, 'Well, why?', an' it was because we were from Cardiff. I was shocked because [we were in] Wales: we're all Welsh – I was really shocked . . . They did prejudge us by our accent, it [created a] boundary. I thought it was a bit strange they judged us by our accent, because we're just the same as them.
>
> (Elmes, 2005: 79)

Lloyd Robson, a Cardiff writer, attempts to 'translate' the spoken accent of city ('Kairdiff' if you want to pro-nounce it in the local accent), in the following way:

> my family from adamsdown & tremorfa; adamsdown when it was part of roath. splott what i get accused of like it's something to be ashamed of by those from the city's west or north, this side of the bridge so not splott at all, but within the call of the splotlands' last orders a pint of dark don't mind if i do thanks very much ta iechyd da 'ware teg down the hatch & up yours butty bach you-arr en shovel it down ya neck then pal.
>
> (Robson, 2003: 95)

From the perspective of fellow writer Niall Griffiths,

> I'm not aware of anybody that does the Cardiff dialect like [Robson]; he gets it well. And he does a lot with it you know, you have to have a really quick wit, it carries some amazing images and visions . . . these dialects are repositories of knowledge so I think the more we keep these dialects alive, the more we keep a little kind of safe of certain types of local knowledge alive . . .
>
> (see Anderson, 2014a)

Griffiths is based in mid-Wales, he writes in this local accent in the following way:

> We pass anavver train travellin in the opposite direction, towards Aberystwyth . . . No doubt bringin maw; maw expectant faces wawkin through tha station, blinking, hoping, lookin faw a certain somefing which they've neva bin able ta find any whir else.
>
> (Griffiths, 2000: 481)

By considering how language is not simply the written word, but also about how one speaks and how one listens, the relations between language and identity become reciprocal in nature, as Valentine et al. state, 'by adopting this approach, we understand language not as constructing identity, nor identity as determining language, but rather we recognise the two as mutually constituted' (2008: 377).

switches in language will occur in a range of places, and in this way our plural worlds become associated with these words, the language mutually informing the place and the behaviour. Using language differently in different places suggests that we have a 'spatial division of language' which tightly coincides with a 'spatial division of identity' (see Anderson, 2004a). Different aspects of our self are articulated in different locations, and often different languages, accents, or words, are used by these different selves in these sites. As Valentine et al. have explored, when individuals are bi- or multi-lingual, this pluralisation of self is clear; as they state, this ability to talk polyvocally, 'opens up . . . a range of possible enactments of the self' (2008: 385, and see below). These multiple selves are performed in different locations, geographically (b)ordering the self and vocabularies that enact them, as Valentine et al. state, 'you are what you speak, and what you speak is *where* you are' (2008: 385). Thus using language differently in different places suggests not only that we have a 'spatial division of language and identity', but also that there is something about these different places that makes a certain accent or choice of words more appropriate in one place than another. In other words, some languages in some geographical contexts are 'out of place'. When this occurs we can begin to diagnose the way that power influences the wor(l)ds of language.

Language is power: 'To dictate definition is to wield . . . power' (Livingstone, 1992: 312)

Language represents and creates relationships [that] include the factors of power, control and domination.

(Carter and Aulette, 2009: 214)

As we have seen in Chapter 5, power is a crucial element in (b)ordering the cultural world. Language influences the power to name valuable traces in the cultural world, as well as the related ability to legitimate that world – by silencing some languages and words, or shouting others from the rooftops. Language therefore plays a key role in defining the meanings and values of the cultural world, and how they are perpetuated.

'[L]anguage', Sakho writes, 'is not simply a neutral medium of communication: it carries historical and cultural biases in favour of the particular power holders of the time'.

(cited in Watson, 2004: 60)

Dominant languages influence who can speak, who is heard, and the nature of the understandings that are produced. We have seen above how we may change our language depending on where we are. We considered that in these cases some languages and words may be out of place in those geographical and cultural contexts. But as we have asked in other chapters of this book, how did these cultural traces become norms, and why are we conforming to them? We may generally agree that English (in some form or other) is the dominant language of the early twenty-first century; this is a product of geographies of Empire (see Chapter 2), and the globalisation of capital by corporations based in the English-speaking world. This dominance often leads to minority cultures learning English (and the associated world views inherent in this language culture) in order to compete is this Anglo-Americanised context. As Brenzinger et al. (1991: 38) explain,

The decision to abandon one's own language always derives from a change in the self-esteem of the speech community. In the cases of language shift one could observe that members, very often the younger generation of minorities, regard their own community as being inferior. Those members frequently try to change their 'negative' social identity by adopting the language (and social identity) of the dominant group.

Speaking your own language, or adopting another's, is a decision of cultural power. Deferring to the dominant language may enable communication and open up new audiences to your point of view, but also admits and perpetuates the inferior status of your mother tongue. As Watson states, 'linguistic dominance . . . is . . . complicit with wider political and economic inequalities: the use of English privileges particular peoples and worldviews, and silences or disadvantages others' (2004: 60). While those inside the Anglo-American universe perhaps see the use of English as the 'natural order', it is not 'natural' or commonsensical to play by Anglo-American rules on an Anglo-American playing field (although cur-rently it may still be a key way to access global markets (see Box 12.4)). However, this Anglo-American dominance is not universal, and is not guaranteed. And, as we have seen in this chapter, the dominance of English is also not without its consequences. The use of English means that we have lost the capacity to straightforwardly explain 'gobray' or 'hiraeth', and through acqui-escence to English as the dominant tongue of the developed world, we are not only silencing other peoples, but also other perspectives on the world. As Abley confirms, 'language reflects and embodies [philosophical] understanding . . . lose the vocabulary, and you lose the [philosophical] relationships too' (Abley, 2003: 5).

BOX 12.4 ENGLISH LANGUAGE AS CONTEMPORARY CULTURAL HEGEMONY

Qn: What language does the The Far Eastern Economic Review call 'Asia's premier language', its 'unifying tongue and its language of opportunity'?
A: English.

(Lavers, 2004)

Four hundred million people speak English as a mother tongue, and over one and a half billion speak it as a foreign language. Some are willing to do anything to acquire it, as Ellis (2004) outlines: 'wealthy South Koreans take their toddlers to plastic surgeons for frenectomies, tongue-lengthening operations intended to help them pronounce English'. Why? Because English is the dominant cultural language; as Lavers (2004) illustrates, in China

English is such a sought-after commodity . . . that its most flamboyant teacher, Li Yang, can fill a football stadium with eager pupils. 'Learning English is a piece of cake!' he yells above a rock'n'roll soundtrack. 'Make 300 million Chinese speak fluent English!' Why? Not because the Chinese love the language, Li explains, 'but because Coca-Cola and Microsoft rule the world'.

As Abley (2003: 84–86) outlines,

When Vladimir Putin took over as president of Russia in the year 2000, one of the first things he did was to create a language committee. He was worried about the state of his language, and he wanted to maintain as much purity as possible. Many of the dangerous, impure words are terms from the brave new Moscow: biznismeni, keeleri. Contract killers, that last word means . . . Other European languages too are struggling to resist the tidal wave [of English]. As the 21st century began, names like Siegfried and Hildegard were rare among German babies. In their place crawled a swarm of little Justins, Jessicas,

Kevins, and Steves. Job ads in one of Germany's leading newspapers appeal for German professionals to fill vacancies like 'Junior Product Manager' and 'Customer Support Representative'. German has admitted 'das Web-Design' and 'die Bypass-Operation', as if it can no longer be bothered to make up its own words. In the country's business circles, a familiarity with English is now essential. Indeed, the nation's largest bank has begun to dispense with bi-lingualism. Board meetings of the Deutsche Bank are held in English only. In June 2000 Rolf Breuer, the chief executive of the bank, was challenged about the absence of German at these new meetings. He replied, 'So what?'.

Language as identity building: nations and cultures

Language is a key means through which we form and articulate our identity. It follows therefore that language can be used by a dominant cultural group in an attempt to construct a collective identity, at or beyond the local level. Chapter 6 explored the ways that nations can be imagined into being, and with language in mind perhaps a national identity can be talked into being. Language can become a practice and world view that individuals share, providing a common 'ground' around which people can gather when they don't share a local neighbourhood. Indeed, the use of language in the construction of national identity is a common endeavour. Benedict Anderson (1983, 2006) argues that language played a central part in the development of nationalism in Europe in the nineteenth century, and that 'the emergence of print languages helped create the conditions necessary for a sense of "national consciousness" to emerge' (Desforges and Jones, 2001: 282). In the US, as Abley (2003: 276) argues, the creation of a shared vocabulary was vital to nation building efforts:

In 1868 a federal commission on Indian [sic] affairs in the United States released a report urging that rapid steps be taken to abolish Indian languages: 'Through sameness of language is produced sameness of sentiment and thought' [it read]. Therefore the Indians'

'barbarous dialect should be blotted out and the English language substituted'.

In the UK, the rise of English as a lingua franca is due in no small part to draconian measures taken by the dominant groups of history to marginalise competing tongues. As Elmes (2005: 78) outlines, in Tudor times, Henry VIII's 'Acts of Union' (1536–1543),

outlawed the language of Welsh . . . its intention [was] 'utterly to extirpe alle and singular sinister usages and customs' [including language]: 'The people of the same dominion have and do daily use a speche nothing like ne consonaut to the naturall mother tonge used within this Realme.'

In the late nineteenth century, marginalisation of the Welsh language continued, as a *Times* editorial of the 1890s put it:

The Welsh language is the curse of Wales. Its prevalence and the ignorance of the English have excluded, and even now exclude, the Welsh people from civilisation, the improvement and the material prosperity of their English neighbours . . . their antiquated semi-barbarous language, in short, shrouds them in darkness.

(Abley, 2003: 255)

In an attempt to impose the English language, in the nineteenth and twentieth centuries, it was

Figure 12.1 Welsh not

Source: St Fagans

'out of place' for children to speak *Cymraeg* (i.e. Welsh) in school. If children did so, they were punished by having to wear a 'Welsh not' sign around their necks (Figure 12.1). 'As crude and undignified as a dunce's hat' (after Elmes, 2005: 80), the 'Welsh not' signs were a constant reminder of the need to talk into being a unified Britain.

The success of these nation building processes, in countries across the world, results in national diaspora seeking out those who speak their mother tongue, regardless of personal affinity or knowledge, in the hope that belonging to this language group (and by extension this nation) will remind them of home. This also occurs for those whose language may be marginalised, but still remains as a key aspect of national identity. As Segrott's (2001) study of Welsh migrants in London demonstrates, 'people often sought out those who shared the same Welsh language background in the city, assuming that if they spoke the same language they might also share the same views or experiences' (in Valentine et al., 2008: 377).

Marginal but mobilised: talking back

Languages embody the intellectual wealth of the people that speak them. Losing any one of them is like dropping a bomb on the Louvre.
(Hale, in Abley, 2003: 126)

We have seen so far in this chapter how the control of language is a key step in any attempt to control the cultural world: if you control

---

BOX 12.5 NEW LANGUAGES, NEW IDENTITIES – CORPORATIVE COLLECTIVES BEYOND THE NATION

We can see, therefore, that language can help to build individual and collective identity, with the latter often grounded at a national scale. However, in the era of globalised capital (see Chapter 7), perhaps another form of supranational identity is emerging, based on a corporatised form of English.

Jean-Paul Nerrière coined the term 'Globish' to refer to the simplified form of English being used by non-native English speakers doing business in Asia (2005). Globish is a version of English without the ambiguity, complexity, and cultural history involved in the standard tongue. Cameron defines Globish as a 'decaffeinated' version of the language (2010). This labelling of Globish is not simply an apt descriptor, but also one that implies a global language of espressos and instant consumption. As Cameron argues, Globish 'meets our practical need for a universal "other tongue" – a simple, neutral, intelligible medium for cross-cultural communication' (2010). In time, Globish may become the official language of trans-national collectives such as the European Union. Accepting standard English as the lingua franca may be resisted by many European political elites due to its associated 'Anglo-Saxon' ideology (see Oltermann, 2013); however Globish may offer a corporate solution. Will we all soon be multi-lingual: fluent in our own language, and the language of a corporate Globish world?

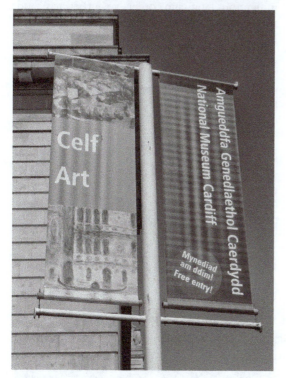

Figure 12.2 Bi-lingual signs

Source: Author

Cymraeg has been a relative success story. From the 1960s onwards, *Cymraeg* has grown from being a marginalised and often ridiculed vocabulary, to a tongue that is now accredited equal status with English in all Welsh government policies and pronouncements.

> place names, at once both material and metaphorical, substantive and symbolic – read, spoken, mapped, catalogued and written in the everyday intimate and official bureaucratic geographies of road signs, street names and addresses – are all about questions of power, culture, location and identity.
>
> (Nash, in Merriman and Jones, 2009: 372)

In the latter half of the twentieth century the status of *Cymraeg*, and the status of the Welsh nation, changed. One of the key signals of this change was the adoption of bi-lingual road signs in Wales (Figure 12.2; see Merriman and Jones (2009)). As Gade (2003: 429) points out, road signs are part of the 'scriptorial' landscape, they are key traces, 'that contribute to the communication of national identity'. Jaworski and Thurlow see such traces as contributing to the 'semiotic' landscape which imposes ideologies onto the cultural world (see Coupland, 2012). Consider for a moment road signs and place names where you live. How do they function to create a sense of

language you can assimilate, marginalise, or even remove different cultural groups. (B)ordering language, and defending language, is therefore vital. Over recent years, there has been a resurgence in some minority languages, and in Wales belonging, history, and identity? Merriman and Jones (2009) outline how *Cymdeithas yr Iaith Gymraeg* (the Welsh Language Society) successfully demonstrated that English-language signs in Wales were 'mundane, ubiquitous and oppressive symbols of anglicisation and of British/English government authority' (2009: 350). In the 1960s, *Cymdeithas yr Iaith Gymraeg* mounted a campaign to reclaim their language and the connections they felt to the land through including indigenous names on road signs. The Society argued that 'Welsh language signs could restore an important link between language and land, re-inscribing the nation's language in local landscapes and providing a highly effective means of defining the national territory' (ibid.: 350).

Incrementally, and through a range of reformatory and radical resistance activities, the Society won its argument. Now, in the twenty-first century, all public signs across Wales are bi-lingual (see Figure 12.3). These signs have not only become representation devices that point the way, but also demonstrate a new direction for the nation of *Cymru*. No longer is *Cymraeg* antiquated and associated with aged cultures and museum artefacts, but is encountered on a daily basis, enabling twenty-first-century mobility and movement. Like Chinese-language signs in Chinatowns across the world, the use of indigenous language confers an authenticity and roots to an area that attracts interest and business, regardless of whether people can read or

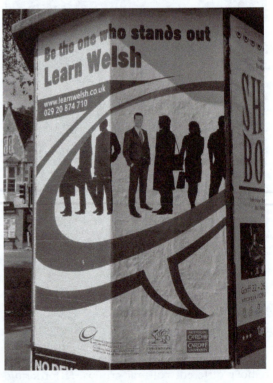

Figure 12.3 Bi-lingualism points the way

Source: Author

understand the script (see Leeman and Modan, 2009). Welsh speakers are no longer marginalised, and perhaps even are treated in some quarters as the true, authentic, communicators of Welsh experience. Welsh speaking (and singing, see www.bbc.co.uk/programmes/p01qmqk1) re-affirms language as a key component of Welsh identity, and contributes to the ongoing re-composition of the (b)orders of Wales in the twenty-first century.

Hybrid languages

Although English is the dominant linguistic culture in the developed world, we have seen that English itself it not a 'pure' language, but has always been re-invented by influences from other languages (see Figure 12.4). Words, phrases, or intonations have either been usurped or particular vernaculars and vocabularies have been assimilated by others. It is easy to frame English as a 'colonising' language, drowning out indigenous words and understandings in favour of its own terms; however, as we have seen in this book, it is impossible for any hegemonic culture to be absolute in its power. Indeed, the case of Globish and *Cymraeg* demonstrates how it is possible to hijack and resist the meanings of English and co-exist with the dominant language culture. Thus due to its impure roots and continually re-composing routes, English is not a one-size-fits-all language, imposing itself everywhere, but is, and always has been, a hybrid or 'creole' language, composed by the meeting or coming together of a variety of lexicons. The case of *Cymraeg* is useful to us again here, showcasing how a local form of hybridisation has occurred on the (b)orders of Wales and England, forging a new language where these two tongues meet. Let me introduce to you: 'Wenglish'.

Figure 12.4 Hybrid languages
Source: Author

Hybrid languages: Wenglish – Whose coat is that jacket?

> The English dialect of south Wales, despite being strongly English in its overall feel, possesses many words with Welsh origins.
>
> (Elmes, 2005: 86)

A 'creole language' is a language of mixed origin; a language that has evolved from the mixture of two or more languages and has become the first language of a group. It can be argued that 'Wenglish' is one local variation of a creole language, emerging from both English and *Cymraeg* (Welsh). Wenglish combines the sentence structures of *Cymraeg*, with the dominant vocabularies of English to create a form of communication that is unique, and dominates the South Wales region. As Penhallurick explains,

> When people say 'there's lovely', it's a translation from a Welsh idiom: the Welsh word is 'dyna' – there's – followed by an adjective . . . Welsh English speakers are [also] well known for tacking 'isn't it' in to sentences: 'We went shopping the other day, isn't it', which again sounds odd. But once again it's a

direct translation of an idiom from the Welsh language [and] because of the way Welsh verbs are constructed, it's quite normal to have that sort of tag at the end of the phrase or a sentence. Then it gets transported into Welsh English.

(cited in Elmes, 2005: 86)

Wenglish phrases were completely alien to me on arrival in South Wales; in England, 'togs', 'scrams' and 'cwtches' mean something different, or have no meaning at all. Now, however, Wenglish peppers my conversation: ach-a-fi! (a Wenglish expression of mild, perhaps sarcastic disgust; for more Wenglish, see www.talktidy. com, and note the 'clean' and 'proper' inference describing this way of talking used in the website name). Cardiff writer Lloyd Robson is vocal in his defence of Wenglish as an important variation of the (now) official *Cymraeg* language. In his words, 'The issue I have fundamentally on *Cymraeg* even though I love it, great language, I refuse to be forced to speak it, or refuse to be forced to feel guilty for not speaking it. I speak English in a way that only occurs in Cardiff, and Cardiff is in Wales, and therefore I speak Welsh. I may not speak the *Cymraeg* language but I do speak Welsh' (see Anderson, 2014a). Do you hybridise your tongues? Have you created your own multi-cultural [insert your place here] English, or multi-cultural *Cymraeg*? How would you feel being told to speak in a particular way, or have to use a particular language? Would this change your identity? Is it more important to have your voice heard in any language, or speak our own language-world into being?

Language: making spaces to be understood

Regardless of the cultural language or dialect we use, as we know from Chapter 3 it is impossible for the language of the word to ever fully represent the language of the world. As Nigel

Thrift puts it: 'the varieties of stability we call "representation" . . . can only cover so much of the world' (2004: 89). Although, as Medeiros points out: 'language is the chisel with which we shape and sculpt and slice the world around us' (2010: 95), our words cannot ever fully 'honour that which lies beyond the scope of discourse' (Bondi, 2005a: 438). Many of our most vital affects and experiences lie beyond language, and there is an 'invisible and yet insurmountable' (b)order (after Medeiros (2010: 95)) separating words from the affective traces that engrave our beings. As a consequence of this we never really know if we are successful in communicating our feelings and ideas to others, as everyone will translate and interpret our carefully chosen words differently. As Medeiros points out, this 'brings us to this horrifying and ultimately mindfuck of a conclusion . . . At best, [our cultural world] can only be felt – but never, ever shared or represented' (2010: 95). How then can we seek to communicate effectively? With whatever language we use, how can we overcome the 'paradox of representation' (Bondi, 2005a)?

As Thrift outlines above, beyond (or before) language's representative function, language also has a performative function (see Chapter 3, and also Anderson (2014b) for how these functions combine to communicate the experience of surfing). This role, as Thrift tells us, 'set[s] up the intersubjective spaces of common actions' (1996: 39).[2] These intersubjective spaces have the potential to be framed as a 'meeting ground' between speaker and listener, a place in which articulations of experience can be shared and understood by both parties. This place will be formed by the speaker through their choice of words and their intended meanings, as well by the silences and ambiguities inherent in the paradox of representation. The listener also has a key role; they interpret these words and silences, making their own meanings through mixing the perceived intentions of the 'author' with their own understandings and ideas. Communication is only successful if this collaborative interpenetration of experiences and representations occurs. Is it only through this 'differentiated solidarity' (after Young, 2000) that we all come to 'know'? Language therefore provides a unique example of the ability of different cultures to engage with one another; only through such engagement can communication, understanding, and toleration come into being.

Conclusion

We have seen in this chapter how language is a vital cultural practice. Languages communicate meanings; they allow new worlds to be imagined, and enable these ideas to become real. Language can dictate belonging and unbelonging, enable identity to be constructed, meaning to be made, and philosophies to be made manifest in the cultural world. As inhabitants of the Isle of Man might put it, 'Gyn Chengey Gyn Cheer' (no language, no identity); language is thus a key mechanism to order and border the cultural world, dictating definition and communicating similarity and difference. Language therefore places us. We make (sense of) the world through the words we use. Choose them carefully.

Notes

1 Indeed sign language is a key form of communication, notably for those with differential hearing abilities, or for babies prior to clear word formation, see also Chapter 13.

2 The existence of this 'intersubjective space of common actions' occurs, perhaps ironically, due to the post-modern insight that we cannot 'extract a representation of the world because we are slap bang in the middle of it' (after Thrift, 1999: 296–297). Due to this 'slap bang in the middle-ness' of our existence, language becomes one medium through which it is possible for an 'author' and their 'audience' to communicate and connect with one another, through not only their shared understanding of the representative function of language, but also through their more holistic connections of experience, emotions, and solidarity.

Suggested reading

The following texts are useful to further explore the issues raised by this chapter.

Abley, M. (2003) *Spoken Here. Travels Among Threatened Languages*. Arrow: London.

Carter, K. and Aulette, J. (2009) Creole in Cape Verde: language, identity and power. *Ethnography*, 10, 213–236.

Merriman, P. and Jones, R. (2009) 'Symbols of Justice': the Welsh Language Society's campaign for bilingual road signs in Wales, 1967–1980. *Journal of Historical Geography*, 35, 350–375.

Valentine, G., Sporton, D., and Nielsen, K. B. (2008) Language use on the move: sites of encounter, identities and belonging. *Transactions of the Institute of British Geographers*, 376–387.

Place and the body

Introduction

As we have seen in this book, traces and places that we assume to be natural, to be the only way to be, are often nothing of the kind. The process of approaching place from the perspective of cultural geography critically disentangles these ongoing compositions of traces and sees them as constructions, rather than givens. In this chapter we turn specifically to what Rich (1986) describes as the 'geography closest in' – the body. This material and somatic site can be considered as a place in the same way as any other: it can be taken and made, given meaning, and be part of a set of trace-chains that affect how we think about and act in the world around us. The place of the body is often assumed to be 'natural' and 'essential' – the body has a biology, gender, sex, and sexuality that are often seen to be as natural as the day we were born. However, as this chapter will explore, the naturalness of these statuses can be questioned as our body's features are (b)ordered in numerous ways. Similarly, the identities and meanings of the places in which are bodies live are constructed too – the geography of the body and the geographies it inhabits are far from being natural – how a body can be in place or out of place is an effect of cultural values and acts of power. This chapter investigates how the body is (b)ordered, and how these (b)orders are changing. It explores how ideas of masculinity, femininity, heterosexuality, and homosexuality are transgressed, and how the

place(s) of the body change as a consequence. We will see how once 'naturalised' and 'essentialised' identities are increasingly crosscut and problematised by the range of other identity and place positions that we have been introduced to in this book. We will see how gender and sexuality are complicated by age, capitalism, religion, and nationality, for example. We will investigate the consequences of these new hybrid traces, how new places are being generated, and how cultural (b)orders are being changed to facilitate or control them. We begin by looking at how gender can (b)order our bodies.

The nature of the (b)ordered body: gendered traces

> Were you raised in a [culture] that assigns gender, as well as sex, at birth? Hint: the answer is yes. Does that [culture] label gender nonconformity 'weird'? Hint: unless you are Native American and were raised knowing about two-spirited people, the answer is also yes.
>
> (McConnell, 2014)

In many cultures, gender in its simplest form refers to the sexual organs you were born with. However, gender involves more than simply our biological assets and capacities; it is associated with what it *means* to be boy or a girl, or a man or a woman: gender involves what it means to be 'masculine' or 'feminine'. Gender, then, is associated with the cultural and geographical aspects of meanings, appearances, practices, and places that make up the 'proper' ways of being a male or a female body. According to Linda McDowell, gender represents: 'differences between women and men's attitudes, behaviour and opportunities that depend upon socially constructed views of femininity and masculinity' (McDowell, 1989: 170).

As de Beauvoir identifies, femininity and masculinity are not something we are born with, but something we become through the cultures in which we are socialised. In her book *The Second Sex*, she claimed women were *not* born, but *made*: 'No biological, physiological or economic fate determines the figure that human being presents in society: it is civilisation as a whole that produces this creative indeterminate between male and eunuch which is described as feminine' (de Beauvoir, 1972: 295).

Tracings of gender on the human body are therefore far from natural, they are (b)ordered into norms according to cultural values and preferences. Such traces come in many forms, from how we dress, to how we move, from how we act, to what we look like. In terms of clothing, for example, skirts are generally considered a feminine clothing item (the example of kilts in Scotland is a notable exception); the use of make-up is likewise a female practice, as is the utilisation of a handbag. In line with these conventions, how would *you* respond to a man holding a handbag and wearing a dress in *your* culture? Would your response depend on where the man was? Why? In the Pacific Islands, specifically Samoa, it is not uncommon for men to wear what we may consider women's clothes. Here it is completely 'normal' for men to dress 'in the manner of a woman' (or *fa'afafine*). *Fa'afafine* are biological males who wear dresses and sarongs, in this culture it is accepted as 'normal' and 'natural' for men to dress this way. We can see in Islamic cultures how specific clothing (b)orders are naturalised, with women wearing the hijab or burka to demonstrate modesty over their physical form in public places (see also Figure 13.1). Similarly masculinity and femininity are 'naturalised' by cultures into accepted body shapes and sizes. Appropriate body shapes for men are generally big, muscular, and hard. Indeed, Johnston argues that the construction of this body shape through exercise is an attempt to 'render the whole [male] body into a phallus' (1996: 330). In contrast, accepted body shapes for women include the absence of such overt musculature. In the West the feminine form is 'appropriate' when slender, svelte, or curvaceous.

Such naturalised body types correspond with specific behaviour and activities that are deemed appropriate for each gender. As Adelman and Ruggi (2008) identify, genders are culturally (b)ordered into particular ways of holding and carrying their body. They argue that in the West females are encultured into,

> mov[ing] their hips emphatically from side to side, thus accentuating the waistline. They take long steps, placing one foot in front of the other, arms moving in such a way as to remain hidden behind the body . . . Males also take long steps, but their arms and legs move in a parallel fashion, meant to accentuate shoulders (strength) rather than hips (sexuality). Thus, corporeal technique is gendered – establishing an asymmetrical femininity and a symmetrical masculinity, which further promotes different readings of the *meanings* of the body.
>
> (Adelma and Ruggi, 2008: 565, emphasis in original)

These accepted ways of carrying the body are also extended into modes of tactility. In the West females are more likely to show tactility towards their friends and family through hugs and kisses, whereas males may only resort to a firm handshake. Within families, fathers are less likely to be tactile with their children than mothers.

Figure 13.1 (B)ordering female bodies

In these examples we can see how the female body can be (b)ordered out of place at different times of the month, or if clothed 'inappropriately'.

Source: Author

However, in terms of sexual practice, it is more culturally acceptable in the West for heterosexual men to be more active than heterosexual women. As Taylor and Jamieson (1997) identify, if women are sexually promiscuous they are likely to be negatively branded (their research study sees promiscuous women defined as 'little scraggies' or 'nookie tramps' by men within a northern UK city – the women's lack of a permanent partner renders them 'placeless' and valueless within this culture). Such (b)ordering of the body is very different in other cultures.

The *fa'afafine* in Samoa are open in their sexual relationships with either gender. 'Natural' (b)orders of 'gay' and 'straight' have little cultural relevance here (there is no specific Samoan term for 'homosexual' for example). In Samoa it is completely 'normal' for *fa'afafine* to have open sexual relations with men and women, without questioning either groups' sexuality.

Our bodies are therefore culturally (b)ordered in terms of what we should wear, how we should carry ourselves, and who we should share the place of our body with. McDowell

(2004) points out that our bodies can also be (b)ordered in terms of how we should perform them. In terms of physical behaviour Western males 'have developed a particular version of a tough, aggressive, sexualised street credibility' (McDowell, 2004: 51). Hart (2008) acknowledges a similar '*mukhayyamji*' style of idealised masculinity amongst males in Palestine. Here males are required to show physical strength, often through aggression and combat.

> Although a true mukhayyamji would not go looking for trouble, should a fight break out, he should be prepared to enter it immediately to save the reputation of himself and his neighbourhood. Describing this responsibility of the young mukhayyamji, sixteen year old Ahmed told me: 'If people know that the shabaab ("lads") in your neighbourhood are strong, others passing through keep their heads down and don't look at the women. They show respect.'
>
> (Hart, 2008: 73)

The '*mukhayyamji*' culture of masculinity thus requires that men be strong and ready for aggressive conflict in order to represent the strength and vigour of their particular place. Here the identity of their neighbourhood is directly connected to the physical characteristics and performance of the men who inhabit it. In the UK, McDowell (2004: 49) cites Willis' classic study of working-class men in the Midlands (1977) which emphasised a similar requirement for physical robustness from the male body.

> These young men's sense of themselves as masculine revolved around a tough aggressive and disaffected stance at school, the disdain of intellectual interests and achievement and the transference of similar behaviour into leisure activities that involved toughness and embodied activities in the streets as well as in sporting activities.

As Willis identified, a range of cultural activities also come to be associated with gendered norms. Men may shun academic interests in favour of aggressive sports. Women may perform well in class, but not be expected to pump iron in the gym (although see below). The sports arena is, therefore, one place in which notions of gender are often emphasised. As Adelman and Ruggi ask, 'just how strong or muscular can or should a woman's body be, and still be "feminine"?' (2008: 567). Similarly what sports can be played by all groups, without calling into question the gender or sexuality of the individuals involved? As Adelman and Ruggi cite, in their research, 'Volleyball [was considered] "so feminine" that, as one player pointed out, it . . . cast a shadow of doubt over the masculinity of the boys and men who practised it, i.e. carrying the stigma of feminization and/or homosexuality' (ibid.: 571).

Immediately from birth, culturally accepted norms of masculinity and femininity are thus (b)ordered onto our bodies. Biology does not produce gender, but cultures do. Particular roles and gender characteristics are constructed through habit, everyday actions, as well as invented ritual traditions. Nawi Ng et al. (2007) identify one of these rituals and its affects on masculine identity in Indonesia (see Box 13.1).

Summary

Bodies therefore have a 'sense of place' that is cultural rather than biological. Culturally constructed (b)orders of the body influence how masculinity and femininity is understood in different places. As Jordan and Weedon (1995: 179–180) point out:

> Our identities as girls and boys, women and men are formed in and through our involvement in social practices, from the family and schooling to culture, sport and the leisure industries. Cultural practices such as the media, marketing, the cinema, sport, literature, art and

BOX 13.1 'IF WE DON'T SMOKE, THEY WILL CALL US FEMININE': SMOKING AND MASCULINITY IN INDONESIA

In Indonesia, boys go through a traditional religious ritual of circumcision at the age of 10–12 years. As Nawi Ng et al. (2007: 798) state,

> Circumcision, which is viewed as a sign of male maturity and adulthood, is celebrated in a village ceremony. During the ceremony, cigarettes are served to the guests, who are mostly teenage boys and friends of the boy being circumcised. Cigarettes are also believed to promote healing of the circumcision wound, which, according to the boys, is a belief shared by their parents and one that has been practiced for many generations.

Through this process, cigarettes have become a symbol of cultural status and masculinity. Cigarettes indicate maturity, attractiveness, success, and potency. As Nawi Ng et al. (2007: 800) report,

> For them, boys have to be brave enough to smoke otherwise they are seen as having an effeminate manner. The smokers stated: 'If we don't follow our peers and smoke, they will call us feminine.' Thus, smoking enabled them to reaffirm their identity as boys.

Such invented rituals and associated cultural status contribute to the gendered smoking rates in Indonesia: 38 per cent of all boys smoke, but only 5.3 per cent of girls. This example illustrates how the place of the body is itself ordered and bordered with culturally distinct ideas of appropriate behaviour for each gender. It is normal and natural for boys to smoke in Indonesia, it is abnormal and 'out of place' for them not to. Such transgression of cultural norms raises questions over their gender: if they do not conform, there must be something wrong with the place of their body.

popular culture construct forms of subjectivity which are mostly gendered. This gendering suggests that certain qualities are appropriate to women and others to men.

These gender (b)orders vary between cultures, but all produce normalised and naturalised roles for males and females. In any culture, 'one form of masculinity rather than others is [always] exalted' (Connell, 1995: 77), and the same is equally true for femininity. These (b)orders effectively limit the ways in which gendered bodies can take and make place.

Gendered (b)orders: gendered places

Gender (b)orders are not simply cultural, they also have geographical affects. Where are the appropriate places for different bodies? Are all genders equally accepted in all places? In Western culture one of the most obvious places that is (b)ordered according to gender is the place of the home. Here it is a cliché that 'a woman's place is in the home'. This naturalised position has in recent decades been adapted and subverted (see below), but it still remains the dominant way in which domestic places are (b)ordered.

Greenberg (1998: 70) cites that many men accept that 'in the traditional home, women hold near total authority' (also cited in Mackay

Yarnal et al., 2004: 688). In this archetype, the male ideal is to go out to work and earn money, whilst the woman is the 'backbone of the family, caring for children and the house' (Miewald and McCann, 2004: 1051). In the place of the home, therefore, dominating power is exercised according to gender, as Miewald and McCann (2004: 1054) identified in the homes of Appalachia,

My mother made all the decisions. My mom was the one that pulled our family through every crisis and everything that went on. My dad had no say whatsoever. He just sat in the rocking chair and rocked which ever child was little.

My dad didn't know how to do anything. [He] didn't even know how to take out the garbage. He didn't know what you were supposed to do with it. He thought you threw the whole can away. It always seemed to be my mother's responsibility. Everything was her responsibility.

The (b)ordering of the body is thus extended into the material geographies our bodies inhabit. Women's bodies may be restricted to domestic places when they may really want to be in the world of business, be Formula 1 drivers, or army generals. Men are similarly (b)ordered into business suits, when they may want to wear stockinged feet and look after their, or other people's, children. Conventionally gendered bodies can therefore work to repress and restrict both genders and the places they can successfully take and make.

Gender (b)orders: patriarchy

Feminists refer to this (b)ordering process, and how it particularly affects women, as *patriarchy*. The term patriarchy refers to 'the system in which men as a group are constructed as superior to women as a group and so assumed to

have authority over them' (McDowell, 1999: 13). In Western societies, patriarchy is deemed an institutionalised process that enhances the idea of the superiority of men and the inferiority of women. Feminists argue that this is achieved through the legal system, the tax and social security system, as well as everyday attitudes and behaviours. Pearson, in his work on surfing in Australia, identifies the process of patriarchy occurring. He identifies a process of misogyny, a continual devalorisation of women and their role, when compared to the superior and exalted role of men.

Some time ago in the letters column some brave soul pointed out that *Tracks* was profoundly sexist – like the surf scene it caters to . . . What could you put instead of 'chicks'? Well, might one suggest 'lady' which is just a little better and of slightly less condescending nature? Point taken. All vessels shall hence forth be called ladies . . . Ed.' (*Tracks* 53: 13)

How do you think it would feel to be walking set of animal functions? . . . The surfing woman is meant to be a handbag, a decorative accessory for the shaggin' wagon like an expensive cassette recorder. For comfort stops who can cook, fuck and make your mates jealous by being as desirable as possible while remaining loyal and clean; a faithful dog who can do the shopping to keep the surfing body beautiful functioning in its holy watery glory. (*Tracks* 2: 51)

(cited in Pearson, 1982: 129)

By referring to women as 'chicks', 'dogs', and 'vessels', one can see that in surf culture femininity is to an extent dehumanised, framed as less important, simply an 'accessory' to what *really* matters. (Note too how faithfulness and loyalty on the part of the female is expected if they are to remain 'clean'.) What really matters in this case is the place of the wave; surf is the place of the male, whilst a woman's place is in the

BOX 13.2 BOYS' TOYS? GENDER ROLES (B)ORDERED

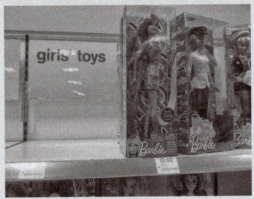

Men at work
Source: Author

Different cultures (b)order gender roles in different ways. None of these (b)orders are as natural as the day we were born, although they often seem to be as we encounter them from our earliest days.

In the Western world, children are encouraged to play with particular types of toy. Although this may appear to be a trivial example, imagine if you had

Boys toys?
Source: Author

never been permitted to play with dolls, or cars, or test-tubes, or footballs? What experiences are introduced to us as children that form our hopes and dreams for the future? In 2013, Goldieblox parodied the Beastie Boys' song *Girls* to encourage children to play with whatever toys they are interested in (see Michaels, 2013; Bates, 2014). Gender (b)orders, therefore, may be made to be broken.

We have seen above how different cultures (b)order the range of clothing types deemed appropriate for different genders. In the UK, messages on clothing also contribute to processes of patriarchy and the formation of gender roles. As Barnett (2011) reported, the retailer Top Man marketed t-shirts with slogans that listed excuses men give for perpetrating domestic violence. Imagine wearing, or dating someone, wearing the following slogans: *'Nice girlfriend: what breed is she?'*; *'I'm so sorry, but: you provoked me/I was drunk'*. Although these clothing lines were withdrawn from sale, their design, manufacture, and short-term availability suggest a particular attitude towards women which falls short of equality between genders, as well as denigrating the control and capacities of men.

margins (in this case, the place of the camper van or the beach). In this way the cultural preferences of bodily order are translated into material geographies: places are taken and made in line with gender roles and relations.

However, as we have seen in previous chapters, the (b)orders that repress and restrict can also offer opportunities and potential. Patriarchy, for example, can restrict women to places that are maternal, family-oriented, and domestic. As peaceable rather than political bodies they belong to the private and are out of place in the public sphere. Nevertheless, Ashe (2006) argues that such apparently restrictive (b)orders have been used by women to make substantial contributions to the public sphere. When women transgress accepted (b)orders and attempt political action they may be framed in a 'feminine' way (as Ashe points out, women campaigners in Northern Ireland have been described as 'pretty and defiant' or 'impossibly glamorous', as well as being defined by the number of children they have brought up). However, these (b)orders also lend credibility and weight to the campaigns they seek to make. As Ashe argues, due to their familial and maternal roles, these women have 'greater moral authority and a *natural* right' to mount campaigns (Ashe, 2006: 164, my emphasis). It is argued that this *cultural* status makes their opinions more significant and resonant than if they were made by a male.

It is important therefore to interrogate how (b)orders of the body can both restrict and enable particular life choices. (B)orders offer both obstacles but also opportunities. How do gender (b)orders confer particular sources of symbolic and cultural capital to particular bodies? How are all genders repressed but also liberated by them? Raising such questions means that issues of gender domination or superiority become complex to answer. Who gets to set the (b)orders of our bodies? Are both genders equally restricted by them? In what places, how, and why?

Changing (b)orders: how bodily ideals are changing

Focusing on bodily (b)orders and the places they create allows us to also interrogate the power relations between males and females, and hetero- and homosexuals. As Panelli (2004: 66) outlines, 'adopting the . . . ideas of [bodily] power relations enables [cultural] geographers to explore the bases and practices of power relations that unevenly affect men's and women's experiences and access to different arenas and spaces'.

Focusing on bodily ideals enables us to see these relative power structures not as taken for granted 'natural' (b)orders, but constructed to benefit certain groups in certain ways. Some cultural groups may be complicit with, or acquiesce to, these (b)orders because they benefit from or agree with them. However it is clear that whatever the relative powers at play, there is always an alternative to these (b)orders. As Blunt and Willis (2003: 93) note, 'To think of [the body] as a social construction has been politically enabling, allowing gender roles and relations to be destabilised and resisted.'

As we have seen, the dominant way Western cultures tend to (b)order our bodies are in line with the dualistic categorisation we have noted elsewhere in this book. We are born as male *or* female in terms of our gender, gay *or* straight in terms of our sexuality. The naturalness of these (b)orders can be easily destabilised and resisted. For example, what about those people who are *trans-* or *inter-*gender? What of those people who are *bi-* or *poly-*sexual? What of people who have used medical advances to change the gender of their birth? These individuals do not conform to the established (b)orders of cultural and geographical classification, they transgress and question the established lines that take and make place. They occupy what we have seen in previous chapters as an ambiguous grey area, in the case of intergender a literal 'no man's land'. Let's consider some examples of bodies that are out of place in terms of these conventional (b)orders (see Figure 13.2).

BOX 13.3 PASSPORT BEYOND (B)ORDERS: THIRD GENDERS IN INDIA AND AUSTRALIA

Although many cultures identify two major genders, in Asia, *hijras* have been part of the cultural world for thousands of years. Hijras – or individuals understood as transgender, intersex, or eunuchs – were originally celebrated in Hindu and Mughal cultures; however British rule in the nineteenth century led to their persecution and criminalisation. In 2014, India re(b)ordered these bodies, ruling that hijras could again be officially recognised as a 'third gender' category by government (see Khaleeli, 2014).

In Australia, individuals also do not have to choose between two gender categories. When arranging their passports, Australians now have a third 'X' box in which they can 'tick' their identity. As Australian senator, Louise Pratt stated, 'X is really quite important because there are people who are indeed genetically ambiguous and were probably arbitrarily assigned as one sex or the other at birth. It's a really important recognition of people's human rights that if they choose to have their sex as "indeterminate", that they can' (BBC, 2011).

Figure 13.2 Hybrid and transgressive genders
Source: Author

Browne (2006) has conducted research on individuals who do not fit into orthodox (b)orders associated with gender and sexuality. Her respondents identified themselves as women, but are not necessarily seen as such by mainstream culture – their size, shape, or surgical transformation means they transgress 'naturalised' categorisations of 'female'. Browne found that these women's cultural otherness also meant they had no geographical place within the mainstream, even in such necessary locations as public toilets. As Browne (2006: 128–129) outlines:

> Where places are segregated into men's and women's spaces, such as toilets, ambiguous bodies can be subject to violence and abuse . . . Attributes, such as large frames, can often be associated with men, constructing the sight of the body as male and consequently 'out-of-place' in women only toilets.

Browne's respondents thus found themselves caught in a liminal or 'in-between' position (a position that we have seen in Chapter 11 also besets youth culture). To the mainstream they were not quite women, but also not quite men. They were somewhere and somebody in-between. Unlike the liminal position of youth culture however, these women had no possibility or desire to 'grow out' of their position, even though the effect of their difference was a chronic displacement from the mainstream. Some of Browne's respondents explain the affects of their difference as follows:

Janet: And you know they [other women] don't look at my face or anything they just look at my build and look at my height and look at my haircut and they just instantly assume that I am some dirty man in the women's toilets. (Janet: focus group)

Lorraine: I can't believe how rude people are though, cos I have been in the toilets with you.

Janet: I dread going to the toilets at [name of nightclub] on a Wednesday night I absolutely dread it, I just think 'fuck I have to go to the toilet'. And I hold it off as long as I can and I just think 'no I have to go'. And I have to get someone to come with me because I just, I just get so many people shouting at me.
(cited in Browne, 2006: 129)

Being born or made into bodies that do not conform to conventional (b)orders therefore means that many individuals have no place within some cultural geographies. It is clear that very real emotional traces are produced by the transgression of bodily ideals. As Davidson (2003) has explored, such transgressions can result in phobias and panic in certain places. In Davidson's words, it can 'destroy one's sense of relatedness to other people, and locatedness in place, alienating the subject from the practices of everyday life' (2003: 3).

> As soon as we recognise that the body and not just the mind is 'a medium of culture' (Bordo, 1993: 65) then the question of how . . . self identities and bodies are maintained and disciplined in and through the agency of a (often hostile) cultural environment becomes crucial.
>
> (Davidson, 2003: 17)

The (b)ordering of bodies and the geographies they inhabit thus reminds us that it is not place itself that is a problem for many people, it is the cultural meanings and associations imbued in them that render some groups 'in' and 'out of place'. To paraphrase Kirby, 'the problem with place isn't just place; it is the fact that there are other people in it – other people who are creating it, determining it, composing it . . . is it surprising then, that place could seem a bit hostile?' (1996: 99).

BOX 13.4 DIS-ABLING (B)ORDERS

Nothing so clearly or inevitably reveals the inner man [sic] than movement and gesture. The moment you move you stand revealed, for good or ill, for what you are.

(Humphrey, cited in Davis et al., 2004: 103)

Just as cultures seek to (b)order gender roles and the appearance of our bodies, they also seek to normalise bodily movements in particular places. As Davis et al. argue, 'body movements associated with illness produce stigma by transgressing the unwritten codes that govern particular public spaces' (2004: 103). As is the case with many cultural (b)orders, we perhaps notice these unwritten codes most explicitly when individuals begin breaking them: what do we think, for example, if someone gambols down the street rather than walking, or someone limps, uses crutches, or a wheelchair on a busy sidewalk (note the important vocabulary used to identify this element of the street, suggesting some activities, and bodies who can perform them, are more appropriate than others)? How do we respond if someone shouts an expletive in public space, or cannot control their limbs or head in ways we have come to consider 'normal'? What does this reveal about our ideas of 'good' or 'appropriate' bodily behaviour?

Imrie (2001) has argued how a range of bodily practices are considered discrepant and deviant in public places – including the use of wheelchairs, guide dogs, canes, etc. In his words,

the physical construction of urban space often (re)produces distinctive spatialities of demarcation and exclusion, from the lack of access to public transport systems to the absence of visual clues or guides in towns to enable vision-impaired people to move with ease. Indeed, the socio-spatial patterns of ableist values [, paces, and practices] are etched across the city in numerous ways, forming a type of architectural apartheid.

(Imrie, 2001: 232).

Can you think about the ways in which different bodies are designed out of public spaces? How could these bodies be designed back in?

We can challenge ourselves to rethink our bodily (b)orders with respect to dis/ability if we consider the case of individuals with Tourette's Syndrome. 'Touretters' (importantly not Tourette's 'sufferers' as perhaps this condition is not always and necessarily an active state, nor should be immediately relegated to a category that is deemed inferior to the non-Tourette's 'norm') often exhibit a range of extraordinary bodily actions. These range from compulsive physical tics (instinctive movements of limbs and head), to 'echolalia' (the repeating of words or phrases), to the more well-known 'coprolalia' (or cursing). Touretters' bodies therefore respond to the cultural world in a range of ways that appear different to the 'norm'; as Davis et al. state: 'Touretters are walking disruptions of doxa' (2004: 109). To many, their bodily behaviours are seen as 'something aberrant, something different, something strange' (Handler, in Davis et al., 2004: 110). Shane Fistell (2012), a Touretter, talks of the cultural and geographical (b)orders that affect him as a consequence of the 'other people' that create, determine, and compose the cultural world:

I think the isolation has been overpowering; it's easier to stay in. It's suffocating, I've lost years . . . I've stayed in this apartment for years . . . Here I am, cooped up like a tiger in a cage. I study people to learn their [body] language, because I'm not familiar with everybody else's gestures; I don't think it's natural to me. I don't understand it. And I thought if I behave as people do I would be more accepted . . . I would be one of you, even if it's my second language.

An individual with a related condition (autism) argues how their experience of the world is not necessarily 'aberrant', but rather a positive, unique, and revealing other 'language' through which their body engages with place (see www.youtube.com/watch?v=JnylM1hl2jc).

Transgressing bodily (b)orders: the process of individualisation

Bodily (b)orders can thus make certain places hostile to some individuals and groups. However, as we have learned in this book, all cultural and geographical (b)orders can be subject to transgression and resistance. Johnston (1996) focuses on how female body-builders are disrupting the accepted norms of femininity and masculinity in relation to both the place of their bodies but also the places in which these bodies are shaped (see Box 13.5).

Bodies can also transgress (b)orders and become individualised through modification. This may be through gender reassignment (see

BOX 13.5 RE(B)ORDERING THE FEMININE BODY

Johnston uses the example of a gym in New Zealand to explore the ways in which traditional gender roles and places interact, but also become perpetuated or transgressed. In this particular gym she notes that,

Women are encouraged to participate in aerobics and circuit training, away from the masculine sexed space of the potentially violent 'Black and Blue Room', a free weights training room. Women dominate the aerobics and circuit room, while the men dominate and actively discourage women from participating in the 'Black and Blue Room'. This particular gym confirms the construction of feminine and masculine spaces, and hence bodies. Within this environment the female body-builder provides a challenge to hegemonic notions of sex and sexuality, as well as participating in dominant discourses which shape feminine bodies.

Jenny: One guy said, 'Oh do you need a hand with that?' And I was carrying it [the bar bell] over to prepare to set up and to do an exercise and I said, 'No I'm alright'. He said, 'Oh are you going to use it?' and I said, 'Yeah', and he had his mouth wide open.

(cited in Johnston, 1996: 328–330).

In this example, Jenny is transgressing both the orthodox (b)ordering of the female body, as well as the conventional (b)orders of the gym itself. Female bodies are 'naturally' (b)ordered as svelte and curvaceous, so to gain a physical muscularity directly resists these ideals and seeks to impose a different cultural and bodily order. Similarly, through her presence in the 'Blue and Black Room' the naturalised order of this site as a place solely for men is challenged. Here the place of the gendered body and the places that it inhabits are being taken and made in ways that resist dominant cultural (b)orders.

above), but can also occur through a range of less invasive actions, including the everyday process of washing our skin and hair, or choosing particular cultural traces to dress our bodies. Clothing decisions may indicate which 'side' we are on in terms of gender or sexuality, and transgressions in these dimensions can be usefully understood through the notion of 'camp'. Camp, as Johnston (2002) points out, generally involves, 'representational excess, heterogeneity and gratuitousness of reference' (Cleto, 1999:

3). Camp is normally associated with homo- or poly-sexual bodies, yet both heterosexual men and women can demonstrate camp characteristics, either temporarily or as a persistent identity trait. Camp therefore not only 'works to undermine accepted values and truth, especially the heterosexual definition of space' (Binnie, 1997: 29, also cited in Johnston, 2002: 127) but also to 'dismantle the static constructions of masculine/feminine and heterosexual/homosexual' (Johnston, 2002: 128). Increasingly the

Source: Author

Source: Anchor: Anonymous

Source: Koi: www.tattoostingo.com/koi-tattoo-designs-and-meaning

Source: Dragon: http://tattooton.com/wp-content/uploads/2013/09/Best-Dragon-Tattoo-for-Men.jpg

Figure 13.3 Tracing the body

influence of camp, gay, and heterosexual characteristics are hybridising traditional identity roles. Metrosexual males may carry 'murses' (male purses), they may pay particular attention to their fashion choices and general appearance, and use hair products or make-up. Similarly all genders and sexualities may dance to certain genres of music in a range of venues. With these complications, does dancing to ABBA in a 'gay bar' necessarily mean you are gay? Would doing so enforce or challenge your body's sexuality in the eyes of others? Why?

Bodies can also be individualised through other forms of 'dress' and practice. Body art is increasingly prevalent in the West, including henna and ink tattoos. Do you have a tattoo? Of what? Where? Can you think of how tattoos have been used to (b)order the body in different places and times? With these questions in mind, tattoos can be thought of as bodily traces that tell us stories; as one individual suggests, 'tattoos map my life' (see Frizzel, 2014). In the same way as graffiti on the streets, tattoos on the body can give us insight into a range of cultural meanings (see Barkham, 2012), they can remind us of significant events (see Locke, 2013), as well as being 'read' experientially (for example, we can ask what is it like to be tattooed, by choice, or against your will?). What do the tattoos in Figure 13.3 mean? Can you think of other ways in which the place of the body is individualised? How do individuals (and how do you) 'carve' and mould this very personal geography?

BOX 13.6 CULTURES (B)ORDERING THE BODY

Traditional bodily (b)orders can also be modified through changes in capitalism. Changes to labour markets through technological developments, communication infrastructures, and globalisation have led to traditional job opportunities and distinctions being eroded. As a consequence, much manufacturing labour has been displaced from the developed economies and the jobs that have taken their place are more associated with bodies that are highly tactile, deferential, and service oriented. These new roles therefore complicate traditional (b)orders of both masculinity and femininity. Although, as McDowell identifies, 'high-touch work' may be 'constructed as "feminised" work in which a docile and . . . conforming body is written into the job description' (2004: 48), the traditional role of 'breadwinner' within a family is accepted as masculine. In this scenario, female bodies have to transgress their (b)orders in order to gain employment, whilst male bodies have to transgress their 'physical' and 'aggressive' ideals in order to conform. In practice, evidence suggests that this process of adaptation may be easier for women than for men, as the following examples illustrate:

In my interviews I found that many of the young men to whom I talked had clear views about the types of work they were prepared to consider, regarding most service sector work as 'women's work' and so beneath their dignity. Even if they were prepared to consider employment in the shops, clubs and fast food outlets that were the main sources of work . . . they often disqualified themselves as potential employees by their appearance (piercings and tattoos as well as inappropriate clothes) and their attitudes during the recruitment process, as employers read the surface signals of bodily demeanour, dress and language as indicators of the underlying qualities they are seeking, or more typically as characteristics they are careful to avoid . . . Many of them [who did get jobs] found it hard to perform the deferential servility required in the service economy.

(McDowell, 2004: 51)

The traditional (b)orders of the body can therefore render some genders out of place in the new geographies of capitalism. Orthodox masculine bodies (strong, pierced, tattooed) are out of place in a new world (b)order requiring deference, tactility, and servility. The mixing and mingling of capitalist and body traces thus destabilises accepted bodily norms.

Similar destabilisation of (b)orders also occurs when other cultural traces merge with the body, for example religion and nationality. As Said (1979) identifies, religions such as Islam are often (b)ordered in the West as violent, oppressive, and patriarchal. However, many in the West are now choosing to convert to Islam, instigating the 'pioneering endeavour' of creating an American or 'European Islamic culture' (Ramadam, 2003: 163). The potentially difficult 'marriage' between Western values and those of Islam prompts a reconsideration of each culture's traces. Mansson McGinty (2007) has looked at this phenomenon, specifically the consequences produced by women in Scandinavia who convert to Islam. In this research, the 'new' cultural geographies constructed through individualisation are clear:

What I find difficult in my life is that since I'm wearing the veil I am questioned. I find it very hard that my intellect is questioned. [Apparently] you can't be intellectual or intelligent and Muslim at the same time. That combination does not exist for most Swedes. In their eyes I have to be somewhat stupid if I, as a Swedish woman, decide to convert to Islam.

(McGinty, 2007: 477).

This may sound very, how can I say it, complacent, but I have always been slim and I have always had a remarkably good-looking body. And that has been a basis for many people in forming a judgment of me. I was only a walking body. And even if you wear jeans and a T-shirt people can tell how you look. With the veil and so-called decent clothes I get judged in a whole different way. Since people can't

see my body they can't think 'wow, what a waist, what a sexy butt and what legs!' Instead people have to look into my eyes. Contrary to what most women believe, it is not so much fun to be so-called 'sexy'. It is not nice when people forget that you actually can have an IQ of 120. Because, in their eyes, if you are sexy you are also stupid. Then I get so damn pissed [whispering].

(McGinty, 2007: 480)

These quotations illustrate how gendered bodies can challenge the relative (b)orders of Islamic and Swedish culture. From a Western perspective, for a white Swedish woman to decide to wear a veil must mean they are somehow abnormal, choosing a servile and apparently 'inferior' position in contrast to the equality that defines Swedish cultural life. However, in their own interpretations of Islamic values, the respondents do not feel as if wearing a veil means deference or inequality. In their reading: 'Islam says that it is not equality through sameness [jämlikhet] we should strive for but equality in the sense of having the same worth and opportunity [jämställdhet]' (ibid.: 478). In practice, the 'transgressive' act of wearing a veil actually means that other aspects of their personality are noticed more (people just don't see them as 'feminine' or 'sexualised' bodies) and thus a new form of equality is attained. As a consequence, this merging of different bodily traces produces new ways to be 'a woman' that transgress the traditional (b)orders of both cultures.

We can see therefore that the individualisation process that we encountered in Chapter 6 is one way through which the dominant place(s) of the body can be destabilised. Through this process a world can be created 'which doesn't have the same borders anymore' (Beck, 2002: 211). The destabilisation of bodily ideals makes it important to chart the reconstruction of (b)orders that are produced in their stead. As the above example illustrates, these Swedish women are changed through their conversion to Islam. Yet Sweden is also changed by having white, Islamic women in it, and Islam is influenced too by its reinterpretation within a twenty-first-century European place. Through such experiments about what it means to be some*body*, what cultural and geographical consequences result from these ongoing compositions?

The consequences of bodily difference: (re)establishing (b)orders

In the final section of this chapter we investigate four consequences of the ongoing composition of bodies and places. We explore how new bodily (b)orders are being constructed but also how traditional traces are being defended from transgression. Through doing so we diagnose the 'natural' place for different bodies within these cultures, and find out whose cultural places these are anyway. In the first two examples we will see how a dominant heterosexual culture has accepted homosexual bodies within the mainstream, but this acceptance has been geographically limited to certain places. Firstly, the gay community is deemed as acceptable and appropriate only in metropolitan areas, they are out of place in rural or non-urban sites. Secondly, the gay community has been segregated into specific areas of the metropolis itself – with 'gay ghettos' being created that at once affirm a gay identity, but also enforce its difference and marginality from the mainstream. In the third example, we explore how certain gendered and ethnic bodies are out of place in Ireland. Finally, we explore how the orthodox (b)orders of Western patriarchy are being re-constructed in one town in China (Box 13.7).

Re-(b)ordering Nebraska as a heterosexual place

In many countries, gay people are conferred the same rights as heterosexual people. However, this cultural acceptance is not universal. In Russia, state authorities take steps to 'protect' minors from 'propaganda' about non-heterosexual lives (see Walker, 2014), whilst in 82 countries homosexuality remains illegal (Williams, 2011). Although homosexuality in legal in America, the 'Defense of Marriage Act' (DOMA) has been introduced in Nebraska to ban same-sex marriage, civil unions, and domestic partnerships. Rasmussen has argued that the threat posed by gay and lesbian difference represents a 'struggle over the very meaning, identity and self-determination of a place called "Nebraska"' (2006: 808). The 'threat' of gay couples taking and making a place in Nebraska led to other communities within the state uniting to repel them. 'Norms' of 'traditional values', 'nuclear families', 'heterosexuality', and 'democracy' were called forth, strengthening the mainstream culture's sense of place. As Rasmussen (2006: 811–812) cites,

> The pro-DOMA forces adopted the strategy of portraying gays and lesbians as a powerful minority that . . . wished to undermine the cultural identity of *their* Nebraska . . . The pro-DOMA campaign portray[ed] advocates of same-sex unions as foreign to Nebraska and its values, a process of representation that drew heavily on stereotypes about cultural belonging . . . [It] was able to capitalize on both the rhetoric of local control and pre-existing stereotypes associating alternative sexualities with metropolitan areas . . . The state natives were represented as rural, rooted and traditional [and] their opponents . . . as cosmopolitan, nomadic outsiders.

As Rasmussen identifies, the 'threat' of gay couples taking and make a place in Nebraska led to other communities within the state uniting to repel them. Stereotypes of belonging and identity were (re)established which placed gays amongst other minority groups in other areas (in this case, the city), and keeping rural places the 'rightful' place for the traditional, straight, majority. As a consequence, gay people had no place in this idea(l) of Nebraska.

According to the mainstream ideal, at least as defined by those in Nebraska, the 'normal' place for the gay community is in the city. Indeed, in many of the major cities of the world there are thriving and vibrant gay areas (perhaps ironically called villages). The idea that is it 'normal' for the gay community to find a sense of place in the city has been explored further by Nash (2006). She argues that although the creation of gay villages or 'ghettos' in many urban areas affords this culture its own place within the mainstream, it also (b)orders this community into particular sites within the city, and is an obstacle to full acceptance by the dominant culture (see also Chapter 10).

> the gay ghetto [can be] regarded as an obstacle rather than a benefit to the liberation of gays and lesbians—a space at odds with and constitutive of a diminished and apolitical homosexual identity . . . Activists argue that exclusively gay spaces wrongly segregate homosexuals from mainstream society in an oppressive and marginal 'ghetto'. Several groups dr[a]w a direct correlation between being forced to socialize in these segregated spaces and what [is] seen as 'inappropriate' homosexual conduct and lifestyles.
>
> (Nash, 2006: 3–5)

Nash argues here that although place is granted for this minority group within the mainstream, their sense of place is still deemed 'inappropriate' by it. Non-heterosexual identities and traces may be accepted as 'normal' by some, but still not as 'natural'. Acceptance has therefore been partial, as a cultural group gays remain both

geographically and culturally marginal. What does this tell us about both cultures? What is the nature of the 'threat' posed by non-heterosexual bodies to the mainstream? Are their biological differences and desires corrupting to the common sense (b)orders of mainstream places? Is partial acceptance a first step towards full 'partnership'?

Pregnant and ethnic bodies: citizenship in Ireland

In Ireland, processes of globalisation and immigration have set in motion the mixing of particular religious, ethnic, and nationalist traces. The dominant cultural group in Ireland (white, Catholic, and perhaps male) has worked to defend its (b)orders from those bodies that are deemed to be out of place there. One example of this re(b)ordering is the 2004 Constitutional Amendment which removes birthright citizenship (i.e. the right to be considered a natural citizen of Ireland) from any child born to immigrant parents in the country. As Cox points out, this Amendment 'tells people in the starkest terms . . . that we do not wish them to know a sense of belonging [in and to Ireland]. It says that . . . genetic connection to the pure Irish race is a necessary prerequisite of Irishness' (Dr Neville Cox, Trinity Law School, Dublin; cited in Tormey, 2007: 70).

According to this Amendment, therefore, the only appropriate tracing that warrants Irish citizenship is blood and race. Being born or growing up in Ireland is not enough, to be Irish your body has to conform to a blood lineage. Tormey (2007) argues that this Amendment renders certain bodies, particularly black, immigrant, and pregnant bodies, as transgressive in Ireland. Due to their inherent visibility, these bodies have become 'matter out of place' (Butler, 1993). As McDowell, Ireland's Minister for Justice, stated at the time, 'anyone with eyes can see the problem' (Tormey, 2007: 69); these bodies are

BOX 13.7 PATRIARCHY RE(B)ORDERED? WOMEN RULE OKAY IN CHONGQINQ, CHINA

In 2007, Chongqing in south-west China planned to build a district based on a local custom which (b)ordered men subservient to their wives. In Chongqing, women maintain dominance over their husbands, enjoying control over the home, work, and financial environment. As Coonen (2007) reports, signs in the district will outline the dominant order: 'Women are never wrong; men can never refuse their needs.' Men who transgress this order are punished by kneeling on a hard board or being forced to do the washing up in local restaurants. The idea of a formalising this modern-day matriarchy has been approved by the local government who wish to use this custom to attract tourists. Does this reversing of patriarchy herald emancipation for women in Chongqing, or does it pander to Western fetishes of domination?

literally marked out as trespassers, they do not fit the physical norm and are thus 'othered' as non-citizens in this place.

Conclusion

Our bodies are the medium through which we encounter the world. The place of the body, and the places it takes and makes, are therefore vital to understand for cultural geographers. As we have seen in this chapter, our bodies can be gendered and sexed, but these subject positions are not only something we are born with, they are also culturally constructed both materially and metaphorically. Our bodies are therefore bound up with a range of cultural meanings that affect how we are expected to act, where we are expected to be, and who we are expected to be like. These bodily (b)orders present us with opportunities and obstacles in our engagement with the world around us.

Through acting in this world our bodies take and make place. Our actions affect the meanings of the world, and our bodies become affected by the world's competing traces. Due to the limits of binary bodily (b)orders (for example of male *or* female, heterosexual *or* homosexual), as well as the competing traces that affect our bodies (for example age, ability, capitalism, religion,

and nationality), an increasing number of people no longer 'fit' conventional (b)orders. How do these 'other' bodies and their traces come to affect the ongoing composition of place?

We can see that destabilising the (b)orders of the body can have a number of effects on place. Dominant (b)orders may be strengthened, leading to the exile of challenging bodies, or they may be loosened for conditional acceptance, albeit with restrictions to the margins. These processes thus resonate to a degree with the control mechanisms outlined in Chapter 10. Challenging bodies may be outlawed, tightly controlled, or accepted through partnership (both in everyday life and in 'civil' union); each culture may even be equally assimilated by the other. How particular bodies secure partnership, where, and why, is thus a key question to answer through the study of the ongoing composition of place.

Suggested reading

The following texts are useful to explore the issues raised by this chapter.

Books

Butler, J. (1993) *Bodies That Matter: On the Discursive Limits of 'Sex'*. Routledge: London.

McDowell, L. (1999) *Gender, Place and Identity: Understanding Feminist Geographies*. University of Minnesota Press: Minneapolis.

Judith Butler offers a relatively advanced insight into the importance of the body as a geographical and cultural place, whilst McDowell gives a more accessible reading of the relations between the body and place, particularly with reference to gender.

Bell, D. and Valentine, G. (eds) (1995) *Mapping Desire: Geographies of Sexualities*. Routledge: London and New York.

Empirical examples of places of the body, gender, and sexuality and how they are constructed and performed differently across the globe are outlined by Bell and Valentine. Examples include places of sex and sexuality, gender in the home, and the body in cyberspace.

Cleto, F. (ed.) (1999) *Camp: Queer Aesthetics and the Performing Subject. A Reader*. Edinburgh University Press: Edinburgh.

Journals

A range of insights into mainstream and alternative sexualities, body places, and gender constructions are explored in the following journals. The paper titles illustrate the subject matter examined therein.

Bondi, L. (1990) Feminism, postmodernism and geography: space for women? *Antipode*, 22 (2), 156–167.

Johnston, L. (1996) Flexing femininity: female bodybuilders refiguring the body. *Gender, Place and Culture*, 3, 327–340.

Binnie, J. (1997) Coming out of geography: towards a queer epistemology? *Environment and Planning D, Society and Space*, 15, 223–237.

Nash, C. (2006) Toronto's gay village (1969–1982): plotting the politics of gay identity. *Canadian Geographer*, 50 (1), 1–16.

Doing cultural geography in practice

Introduction

This book has begun from the premise that we are part of the cultural world. All our ideas and actions combine to produce places. This production occurs both in our everyday activities, but also in the process of us investigating cultures and their locations – in *doing* cultural geography. In their own ways, our research encounters take and make place. They leave traces which mark our presence in a place during our research, then absence when we leave. The nature of the traces we leave says something about our research methods, and says something about us. What sort of research methods should we use to investigate cultural geography, and what sort of traces should we leave?

In this chapter we investigate some of the key methods that can be used to approach place from the perspective of cultural geography. As we know, this approach looks at places as ongoing compositions of traces. These traces are both material (for example 'things' such as buildings, signs, or statues), but also non-material (i.e. events, performances, or emotions). As a consequence of this approach cultural geographers need to employ research methods that can explore *both* types of traces. In essence, therefore, we need to use methods that can investigate both the representational and the non-representational in cultural geography; to echo the words of Greenhough (2004) we need to use methods that can explore both the language of the word

and the language of the world (see Chapter 3). This cultural geography approach to place – this 'more than representational' cultural geography (see Lorimer, 2005) – seeks to include both theories and things, but also emotions and feelings in its study. In order to do so, a range of different methodologies is required. In this chapter we outline some of the key methods that can do just this. The chapter begins by exploring some of the key issues to be considered when conducting any method, including access, operation, and ethics. It continues by outlining some of the methods used by cultural geographers to understand the language of the 'word', namely interviews and the analysis of cultural traces. The chapter goes on to explore methods that go further into the language of the world, most notably ethnography, mobile methods, and the use of film, before outlining how all these methods often combine through case studies to give us cultural geography's approach to place.

Positioning methods

The need to choose appropriate methods for any research project is vital. When choosing the best methods for your work it is important to remember the illusion of the 'god trick'. As Haraway (1988) informed us in Chapter 2, the 'god trick' is the fantasy that fools us into thinking that we can stand apart from the world we are studying and gain a universal and detached view of it.

However, we are not gods. We are thoroughly attached to the cultural worlds we study, and as a consequence can only gain a partial and positioned view of it. As Cook and Crang (1995: 7) note, 'rather than claiming some sort of Archimedean point from which the world can be critiqued, the researcher's viewpoint is largely a product of social relations both within the academy and between it and the world at large'.

When positioning ourselves and our methodological choice we therefore need to remember that any research encounter does not simply uncover the unmediated world of 'others', but it is facilitated and limited by our *interrelation* with this world (following Hastrup, 1992). Our relationship with the cultural groups and places we study will significantly affect the type of access, operation, and ethics we adopt to implement our methods, and the quality of the insight we get into 'their' world. For example, what is the relationship between you and the cultural geographies you wish to study? As a student, and part of the academic community broadly speaking, what orders and borders connect but also separate you from the cultural places you are interested in? Are you an 'insider' to these places, or a strange and alien 'other'? What implications do these relations have? How should you present yourself to the groups you wish to explore? How can you cross (b)orders and gain access in ways that will avoid alienating you from 'them'? How will you gain their understanding, and vice versa?

The (b)orders that connect and separate us from different cultural geographies mean that issues of access, trust, and ethics are important to consider for research methodology. How, for example, do we begin to gain *access* to a particular culture? Access arrangements are of vital importance; without them any research project is doomed before it begins. In the first instance, gaining access is made easier by doing background exploration on the nature of the culture under study (this can often be achieved through the analysis of the traces produced by that culture, and this is discussed further below).

By using the Internet, documentary material, contact directories, and other publicly available archives, key issues can be ascertained such as the key people and places to that culture's geography. But how do you present yourself to this culture and its places? Should you sneak under their (b)orders and study this culture secretly, or should you gain permission to access their lives? What do you think? In what cases would covert study be appropriate? Why would consent be important?

In some cases, secret investigation can be effective, and perhaps the only option available for research. Covert research can be justifiable on grounds of safety (when revealing your student status may endanger you or others, perhaps when studying gangs or turf wars for example), or when the nature of the culture you are studying means overt access would be impossible (for example if studying illegal activities). However, in the vast majority of cases, covert research is difficult to justify. Covert research not only limits the types of circumstance you can readily gain access to without causing suspicion, but also makes it difficult to maintain honest and open relations with those you wish to study. At some stage during covert research it is likely that your true identity will be revealed. At this stage, such transgression often results in strong feelings of antipathy not only towards you, but also to the broader research community of which you are a part. Will the outcomes of your research project outweigh the damage caused by such antipathy?

Thus it is crucially important to consider the traces you may leave during your research, even when deciding how to present yourself to those under study. Being open and honest about the nature of your research, what it will be used for, and what you want from your respondents is often not only the most *effective* course of action to gain the best knowledge of their world, but also the most *ethical*. As Hay (2003: 39) defines, 'ethical behaviour protects the rights of individuals, communities and environments involved in, or affected by, our research'. Being overt is often the most ethical course as it protects the inter- ests of those we may wish to study, giving them the opportunity to 'sign up' to research knowing all the 'small print' involved in their participa- tion. Will they, for example, have the option of being anonymised in your work, to speak off the record, or to read through your finished project before it is made public? In what circum- stances would such opportunities be appropriate, or not?

Ethical behaviour when gaining access to the culture and places under study helps the successful operation of your research project. Acting professionally, courteously, and with respect not only increases the chances of people responding in kind, but also reflects well on the broader academic community. From the onset of recruitment (through letter, email, or personal communication), through operationalising your method, then leaving the field, rapport and good relations ensure effective and ethical research.

However, as we have seen with the examples of covert research, the best ethical approach is not the same in all cases. There is no one 'right' or 'wrong' course of action and good practice will change with every research project. With this in mind, it is up to all cultural geographers to be *reflexive* about their actions in place. Critically thinking about your behaviour in all aspects of your research – considering not only how you would feel being contacted in this way or that way, but also putting yourself in your respondents' shoes – these are often good ways of reflecting on how you should behave. In many universities there are ethical guidelines to help you make the best possible decision for your research project. These guidelines do not ban or outlaw particular types of research practice, but encourage you to reflect on the possible outcomes involved in one course of action or another. Why did you make the decision that you did? Would another action be more appropriate? If you can robustly justify your potential actions from such questioning then it is likely your ethi- cal imagination is properly switched on. At this

stage you are ready to choose the methods that are correct for your study.

Choosing methods

> Choosing a research method is not just a case of picking one that seems the easiest but picking the most appropriate relative to the knowledge that you require.
>
> (Kitchin and Tate, 2000: 211)

Choosing the appropriate method is fundamen- tally important to a successful research project. As Kitchin and Tate point out, choosing methods is not a matter of choosing the easiest or quick- est, rather it is an issue of choosing the most appropriate method to access the insights and knowledge that matter to you. As we have seen in this book, our understandings and construc- tions of the cultural world are not 'natural' in form and structure; they are generated by differ- ent cultural groups in different cultural places. In order to conduct effective research we therefore need to employ methods that can access these different constructions and understandings. It is vital that our methods are capable of getting at the meanings, interpretations, and belief systems of particular cultural groups, as well as the emo- tions and feelings that may motivate or move groups to take and make place. In these cases, *qualitative methods* are often the most appropri- ate to employ.

Qualitative methods are a group of techniques which help researchers gain access to the mean- ings ascribed to cultural traces and places. As Hakim (1987: 26) outlines, qualitative research offers insight into:

individuals' own accounts of their attitudes, motivations and behaviour. It offers richly descriptive reports of individuals' percep- tions, attitudes, beliefs, views and feelings, the meanings and interpretations given to events and things, as well as their behaviour; [it]

displays how these are put together, more or less coherently and consciously, into frameworks which make sense of their experiences and illuminates the motivations which connect attitudes and behaviour, or how conflicting attitudes and motivations are resolved in particular choices.

Qualitative methods therefore offer access to meanings, beliefs, and attitudes of individuals and groups. They can be effective at accessing 'thought-through' interpretations and representations of our world (through interviews); they can give insights into everyday or impulsive interactions (through ethnography); as well as interpreting material traces (through the analysis and decoding of cultural traces). Qualitative methods are therefore effective at accessing not only the language of the word, but also the language of the world. This chapter continues by outlining in more detail specific methodologies that can do just this.

Methodologies and the language of the word: talking about and reading cultural traces

As Crang (2003) identifies, in recent decades the most popular methods in cultural geography have been those that attempt to access the language of the word. The popularity of the verbal and written word in methodology is a direct consequence of the dominance of the representational approach to cultural geography. As outlined in Chapter 3, this approach focuses on the meanings inscribed by cultural groups on a whole manner of traces, investigating how these are produced, transgressed, and resisted. It assumes that meaning is constructed by a self-conscious, intentional process, resulting in rationally thought-through comments, texts, and material products. As a consequence, through the correct method traces can be decoded by the astute scholar into a coherent understanding of

the cultural geographies out there. The specific methodologies employed to access the language of the word include *interviews* and the analysis of cultural 'texts'.

As Ackroyd and Hughes define, interviews are 'encounters between a researcher and a respondent in which the latter is asked a series of questions relevant to the subject of research' (1983: 66). In many senses, therefore, interviews are a variation of 'normal' interactions between people, they are dialogues that transfer knowledge and information, predominantly through verbal exchange. Interviews can range from a systematic interrogation akin to a cross-examination by authorities, through to informal 'conversations with a purpose' (after Eyles, 1988). In whatever form they take, interviews seek to discern the experiences, thoughts, and feelings of those under study.

Interviews are therefore primarily used to access what people *say* about their cultural geography. They are useful in finding out how people interpret their worlds, the meanings they give to cultural traces either through reasoned analysis or less coherent ramblings. Interviews are self-reports of experiences, opinions, and feelings, and the success of this method relies not only on the individual's ability to comment on their own culture, but also the interviewer's ability to ask the right questions to access this knowledge. As Longhurst (2003) identifies, it is important that the researcher enjoys a degree of literacy on the topic under investigation before conducting an interview. Although this may seem counter-intuitive – you are asking questions of an other because you *don't* know about their culture – background literacy is vital so questions are not formulated that may prove embarrassing (to yourself or your respondent) or waste time (for example, are you asking the right questions to the right person? Do you *need* to ask them questions – can the information required be found through a website or other data source?).

From this background literacy a range of themes can be generated to use in your interview.

These themes can be employed in a number of ways. You may wish to fashion a range of specific questions to guide the interview from a beginning, through a middle, to its end. This *'structured'* interview format (after May, 1993) is often useful for those who may be inexperienced in or nervous about conducting interviews, providing for them a path to follow throughout the conversation's duration. This format is also useful if multiple interviews are going to be carried out and the results systematically compared (in the style perhaps of a mass-produced survey). In many cases, however, the structured format precludes the realisation of the greatest strengths of the interview method: the opportunity to build rapport between researcher and respondent, and the discovery of the unexpected through improvised conversation. In order to realise these benefits it is often the case that a more semi-structured approach is adopted.

Semi-structured interviews use the general themes obtained through background literacy to identify key questions or areas that the interview wishes to cover. From this, only a small number of questions will be asked in a strictly chronological order, perhaps questions that seek to establish rapport and open conversation between the involved parties. From these, the interviewer follows the responses in a more improvised way, endeavouring to gain all the information sought, but not necessarily in the order envisaged at the outset. In this way, semi-structured interviews allow the potential for a much broader discussion to occur than the researcher may have anticipated, with new ideas and avenues for exploration being discovered along the way (it also gives the interviewer the opportunity to jettison questions as they prove redundant or inappropriate). The semi-structured approach therefore requires a degree of confidence on the part of the interviewer, but through establishing rapport through conversation this format often results in rich and highly detailed accounts of individuals' experience and life.

The third major variety of research interview is the *unstructured interview*. These dialogues

have a general aim – perhaps to talk about a life history, a campaign, or a culture – but have no specific questions set and are fully improvisational. These interviews are often selected when the interviewer and interviewee already enjoy a degree of rapport so that trust and conversation are easily generated and maintained. In many cases such interviews will be conducted following an extended period of ethnography (see below, and, for examples, see Anderson, 2004a). In whatever form employed – be it structured, semi-, or un-structured – interviews generate rich, detailed, and textured insight into cultural geographies. They can be tailored to take into account the positioning of the individuals concerned, and thus make it possible for them to use their own words to give accounts of their experiences. These accounts depend on a good rapport being established between researcher and respondent, the use of appropriate questions, but also the choice of an appropriate location for the interview.

Place and context in interviews

We have seen throughout this book that place matters. In this chapter more specifically we have seen that our own positionality in relation to method matters too – our metaphorical place influences how we connect to our research respondents. However, place also matters for methodology in a material sense as well. As every place is subject to different orders and borders, interview participants are subject to varying senses of belonging, comfort, or ease in a range of different sites. The choice of location of an interview can therefore make a crucial difference to the effectiveness of methodology. Consider, for example, the quality of information you may gain access to if interviewing women in a male-dominated context, or homosexuals in a homophobic environment. Will your respondents feel like sharing their accounts when they feel threatened or out of place, regardless of your sensitive questioning? The place of interview should therefore be chosen for its capacity to put your respondents at ease and to ensure both your and their safety. The location may even be chosen for its ability to stimulate and prompt discussion about the issues you are interested in. For example, talking to environmental campaigners about their relationship to the non-human world may be more effective in a wild location rather than an office or living room; asking residents of Belfast to walk down their own street and comment on its murals may be effective in raising insights about their sense of belonging. Conversely, it may be inappropriate to talk to an alcoholic about their cultural geographies in a bar or public house. The place of method therefore deserves as much reflexive consideration as issues of access or recruitment.

The potential of interviews can therefore be realised through being sensitive to the geographical context in which they are conducted. But context also needs to be taken into account in terms of the type of responses given. In the first instance this involves listening. Although interviews directly involve asking questions, their potential is only realised if answers are produced. Listening to responses and judging their meaning are key to conducting an effective interview. Are responses ironic or satirical? Are they meant as a joke, or are they deadly serious? Being sensitive to these responses allows the interviewer to judge the type of question that may be appropriate in following up this exchange. Does the response require clarification, does it suggest another question, or does it say it all? Of course, the context of responses (be they the geographical place of the interview, or the meanings exchanged during dialogue) may be clear at the time the research is undertaken, but may be lost when the notes of the interview or its recording (for example through audio-digital capture) are played back. How will you ensure that the influences of this geographical location, or the nuances of language understood in conversation, will not be lost when they are transcribed onto a two-dimensional screen or into an essay?

The analysis of cultural texts

The transcription of interviews, along with a detailed account of the contexts in which these research techniques are conducted, is one example of a variety of texts that can be used to access the language of the 'word'. Field diaries, policy documents, graffiti, web-pages, campaign material, even documentaries or films can also be seen as texts that can be read to understand cultural geographies. All these texts, in a similar way to other material traces outlined in this book (e.g. statues, buildings, parks, and places in general), can be interpreted and decoded to gain insight into different cultures. As Cosgrove cites,

> the kind of evidence that [we] now use for interpreting the symbolism of cultural landscapes is much broader than it has been in the past. Material evidence in the field and cartographic, oral, archival and other documentary sources all remain valuable. But often we find the evidence of cultural products themselves – paintings, poems, novels, folk tales, music, film and song can provide as firm a handle on the meanings that places and landscapes possess, express and evoke as do more conventional 'factual' sources.
>
> (in Gregory and Walford, 1989: 127)

In this way, art, books, media, and museum archives can be used to interpret the worlds of exotic or mundane places. Even films and adverts can be employed to get at cultural understandings (as Aitkin points out, in recent years the prospect of studying cultural geography 'in armchair with popcorn' has become as popular as fieldwork study in the traditional sense (1997: 197)). One 'key text' that cultural geographers have utilised with increasingly popularity is the photographic image. As Lombard (2013: 23) notes, geography has always been a visual culture, and knowing the world has often been about 'seeing the world' (after Rose, 1996: 281). With this in mind taking a photographic record of a cultural activity under study, and more particularly asking our respondents to take photographs of their practices, is a key way of seeing the world through the eyes of a particular culture. More than simply a way of providing a visual record of a particular event or place, photographs can be used to encourage or 'push' reflection and thinking (after Pyyry, 2013: 2). Used alongside a semi-structured interview, looking through photos may elicit thought and comments by both researcher and respondent, and allow a form of interpretation, explanation, and meaning-making that may not have occurred without the photographic prompts. Respondents can be asked why they took the photos they did, if the process was practical in their cultural places, what the act of photo-taking felt like. In this way, a 'photo-voice' can be given to respondents (after Warren, 2005, and Scarles, 2010), and light can be shed on the cultural context of the chosen activities. In these ways, as Radley and Taylor state, photographs offers a 'more powerful tool for eliciting (respondents) experiences than would an interview alone' (2003: 79).

As we have seen throughout this book, there are many ways in which images, and texts of all types, can be read by different groups. It is vital, therefore, to take into account the context (both geographical and cultural) that existed when these texts were produced. How were meanings conveyed in these contexts, and what was their significance? It is only through critically reflecting on the values of, for example, a swastika or a cross in different cultures in different places, that we can appropriately read these cultural texts. As Jackson points out, 'It is an impoverished view of culture that stresses text, sign or discourse to the exclusion of context, action and structure. Meanings must always be related to the material world from which they derive' (1989: 195).

With this in mind, Doel (2003: 509) outlines five key questions that can readily be used to critically reflect on any cultural text:

1. How, why and for whom has the text been constructed?
2. What codes, values, dispositions, habits, stereotypes, and associations does it draw upon?
3. What kind of personal and group identities does it promote? And how do they relate to other identities?
4. More importantly, what kind of work does it do? And who benefits?
5. How might it be modified, transformed, or deconstructed? How could this social space be inhabited differently?

Doel's final question brings to mind the range of ways in which different cultural texts and traces can be interpreted by different groups. Are these traces seen as natural, normal, or novel (see Chapter 5)? In what ways are they recuperated, commodified, or transgressed? What strategies are adopted to resist them, and what trace-chains are set in motion as a consequence (see Figure 14.1)?

Summary

The use of interviews and the analysis of cultural texts are key ways in which cultural geographers attempt to access the language of the word. As we have seen in Chapter 3, these representational approaches tend to draw our attention towards rationally thought-through explanations and intentionally constructed traces. Research becomes focused on individuals' considered responses to questioning, and on traces that are planned, edited, and modified before their completion. This research approach remains vital as it sheds light on key aspects of places and traces. However, it is also partial as it tends to draw attention away from the everyday impulses, habitual practices, and un-thought-through actions that go together to make up our world. There is, therefore, another language which produces our cultural geographies. This other language is hinted at in interviews where communication before or beyond the verbal occurs, perhaps through 'the blink of the eye and the shrug of a shoulder' (Herbert, 2000: 554). This other communication is the language of the world and a different set of methods have been adopted by cultural geographers in order to access it.

Figure 14.1 Interpret these traces

What do these signs in Figure 14.1 mean to you?

Source: Author

Methodologies and the language of the world: ethnography and embodied practice

There is an urgent need to supplement the familiar repertoire of . . . methods that rely on generating talk and text with experimental practices that amplify other sensory, bodily and affective registers.

(Whatmore, 2006: 606–607)

Up until the turn of the century many research projects tended to concentrate on the language of the word. As Latham points out, for this form of study 'what really mattered was talk' (2003: 1999). However, as this book has argued, cultural traces are not simply associated with words or things. Traces also include the complexities of cultural practice, the feelings that we have before or beyond verbal communication, the blushes, goosebumps, and shivers, as well as the emotions, impulses, and movements that affect us every day. Traces are not only generated by thought-through and intentioned activity, but also by mundane and habitual acts whose conscious motivation may have been forgotten, if it ever existed in the first place. By using methods that seek to access only the language of the word these other languages are either ignored, silenced, or (mis)translated into a form of communication that often does not do them justice. Interest in languages before and beyond the representational has thus led to other research methods becoming popular within cultural geography. A key method used is ethnography.

Ethnography is a research method that seeks to not only explore what people *say*, but also what they *do*. It seeks to investigate the constitution and preferences of a cultural group through engaging with their practices, emotions, and senses of place. As Cook and Crang (1995: 4) state, 'The basic purpose in using these methods is to understand parts of the world as they are experienced and understood in the everyday lives of people who actually "live them out".'

Thus ethnography is employed to 'understand the world and ways of life of actual people from the "inside", in the contexts of everyday, lived experiences' (Cook, in Flowerdew and Martin, 1997: 127). In order to appropriately understand these cultural experiences ethnography involves a number of key elements:

In its most characteristic form it involves the ethnographer participating, overtly or covertly, in people's daily lives for an extended period of time, watching what happens, listening to what is said, asking questions – in fact, collecting whatever data are available to throw light on the issues that are the focus of the research.

(Hammersley and Atkinson, 1995: 1)

As Hammersley and Atkinson outline, ethnography involves participation in the cultural geographies under study. As this chapter has already acknowledged, we are all involved in the cultural world; however, through ethnography, our involvement in our respondent's world is explicit and open. Ethnographers gain their insight through becoming part of the cultural group under study, spending time in their places and becoming part of their world. Through sharing in their life, participation allows ethnographers to acquire a 'native competence' about the culture they are interested in (after Collins, in Bell and Roberts, 1984) and through this participation comes comprehension. As Herbert outlines, 'ethnography provides unreplicable insight into the processes and meanings that sustain and motivate social groups' (2000: 550). As noted above, ethnography can be undertaken overtly or covertly (notwithstanding the ethical dilemmas involved in each approach, see above). Ethnography also requires an extended period of time for research, especially when compared to interviews or the analysis of cultural traces. Getting under the skin of a culture, understanding its meanings, practices, and unstated rituals cannot be achieved in an afternoon. Although there is no minimum limit – time lengths will depend on the researcher's 'literacy' of the culture and the representativeness of the time within it – ethnography often involves extended commitment (see Anderson, 2002).

As Hammersley and Atkinson suggest, ethnography involves both participation and observation. In some cases ethnographers will adopt a 'complete participant' position (where researchers get involved in all aspects of the cultural geographies being investigated, living their life from the inside) or a 'complete observer' position (for example as a legal observer at a rally or demonstration) (after Gold, cited in Burgess, 1984: 80). Yet it is more common for researcher to adopt a mixed position, both participating and observing during their study. As Cook explains, ethnography commonly involves:

> researchers moving between participating in a community – by deliberately immersing themselves in its everyday rhythms and routines, developing relationships with people who can show and tell them what is 'going on' there, and writing accounts of how these relationships developed and what was learned from them – and observing a community – by sitting back and watching activities which unfold in front of their eyes, recording their impressions of these activities in field notes, tallies, drawings, photographs and other forms of material evidence.
>
> (Cook, in Flowerdew and Martin, 1997: 127)

Ethnography thus involves both participation and observation in order to understand a culture. However, because research is the product of the *interrelation* between the academy and the world at large, this method often presents a number of difficulties for the researcher. When doing ethnography, researchers live across the (b)orders between the academy and their study places. Although they remain part of the academic community, they are distanced from it

due to a new positioning in their ethnographic world. Similarly, although they are part of the group under study, their observation and questioning of it means they are always somehow *dislocated* from this world. Ethnography therefore places the individual researcher in a difficult position: which side are they really on? Will they 'go native', leaving their research project to become a permanent part of the culture under study? What responsibilities do they have to their relationships in each place? How possible and desirable is it to leave a new world when the research project is complete?

Doing ethnography therefore makes explicit the emotional, personal, and very real attachments research creates between it and the world at large. These attachments become embodied within individual researchers, their own fully human nature making the research project both an intellectual as well as an emotional event. Ethnography emphasises that scholars are not dispassionate and disembodied, rather, as McDowell notes, they are 'touched by relations of power [and] feelings of jealousy, envy, doubt, love, [and] possession [amongst others]' (1994: 241). These embodied emotions cannot be hidden or ignored by the ethnographer, rather they are used and harnessed to give new insight into what the experience of that cultural world is like.

Moving methodologies

With cultural geographers increasingly sensitive to issues of mobility (see Chapter 8), it is vital that methods are also capable of capturing and understanding this aspect of the cultural world. As Büscher and Urry confirm, 'in order to deal with [a world of mobility], it is necessary to develop research methods that are "on the move"' (2009: 103–104). This need to capture mobility has led to geographers using ethnography to 'follow' their object of study (after Marcus, 1995) and 'capture, track, simulate, mimic, parallel and "go along with" the kinds of moving systems and experiences that seem to characterise the contemporary world' (Büscher et al., 2011: 7, cited in Merriman, 2014: 168). Tracking mobility has been undertaken in a number of ways. For example, geographers have begun doing interviews 'on the move' through processes such as 'talking whilst walking', allowing respondents, interviews, and the places through which they are mobile to all play a role in prompting conversation (see Anderson, 2004b; and Reed, 2002). Mobile ethnography is also possible through methods such as the 'go along' (see Kusenbach, 2003). Anderson and Jones (2009) demonstrate how moving with youth groups as they bike and hike across the city enables useful insights into how 'neo-tribes' take and make different places. In some cases, however, it is not always possible to 'follow' in real time the person or research encounter you are interested in. In these cases, modern technologies such as GPS-tracking or film recording can be used to gain a mediated experience of a particular cultural activity. It is possible, for example, to mount a small digital camera on a bicycle or car in order to gain a different understanding of the nature of mobility (or immobility) in urban traffic. In such cases, 'mobile video ethnography can contribute to research . . . as a way of "feeling there" when you can't be there . . . as a way of apprehending fleeting moments of mobile experience; and . . . as a tool to extend sensory vocabularies' (Spinney, 2011: 161). These technologies not only help to depict and describe acts of mobility in a similar way to photo elicitation (see above), they can also 'replay' respondents' experiences, as well as 'pause' moments that can be followed up in interviews later. In this way, as Spinney attests, 'video can help participants articulate their experiences of fleeting sensations because it allows them to be slowed down and replayed in a non-linear manner' (2011: 171).

Field diaries

In this way, the use of film can help to capture not only an increasingly mobile world, but also the languages of that world that are not limited to words or talk. Films can therefore help us to gain a different insight into the language of the world as they are 'multisensorial, capturing sound, image, movement, gesture, time and place' (Garrett, 2012: 531). Another way of capturing these multisensorial experiences is through a written or filmed 'field diary'. As Cook and Crang (1995: 29) outline, field diaries are

> some kind of record to how the research progresses, day by day, chart[ing] how the researcher comes to certain (mis)understandings. Diaries should represent the doubts, fears, concerns, feelings, and so on that the researcher has at all stages of her/his work. How your understandings are affected by particular perspectives; your developing positionality in the community; power relations which can be discerned in this, how your expectations and motives are played out as the research progresses; what you divulge, and why and to whom and how they appear to react to this; how various aspects of the research encounter make you 'feel' (e.g. swings) and how this affects what you do; what you dream about; what rumours have come back to you about yourself and the reasons for your presence in the community.

Field diaries are often used to keep an account of experiences within the cultural geographies under study. From writing scrambled notes on scraps of paper, through to longer, more reflective pieces, the field diary comes to capture the flavour and sense of being in a cultural place. When reading (or watching) it back, it offers a window into that three-dimensional context, offering a portal to the resonant emotions conjured up in that place. In this way, ethnography comes close to auto-ethnography – using the researcher's own experiences and life in the place of the other to shed light on this world. Such auto-ethnography is not insular and narcissistic, but explicitly acknowledges the expanded field of academic enquiry, and the researcher's inevitable positioning within it (see Okely and Callaway, 1992; Ellis, 2004). As Cook outlines, auto-ethnography is not about the researcher themselves, but about the interrelations between worlds that we all occupy,

> What have you 'learned' from reading it [my diary]? Not much about 'me', I hope. It's not a me-me-me-me-me-me-me-type narrative. Is it? I think it's an it-me-them-you-here-me-that-you-there-her-us-then-so- . . . narrative. It's an 'expanded field' thing. And you're in it too. Aren't you?
>
> (Cook, 2001: 120)

Thus mobile ethnographies and field diaries can harness the researcher's own experiences in order to understand cultural geographies, but it is not only their experience that is sought through these methods. As in the case of photo elicitation, field diaries can be written or filmed by respondents to access both their impulsive and reflective responses to experiences in cultural places. Through the use of these ethnographic and embodied techniques, glimpses of the language of the world are brought into geographic inquiry. The thought-through is combined with the impulsive and intuitive, and a bricolage of languages is employed to understand the lived experience of places. By attempting to access this lived experience, these methods seek to convey not just what it *means* to live within particular (b)orders, but also what it is *like* to live there. In this way, cultural geography can include our bodies in our writing and responses; it can engage not only our mind, but perhaps also our heart and soul too.

Communicating the field through film

We have seen how film technology can be used to 'follow' mobile cultures. However, many geographers are also using this technology alongside the written and spoken word to not only record the moment of their research encounter, but also convey its insights (see Anderson, 2013a, and for some short examples, www.placehacking.co.uk/documentaries/; http://alphaecho.co.uk/soundspace-1.html; and www.spatialmanifesto.com/teaching-projects/culture-place-space). As digital video technology and software have become widely available, there are new opportunities to use film to open up the range of voices that can be heard in our research, as well as the range of audiences who wish to hear them (see Fuller, 2008; Fuller and Askins, 2010; Ward, 2006). Using film offers the possibility to integrate lay voices, informal languages, music, and emotive commentary into a world of understanding that is often quite different to that produced in traditional forms of academic reportage. As Youngs notes, 'film . . . can provide data on at least two levels of consciousness – tangible and intangible . . . personally, it is the latter that is most significant' (in Garrett, 2012: 533). Here Youngs refers to the capacity of film to resonantly affect its audience – to move them, inspire them, or mobilise them politically. These aims re-emphasise the key role of cultural geography – not simply to record or document events, but to engage our audience and prompt them to take a position. Using film, or indeed using any creative media, to communicate can help cultural geographers to articulate the world as they, and their respondents, see it, articulating in intellectual and emotional ways what has been understood. As Denison and Markula state, 'we [shouldn't] be afraid to depart from traditional practices of representation and experiment more freely with form, content, and style' (2003: 9). In this situation, what means and methods of communication will you use to convey your research?

In practice, methodologies that seek to access the language of the word and those that explore the language of the world often come together in *case studies*. Case studies are empirical enquiries focusing on one particular place and its (b)orders. These cases are not meant to be representative of broader phenomena, rather illustrative of the practices and traces that come together to constitute the cultural world. It is in case studies that the *primary* source of the *world* is brought together with the *secondary* source of the *word*. In this way the research questions set are explored 'with the widest array of conceptual and methodological tools that we possess and they demand' (after Trow, in Burgess, 1984: 163). In case studies, interviews, 'textual' analysis, and ethnography are often combined. The outcomes of different methods are *triangulated,* they are compared and contrasted in order to reinforce or contest a particular perspective, response or meaning (after Denzin, 1970).

Conclusion

This combination of methodologies that seek to access the language of the word and the language of the world refocus our attention on the inevitable fusion of the representational and the non- or more-than-representational (see Chapter 3). Our research will always rely on insights that are before or beyond the spoken and written word, as well as the reflective and thought-through responses on these impulses and experiences. As our projects are a product of the *interrelation* between the academy and the world of our respondents, it is appropriate that both parties' responses are integrated into our write-ups (or presentations, films, artwork, etc.). However partial and momentary, the interpretative claims of respondents and researchers are equally important in order to ensure as broad and transparent a view as possible on the cultural geographies under study. As outlined in Chapter 3, our work as cultural geographers and our

lives as people are more than simply experiencing and describing events. Our purpose must be about sharing what these events mean to us on an individual level, finding out what others say they mean to them, and allowing our audience to take in these positions and work out their own impulsive and thought-through opinions on them. Such critical engagement with the world and its traces is the aim of our cultural geography approach to place (see Routledge, 1996).

Suggested reading

The following texts are useful to further explore the issues raised by this chapter. There are some excellent textbooks on qualitative methods, both in general (e.g. Kitchin and Tate) and for cultural geography in particular (e.g. Shurmer-Smith). Key books and journal papers elaborate on specific methods in detail.

Books

Clifford, N. and Valentine, G. (eds) (2003) *Key Methods in Geography.* Sage: London.

Denison, J. and Markula, P. (2003) *Moving Writing. Crafting Movement in Sports Research.* Peter Lang: New York.

Denzin, N. and Lincoln, Y. (eds) (2000) *The Handbook of Qualitative Research.* Sage: Cambridge.

Ellis, C. (2004) *The Ethnographic I. A Methodological Novel about Ethnography.* Alta Mira: Walnut Creek.

Flowerdew, R. and Martin, D. (1997) *Methods in Human Geography. A Guide for Students Doing a Research Project.* Longman: Harlow.

Hammersley, M. and Atkinson, P. (1995) *Ethnography: Principles into Practice.* Routledge: London.

Kitchin, R. and Tate, N. (2000) *Conducting Research in Human Geography: Theory, Methodology and Practice.* Prentice Hall: Harlow.

Pink, S. (2001) *Doing Visual Ethnography.* Sage: London.

Pink, S. (2006) *The Future of Visual Anthropology: Engaging the Senses.* Routledge: London.

Shurmer-Smith, P. (ed.) (2002) *Doing Cultural Geography.* Sage: London.

Papers

Anderson, J. (2013a) Active learning through student film: a case study of cultural geography. *Journal of Geography in Higher Education,* 37 (3), 385–398.

Anderson, J. and Jones, K. (2009) The difference that place makes to methodology: uncovering the 'lived space' of young people's spatial practices. *Children's Geographies,* 7 (3), 291–303.

Crang, M. (2003) Qualitative methods: touchy, feely, looksee? *Progress in Human Geography,* 27 (4), 494–504.

Garrett, B. (2012) Videographic geographies: using digital video for geographic research. *Progress in Human Geography,* 35 (4), 521–541.

Herbert, S. (2000) For ethnography. *Progress in Human Geography,* 24 (4), 550–568.

PART 7 CONCLUSIONS

A cultural geography approach to place

Towards a conclusion

This book has introduced how cultural geography approaches place. In this final chapter we reiterate the key premise on which this approach is built – how places can be understood as an ongoing composition of traces. We then outline one example, the case of the World Trade Center, New York, USA, to illustrate this argument for a last time.

The cultural geography approach to place adopted by this book is built on the premise that you and I are in the cultural world. We live in it, we survive it, and we contribute to it, in all our actions. Cultural geography is interested in how and why we contribute to this cultural world in the way we do. It is interested in the whole range of human practices that comes together to create the cultural world. As culture is what humans do, cultural geography focuses on all the things, ideas, practices, and emotions that constitute culture. In this light culture can include all aspects of society, politics, and the economy, and can be categorised by a range of different (and sometimes overlapping) activities. Culture can be ethnic, youth, or straight; capitalist, environmental, or national; through our own actions, culture can also come to be new and exciting combinations of these broad categories. Whatever culture we focus on, this book argues that all cultural activity has vital and inevitable geographical consequences. As we have seen in Chapter 1, our cultural life '*does not take place in a vacuum*'. Thus a key component of our cultural actions is that they *take place*. Straightforwardly, our actions take place in the sense that they occurred, they happened; but more importantly they also take place in a geographical sense too. Our actions aggressively, passively, intentionally, or otherwise, take and make places. Through doing so our actions come to define and claim place in small but significant ways. As actors in the cultural world we go about taking and making place through our everyday actions. Although we may say that practice makes perfect, we may also say that our practices make and take place.

Our actions and practices that we make in the cultural world, and their consequences, this book understands as traces. These traces are most commonly considered as *material* in nature (material traces include 'things' such as buildings, signs, statues, graffiti, i.e. discernible marks on physical surroundings), but they can also be *non-material* (including activities, events, performances, or emotions). Cultures therefore leave visible traces in places, but we can also sense our traces in other ways (we can hear them, smell them, even taste them or feel them). Cultural traces can therefore be durable due to their solidity and substance as things, whilst they can also leave indelible marks on our memory or mind.

Due to the constant dynamism of cultural life, places and their associated traces are not static in nature. Rather they are lively, vibrant, and open to change. Traces may originate from cultural activity occurring in our proximate locale – from actions occurring in our neighbourhood, town, or city, but traces may also derive from cultural activity occurring elsewhere – from decisions made in places on the other side of the globe. Traces may also be contemporary – from actions in the present – but they may also spring up from places-past as historical activities are remembered, re-enacted, or revived in the here and now. Coupled to this, human traces may also interact with those generated by non-humans. Animals, weather events, viruses, and climate changes all come together to affect the form, shape, and structure of the ongoing composition that is place. Cultural places, therefore, are not straightforward or easy to read processes. They are at once unique, but they are defined by trace-chains that connect the contemporary to the past, the local to the global, and the human to the non-human.

Thus through a variety of traces, places are taken and made. However not all traces affect place to the same extent. Some traces have the ability to transform the identity and definition of place in important ways, whilst some do not. Those traces that can transform place are known as *dominating traces* and these enjoy the power to impose their own ideas, value systems, and beliefs onto places – they have the ability to impose their own *cultural order*. This cultural order transforms a complex, chaotic, and up for grabs reality into a place that stands for particular ideas, authorities, and identities. It is this process that generates a recognisable culture, it is through these acts of dominating power that groups publicise, perpetuate, and strengthen their identity. Of course, there is a reciprocal relationship between cultural and geographical action: cultural practice inevitably involves geographical consequences, and vice versa. In this way, cultural ordering goes hand in hand with *geographical bordering*. It is in physical places that cultures are made manifest, with walls and wires, fences and frontiers, marking out the extent of

a specific cultural realm. These *(b)ordering* processes mark out the organisation and limits of a place, as well as the cultural group who claim it. It marks out the geographical *here* and *there*, the cultural *them* and *us*. It defines who and what belongs in this location, which groups enjoys a sense of place, and, most importantly, which groups do not.

A cultural geography approach to place thus investigates the relationships between people and places. It looks at how the ideas and prejudices about a place define its (b)orders, and who and what is granted belonging as a consequence. It investigates how these politics and prejudices mark out some traces as 'normal' or 'natural', and how others are deemed as 'novel' and threatening. It analyses how different cultural groups use their power to *transgress* or *resist* these (b)orders, and attempt to get their own meanings and interpretations of traces accepted into the dominant reading of a place. A cultural geography approach sees transgression as destabilising (b)orders, as actions that turn lines into question marks. As (b)order crossings occur, conventional readings of place become disrupted. Whose place is this now, following these transgressions? Is it (still) ours? Do we belong? How should we behave? In many instances new traces change the identity of place so much so that it seems strange to us. This strangeness may affect us and our reactions to it – we may feel threatened, liberated, or excited; we may seek to perpetuate the revolution, or re-establish old (b)orders. In part, therefore, cultural geography is about locating ourselves both materially and metaphorically in the ongoing composition of place.

Places are thus the medium and the message of cultural life. They are where cultures, communities, and people root themselves and give themselves definition. Places then are saturated with cultural meaning. They are crucial for understanding who we are and where we fit in to the culture and geography of our lives. Cultural geographers analyse and interrogate all the agents, activities, ideas, and contexts that combine together to leave traces in places. A cultural geography approach to place swims in the context of places, exploring their traces and how these come to be constructed as (b)orders. A cultural geography approach is representational in nature – it looks at what traces and place stand for, but it is also more-than-representational – it explores the emotional affects of these traces and (b)orders on our sense of place. It looks at the materialities and meanings, the artefacts and affects, and the identities and intensities that come to define places as ongoing compositions of traces.

A cultural geography approach to place: identity and intensity

Case: The World Trade Center, New York City, New York

When I first started working in the World Trade Center, my first day here I was like 'Wow!' I was overwhelmed, I was like, 'This is really New York City, this is really what Wall Street is about!'

(Joanne Capestro, 87th Floor, North Tower, in Armitage, 2006)

Figure 15.1 The World Trade Center

Source: Shayna Marchese www.flickr.com/photos/87538314@N00/218783201/

New York. Skyscrapers. No-one ever had to ask what they stood for. They stood for New York. They stood for America. 'Build me the biggest towers in the world' the architect of the World Trade Center was told in the 1960s, and he did. And they stood for America. That's why they were destroyed.

(Narrator, Sim, 2004)

This was New York City's darkest hour. The attacks on the World Trade Center on September 11th 2001 were a direct attack on the two biggest symbols of the capitalist system.

(Stiglitz, 2005)

We turn to the case of the World Trade Center (see Figure 15.1) to outline our cultural geography approach to place for a final time. The above quotations outline three different but interrelated ways in which a cultural place – in this case the World Trade Center, New York City – is constituted by an ongoing composition of traces. In the first quotation, Joanne Capestro, an office worker who worked on the 87th floor of the North Tower, explains what the World Trade Center means to her. To her it represents Wall Street, the financial heart of New York City. To her this building, in this street, embodies what New York as a city means – it means money, and this means power. She is overwhelmed by the fact that she is working in this place, it means she is living her (American) dream.

In the second quotation, the idea that the World Trade Center can be read as a defining symbol of New York City is reinforced. Indeed, this quotation suggests that this building was constructed with this specific intention, to become an icon that represents the power and authority of the city across the skyline. It also suggests that this symbolic meaning was not limited to those within the American nation, but also had resonance far beyond its shores. Indeed the success of this symbol and its resonance across the globe had an unforeseen and

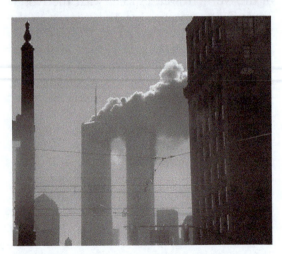

Figure 15.2 9/11: the attacks on the World Trade Center

Source: Macten www.flickr.com/photos/ macten/237364092/in/set-72157594274390075/

tragic consequence. The representational value of the twin towers made them a target for attack (see Figure 15.2).

The third quotation, from Joseph Stiglitz, a Nobel Prize winning economist, refers directly to this attack. For Stiglitz, the World Trade Center stood not simply as a symbol of New York, but of America. An America that embodied the values and objectives of one form of cultural organisation in particular: the culture of capitalism. For him, it was the values of capitalism that were attacked as the building was hit by two hijacked passenger jets on 11 September 2001.

In one sense, therefore, the World Trade Center was just a building. But for cultural geographers it's not *just* a building. It's a building that represented a street, a city, a country, and an economic mode of organisation, in sum: the World Trade Center was a composition of cultural traces. In one sense it was a material trace, made of bricks, mortar, and glass. In another it was an amalgam of non-material traces, of dreams and ambitions, of hopes and greed. The World Trade Center was constituted by traces that derive in the local area, from its workers that commuted to work up in the sky, but also by traces that came from elsewhere – it was the *world* trade centre after all, one node in a network of trace-chains that inspired both celebration of and resistance against its dominant (b)orders. As the above quotations suggest, the dominant cultural ordering of the World Trade Center was capitalist in nature. It became an icon of corporate capitalism that, as we have seen in Chapter 7, is both culturally and geographically transformative to many other cultures across the globe. The geographical and cultural hegemony of capitalism means that it is often seen as the best and natural way to (b)order the world. However, there are many ways to read all cultural traces, and although capitalism's ability to transform is often dominant, it is never absolute. The attack on the twin towers illustrates this point in a tragic and profound way: to some capitalism is liberating and offers opportunity, to others it is exploitative and offers invasion. The ongoing composition of the site of the World Trade Center will always have these dominating and resisting traces bound up within it.

> Ground Zero. On September 10th, this was 16 acres of real estate. A day later, it was sacred ground. Sacred to the people who were killed here. Sacred to an idea of America that was attacked. Nearly 3000 people had been killed because they went to work in an icon . . . Icons had become dangerous. After 9/11 how would you go about building another one?
>
> (Narrator, in Sim, 2004)

By taking a cultural geography approach to place we know that we are part of the cultural world. As a consequence we know that our traces take and make place, and these traces come to define both ourselves and our place – both materially and metaphorically. With this in mind, thinking how places could be, and should be, becomes a difficult and responsible project. What places do we want our actions to produce? What do we want our places to stand for, and what do we want them to say (about us)? In the case of the World Trade Center, on Ground Zero, a new place could be born. Ellie Hartz, who lost her husband John in the attacks, wanted this new birth to make her feel that 'we triumphed' over 9/11 (in Armitage, 2006). Hartz doesn't make it clear if, by this 'we', she means her family, her city, her culture, or humankind as a whole. Ed Hayes, lawyer to the architect hired to jointly design the future of Ground Zero, wants a place that resists the values of corporate capitalism, as he puts it:

> I don't want my mother walking down the street dwarfed by a tower that is a monument to corporate interests. Look: they are going to work and there is a guy at the top of that tower who they work for who doesn't give a goddamn whether they live or die. And

these towers *cannot* reflect that theory of life. They have to reflect something that values my mother as much as the guy that's at the top of the tower.

> (in Sim, 2004)

How would the new place of Ground Zero be ordered and bordered? What cultures would it unite and which communities would it divide? What cultures would find a home and place of belonging at this site, and which people and activities would be left dislocated and out of place? If you had dominating power, what would you do?

> Every building bears the scars of not only the emotional trauma that is so engraved into these few city blocks, but of the battles that have been fought since 11 September 2001 to agree on anything being built at all. An untold amount of passion, energy and care has gone into the creation of every single piece of the 9/11 . . . complex – but these faceted forms are buffeted not just by the anguish and loss of that day, but by the compromise of competing interests that have waged war across the site ever since.
>
> (Wainwright, 2014)

In practice, the taking and making of Ground Zero illustrates the key assumptions underpinning our cultural geography approach to place. The place of Ground Zero was designed by two 'trace-makers', Polish architect Daniel Liebeskind – winner of an international competition to plan the future of the site – and David Childs, an architect appointed by Larry Silverstein, owner of the site. The different values and ideas of these two architects resulted in the designs for Ground Zero being keenly contested.

Liebeskind, an immigrant who sailed into New York harbour as a child past the inspiring vision of the Statue of Liberty, wanted Ground Zero to be a place where a story of freedom and hope could be told. Liebeskind wanted a tower

BOX 15.1 THE FREEDOM TOWER, ONE WORLD TRADE CENTER

How – and to whom – shall it speak?

(Watts, 2009: 415)

Marcus Robinson was an artist commissioned to record the construction of the new World Trade Center. In his words, the architects, designers, and construction workers are 'healing a scar in the bedrock of the city, in its skyline, and in many ways what they are doing is part of a much greater act of rebuilding and healing' (2013). You can see Robinson's record of this ongoing composition here: http://m.guardian.co.uk/world/video/2013/may/11/one-world-trade-centre-build-time-lapse-video. Healing this 'scar' in the city of New York entailed designing and re-building a city icon, but also perhaps a national identity, and a message to the world (in the same way that Trafalgar Square did in London in the eighteenth century). What is this new identity, and this message to the future? How might we read this message in 200 years time? As Watts (2009: 416) asks, 'How will the site speak across time to those who wish to situate themselves in relation to the tragedy? How may we infuse the site with features that inspire reflection, dialogue, and action?'

The Freedom Tower

Source: Kimberley Peters

Source: 2005 Skidmore, Owings & Merrill, LLP/ dbox www.renewnyc.com/plan_des_dev/wtc_site/ new_design_plans/Freedom_Tower/default.asp to echo the shape of the Statue of Liberty and sit in complement with the established skyline of the city. In his words,

> The sky [should] become home once again to a towering spire that speaks to our vitality in the face of danger and our optimism in the aftermath of tragedy. Man or woman doesn't live by function alone. There has to be something that raises our hearts and eyes above all the prosaic daily needs.
>
> (in Sim, 2004)

Like Hartz, Liebeskind wasn't specific about which cultures should feel optimism as a result of his designs, and which places would be revitalised in the face of danger. Childs however had different views. Alongside Silverstein, Childs wanted to maximise the rental area of the new site, which occupied a prime location in downtown New York. He also wanted to build as high a building as possible to facilitate satellite and aerial communication. As Roland Betts, Chair of the Ground Zero Site Committee put it,

> Larry Silverstein wanted a tall tower . . . why? To make more money, more rental square feet . . . I said to Silverstein, 'You see this as your property. We see this as belonging to the citizens of New York and the citizens of the world. The process we're running is to engage with all these voices.'
>
> (in Sim, 2004)

Liebeskind had the blessing of the Mayor of New York, George Pataki, but Childs was on the payroll of Silverstein, the owner of the site. Who did this place belong to? Would the bought and sold legal traces owned by Silverstein be stronger that those senses of belonging felt by everyday citizens? It was Betts' job to broker a partnership between the two camps. In practice, Childs gained overall control of how the replacement for the World Trade Center was

(b)ordered, and Liebeskind offered his collaboration. On 19 December 2003, the final designs for the Freedom Tower were unveiled (see Box 15.1).

In the designs for the redevelopment of the World Trade Center Childs' influence was clear. This building would be the tallest in the world, rising from Ground Zero to a height of 1776 feet tall. Yet Liebeskind's collaboration could also be discerned – the symbolic height of the building was no accident, it referred to the year of the United States' declaration of independence. In practice, therefore, the World Trade Center would become a place that stood for contested and contradictory ideas. It was America that would feel revitalised in the face of danger, and it was capitalist culture that would benefit from the building's vast dimensions. However, it remains to be seen how future cultures will read the value and symbol of this material trace. Will the place of the World Trade Center be seen as a triumph over adversity? If so, for which groups, in which places? Will it be a monument to a nation, or corporate interests, or both? Will it signal hope and vitality in the face of danger, and how will it link the past to the future, and this location to sites elsewhere? Will it bring about a revolution, or re-establish old (b)orders? In short, what sort of cultural world will the World Trade Center stand for? Whose place is this anyway (see Box 15.2)?

> Our moments sculpted out of time and space and each other are always to be found in an eternal art gallery at the end of each day. The art gallery that houses all of our retrospectives is the univers[al] [city].
>
> (Levitch, in Watts, 2009: 416)

> New York is a [universal] city that is carved out as a composition. That is what is so interesting about [this place], it's about the composition of the whole city, almost like an organic work of art that is ongoing.
>
> (Liebeskind, in Sim, 2004)

BOX 15.2 ENCOUNTERING THE TRACES OF THE WORLD TRADE CENTER: AFTER-SHOCKS AND AFFECTS

Whose place is the World Trade Center? According to Moore (2011), the redeveloped site 'is very New York', by which he means 'that everyone has [had] an opinion and an interest, which [had to] face down the opinions and interests of others'. In many ways, the spatial struggle for power at the Ground Zero site is emblematic of the conflicting opinions and practices which seek to compose any geographical place. We have seen how places on the ground connect closely to our ideas, feelings, and practices, and help to strongly position us in the cultural world. Our position, both culturally and geographically, is formed through these relations, and we have to argue them, practise them, and often fight for them, in order to retain our sense of place. Geographical places are the sites where these cultural ideas are disputed, and thus in the case of the World Trade Center the 'jostle of buildings reflects this jostle for power' (Moore, 2011).

'1 World Trade Center' thus tells many tales. As Moore recounts, 'this place is indeed a product of New York's version of democracy, in which no one has the ability to assume complete control . . . [the new site] looks assertive and confident, like the towers of corporate America anywhere, but it has contradictions. It is unsure if it is a symbol or a piece of commercial real estate' (2011). As we know, however, a cultural geography approach to place involves more than an investigation into what a place may mean, it also includes the 'particular affective capacities' generated by that place (after Thein, 2005: 14). Have you visited the new World Trade Center?[1] What was it like for you?

> Most attempts to re-create history fail to understand that history is not a documentation as much as it is a sensation . . . It is something to be felt. How does it feel to me?
>
> (Levitch, in Watts, 2009: 417)

Architectural commentator Oliver Wainwright (2014) records his own views of the museum:

> Everywhere you look there are steel plates of immense thickness, effortlessly curled like pencil shavings, peeled back like the discarded skin of a banana. Raised on plinths or mounted on the walls and dramatically spotlit, these gut-wrenching fragments are treated like . . . sculptures, elevated into art objects with ghoulish glee, in a way that's not entirely easy to stomach.

According to the museum architect Craig Dykers the mirrored exterior is a more joyous affair; he states how the reflective walls are 'proving a popular place for selfies. If we can get someone to smile or have a giggle at a place of such sorrow, we've done our job' (in ibid.). Like any place, however, (b)orders are in the process of being made. Although it's encouraged to take selfies here, what is this place really for? Is it a place for business, for memorial, or for tourism? What behaviours are out of place here, and why (see Reuters, 2014)?

Our cultural geography approach to place thus focuses on this range of trace-making. It focuses on the materialities and meanings, the artefacts and affects, and the identities and intensities that come to define places as ongoing compositions of traces. This interpretation of culture is expansive, and, as we have seen in this book, cultural geography covers a wide range of areas. We have seen how different cultures naturalise 'nature' in different ways and to different ends. How a range of (b)ordering tactics can seek to eliminate some ethnicities but partner others. How traces such as patriotism and religion combine to strengthen but also complicate our sense of place. How adult and youth cultures interact to take and make the places around them, and how traces can be written on the body through our gender and sexuality. This interpretation of culture thus may be expansive, but it does not preclude our ability to engage with the 'so what?' question.

The 'so what?' question

Understanding places as unique compositions of traces can make it seem difficult to pronounce on the rights and wrongs of particular cultures and their geographical affects. As Duncan and Duncan identify, 'it is like acknowledging that everyone is utterly unique. At one level this is true . . . but ultimately [this is a] paralysing perspective' (2004: 398). Such a broad interpretation of culture is thus open to the following critique:

> Culture has become central by becoming vacuous . . . [It] returns to its comfortable niche as a cataloguer of difference.
> (Mitchell, cited in Nash, 2002: 322)

This critique suggests that even though new branches of cultural geography have responded to the external pressures set in motion by the globalising and increasingly politicised climate of the twenty-first century, and the internal debates prompted by new academic ideas and approaches, the discipline has not ventured far from the apolitical empiricism of Carl Sauer. In other words, looking at cultures and their traces is interesting, but *so what?* What does studying the difference that place makes actually make to the cultural world?

The cultural geography approach to place is a response to this critique. By asking what cultural traces dominate a particular place and how different actions may resist them, by investigating what alternatives these traces stand for and what trace-chains are set in motion by their practice, we return to our original premise that we are part of the cultural world. All our ideas and actions combine to produce new generations of places. As a consequence, it is not enough to chart the different cultural traces that come together to form places, we must see ourselves as part of these places too. It is not enough to decide that, for us, capitalism is good (or bad), or that all ethnicities have a right to exist in our neighbourhood, city, or nation (or not). We must choose actions that support and endorse the traces that we believe in, or at least not acquiesce to those that go against our values and cultural geographies. Thus to undertake cultural geography is not to engage in 'cultural relativism' (see Schweder, 2000), because for each of us 'not all ideas or uses of [place] are equally defensible' (after Cronon, 1995: Foreword). To conclude, we must use a cultural geography approach to help us better understand our place in the world around us, and use our thought-through and impulsive responses to this understanding to help us act better in our world.

Note

1 (If not, why not take a virtual tour of the 9/11 museum, found here: www.guardian.co.uk/world/gallery/2012/apr/30/one-world-trade-centre-new-york-in-pictures#/?picture=389392745&index=16).

Bibliography

Aaltola, M. (2005) The international airport: the hub-and-spoke pedagogy of the American empire. *Global Networks,* 5 (3), 261–278.

Abbey, E. (1992) *Desert Solitaire: A Season in the Wilderness.* Robin Clark: London.

ABC (2012) Paying to maintain the status quo. *Media Watch,* 6 August. Available at: www.abc.net.au/mediawatch/transcripts/s3561819.htm (accessed 10 November 2014).

ABC News (2007) Thai national anthem could stop traffic. 24 November. Available at: www.abc.net.au/news/stories/2007/11/24/2100057.htm (accessed 10 November 2014).

Abley, M. (2003) *Spoken Here. Travels Among Threatened Languages.* Arrow: London.

Abse, D. (2003) Double footsteps. In Abse, D., Finch, P., O'Reilly, K., Lewis, G., Brito, L., Clarke, G., Davies, G., Schiller Levey, S., Williams, J., and Robson, L. (eds) *Cardiff Central: Ten Writers Return to the Welsh Capital.* Gomer: Llandysul, 1–12.

Ackroyd, S. and Hughes, J. A. (1981) *Data Collection in Context.* Longman: New York.

Addley, E. (2007) Blood on the tracks. *The Guardian,* 20 January. Available at: www.guardian.co.uk/uk/2007/jan/20/ukcrime.prisonsandprobation (accessed 10 November 2014).

Addley, E. (2011) Occupy movement: from local action to a global howl of protest. *The Guardian,* 17 October. Available at: www.guardian.co.uk/world/2011/oct/17/occupy-movement-global-protest (accessed 10 November 2014).

Adelman, M. and Ruggi, L. (2008) The beautiful and the abject: gender, identity and constructions of the body in contemporary Brazilian culture. *Current Sociology,* 56 (4), 555–86.

Adey, P. (2007) 'May I have your attention': airport geographies of spectatorship, position, and (im)mobility. *Environment & Planning D. Society & Space,* 25 (3), 515–536.

Adey. P. (2008) Airports, mobility and the calculative architecture of affective control. *Geoforum,* 39, 438–451.

Adey, P. (2010) *Mobility.* Routledge, London & New York.

Adey, P. (2014) *Air: Nature and Culture.* Reaktion: London

Adey, P., Bissell, D., McCormack, D., and Merriman, P. (2012) Profiling the passenger: mobilities, identities, embodiments. *Cultural Geographies,* 19, 169–193.

Adey, P., Budd, L., and Hubbard, P. (2007) Flying lessons: exploring the social and cultural geographies of global air travel. *Progress in Human Geography,* 31 (6), 773–791.

Adorno, T. (1991) *The Culture Industry.* Routledge Classics: London.

Agamben, G. (2005) *State of Exception.* University of Chicago Press: Chicago.

Agnew, J. (1993) Representing space: space, scale and culture in social science. In Duncan, J. and Ley, D. (eds) *Place/Culture/Representation.* Routledge: London.

Agnew, J. (2005) Space: place. In Cloke, P. and Johnston, R. (eds) *Spaces of Geographical Thought*. Sage: London, 81–96.

Agnew, J. and Duncan, J. (eds) (1989) *The Power of Place: Bringing Together Geographical and Sociological Imaginations*. Unwin Hyman: London.

Agnew, J., Livingstone, D., and Rogers, A. (eds) (1996) *Human Geography: An Essential Anthology*. Blackwell: Oxford.

Aitkin, S. (1997) Analysis of texts: armchair theory and couch-potato geography. In Flowerdew, R. and Martin, D. (eds) *Methods in Human Geography: A Guide for Students Doing a Research Project*. Longman: Harlow, 197–212.

Alam, F. (2004) Muslim boxing hero who unites us all. *The Observer*, 29 August, p. 1.

Aldridge, A. (1994) The construction of rational consumption in 'Which' magazine: the more blobs the better. *Sociology*, 28 (4), 899–912.

Amin, A. (2004) Regions unbound: towards a new politics of place. *Geografiska Annaler*, 86 B 1, 33–44.

Amit, V. (ed.) (2000) *Constructing the Field: European Association of Social Anthropologists*. Routledge: London and New York.

Amos, A. and Bostock. Y. (2006) Young people, smoking and gender: a qualitative exploration. *Health Education Research*, 22, 770–781.

Anderson, B. (1983; 2006) *Imagined Communities: Reflections on the Origin and Spread of Nationalism*. London: Verso.

Anderson, B. and Harrison, P. (2010) *Taking-Place. Non Representational Theories and Geography*. London: Ashgate.

Anderson, J. (1992) *Questions of National Identity, Social Sciences Foundation Course D103*. BBC TV: London

Anderson, J. (2002) Researching environmental resistance: working through Secondspace and Thirdspace approaches. *Qualitative Research*, 2 (3), 301–321.

Anderson, J. (2004a) Spatial politics in practice, the style and substance of environmental direct action. *Qualitative Research*, 2, 301–321.

Anderson, J. (2004b) Talking whilst walking: a geographical archaeology of knowledge. *Area*, 36 (3), 254–261.

Anderson, J. (2009) Transient convergence and relational sensibility: beyond the modern constitution of nature. *Emotion, Space, and Society*, 2, 120–127.

Anderson, J. (2010) The surfed wave as transient convergence. *You Are Here: The Journal of Creative Geography. Special Issue: Spatial Imaginaries and Imaginary Spaces*, XIII, 31.

Anderson, J. (2012) Relational places: the surfed wave as assemblage and convergence. *Environment & Planning D: Society & Space*, 30 (4), 570–587.

Anderson, J. (2013a) Active learning through student film: a case study of cultural geography. *Journal of Geography in Higher Education*. DOI:10.1080/03098265.2013. 792041.

Anderson, J. (2013b) Mapping the consequences of mobility: reclaiming jet lag as the state of travel disorientation. *Mobilities*. DOI: 10.1080/17450101.2013.806392.

Anderson, J. (2014a) *Page and Place: Ongoing Compositions of Plot*. Rodopi: Amsterdam.

Anderson, J. (2014b) Exploring the space between words and meaning: understanding the relational sensibility of surf spaces. *Emotion, Space & Society*, 10, 27–34.

Anderson, J. (no date) *Spatial Manifesto*. Available at: www.spatialmanifesto.com/teaching-projects/culture-place-space (accessed 10 November 2014).

Anderson, J. and Erskine, K. (2014) Tropophilia: a study of people, place and lifestyle travel. *Mobilities*, 9 (1), 130–145.

Anderson, J. and Jones, K. (2009) The difference that place makes to methodology: uncovering the 'lived space' of young people's spatial practices. *Children's Geographies*, 7 (3), 291–303.

Anderson, J. and Peters, K. (eds) (2014) *Water Worlds: Human Geographies of the Ocean*. Ashgate: London.

Anderson, K. (1999) Introduction. In Anderson, K. and Gale, F. (eds) *Cultural Geographies*, 2nd edn. Longman: Mebourne, 1–17.

Anderson, K. (2000) Thinking 'postnationally': dialogue across multicultural, indigenous, and settler spaces. *Annals of the Association of American Geographers*, 90, 381–391.

Anderson, K. (2008) Race in post universalist perspective. Institute for Culture & Society Pre Print Journal Articles. 1–21 [Anderson, K. (2008) *Cultural Geographies*, 15 (2), 155–171.]

Anderson, K. and Gale, F. (eds) (1999) *Cultural Geographies*, 2nd edn. Longman: Melbourne.

Anderson, K. and Jacobs, J. (1997) From urban aborigines to aboriginality and the city: one path through the history of Australian cultural geography. *Australian Geographical Studies*, 35 (1), 12–22.

Anderson, K. and Smith, S. (2001) Editorial: emotional geographies. *Transactions of the Institute of British Geographers*, 26 (1), 7–10.

Anderson, K., Domosh, M., Pile, S. and Thrift, N. (eds) (2003) *Handbook of Cultural Geography*. Sage: London.

Anon. (2006) The writing on the wall. *The Guardian*, Friday 24 March. Available at: http://arts.guardian.co.uk/features/story/0,,1738453,00.html (accessed 10 November 2014).

Anon (2007) Banksy's graffiti masterpiece 'comes with free house', *24 Dash*, 10 February. Available at: www.24dash.com/news/Housing/2007-02-10-Banksys-graffitimasterpiece-comes-with-free-house (accessed 10 November 2014).

Anzaldua, G. (1987) *Borderlands: La Frontera*. Aunt Lute Books: San Francisco.

Appadurai, A. (1990) Disjuncture and difference in the global cultural economy. In Featherstone, M. (ed.) *Global Culture: Nationalism, Globalization and Modernity (Special Issue of Theory, Culture and Society)*. London: Sage.

Appadurai, A. (1996) *Modernity at Large: Cultural Dimensions of Globalization*. University of Minnesota Press: Minneapolis.

Armitage, S. (2006) *Out of the Blue*. Channel 5, 11 September.

Ashe, F. (2006) The McCartney sisters' search for justice: gender and political protest in Northern Ireland. *Politics*, 26 (3), 161–167.

Associated Press (2011) Saudi woman to be lashed for defying driving ban. *The Guardian*, 28 September. Available at: www.guardian.co.uk/world/2011/sep/28/saudi-woman-lashed-defying-driving-ban (accessed 10 November 2014).

Asthana, A. (2004) How graffiti artists are cleaning up. *The Observer*, 15 August. Available at: http://arts.guardian.co.uk/features/story/0,,1283459,00.html (accessed 10 November 2014).

Attfield, R. (1991) *The Ethics of Environmental Concern*, 2nd edn. University of Georgia Press: Athens, GA, and London.

Aufheben, J. (1996) Review: senseless acts of beauty. *Green Anarchist*, 34. Available at: www.geocities.com/aufheben2/5.html (accessed 10 November 2014).

Auge, M. (2008) *Non-places. Introduction to an Anthropology of Supermodernity*. Verso: London.

Ayres, C. (2005) Rappers are lovin' it as burger chain buys a slice of their act. *The Times*, 31 March.

Bacon, F. (1604) *Valerius Terminus: Of the Interpretation of Nature*. Available at: http://ebooks.adelaide.edu.au/b/bacon/francis/valerius/ (accessed 10 November 2014).

Bakhtin, M. (1968; 1984) *Rabelais and His World*. MIT Press: Cambridge, MA; Indiana University Press: Bloomington.

Bancroft, A. (1999) Gypsies to the camps!: exclusion and marginalisation of Roma in the Czech Republic. *Sociological Research Online*, 4 (3). Available at: www.socresonline.org.uk/4/3/contents.html (accessed 10 November 2014).

Bandini, P. (2013) Richie Incognito ignited

very public row in Miami and crossed the line. *The Guardian,* 18 June. Available at: www.theguardian.com/sport/blog/2013/nov/16/miami-dolphins-incognito-racial-slur (accessed 10 November 2014).

Banksy (2001) *Banging Your Head Against a Brick Wall.* Banksy: No Place.

Banksy (2005) *Wall and Peace.* Century: London.

Barnes, T. (2005) Culture: economy. In Cloke, P. and Johnston, R. (eds) *Spaces of Geographical Thought.* Sage: London, 61–80.

Barford, V. (2013) The secret colonies of graffiti artists and skateboarders. *BBC News Website,* 20 May. Available at: www.bbc.co.uk/news/magazine-22551669 (accessed 10 November 2014).

Barkham, P. (2012) Tattoos: the hidden meanings. *The Guardian,* 26 June. Available at: www.guardian.co.uk/fashion/2012/jun/26/tattoos-hidden-meanings (accessed 10 November 2014).

Barnett, L. (2011) What was Topman thinking with these T-shirts? *The Guardian,* 14 September. Available at: www.theguardian.com/lifeandstyle/the-womens-blog-with-jane-martinson/2011/sep/14/topman-sexist-t-shirts (accessed 10 November 2014).

Barnett, L. (2013) Katharina Fritsch on her Fourth Plinth cockerel sculpture: 'I didn't want to make fun – but I was invited'. *The Guardian,* 24 July. Available at: www.theguardian.com/artanddesign/2013/jul/24/katharina-fritsch-fourth-plinth-cockerel-sculpture (accessed 10 November 2014).

Barry, M. (2006) *The Corporation Feature Transcript.* Available at: www.thecorporation.com/media/Transcript_finalpt2%20copy.pdf 12.30.27 pp10 (accessed 10 November 2014).

Barthes, R. (1973) *Mythologies.* Paladin: London.

Basok, T. (2002) *Tortillas and Tomatoes: Transmigrant Mexican Harvesters in Canada.* Queen's University Press: Montreal.

Bates, L. (2014) Little girls deserve better than to be told to make themselves sexy. *The Guardian,* 10 January. Available at: www.theguardian.com/lifeandstyle/womens-blog/2014/jan/10/little-girls-deserve-better-toys-sexism (accessed 10 November 2014).

Battista, K., LaBelle, B., Penner, B., Pile, S., and Rendell, J. (2005) Exploring 'an area of outstanding unnatural beauty': a treasure hunt around King's Cross, London. *Cultural Geographies,* 12, 429–462.

Bauman, Z. (1992) *Intimations of Postmodernity.* Routledge: London.

Bauman, Z. (1995) *Life in Fragments.* Blackwell: Oxford.

Bauman, Z. (2000) *Liquid Modernity.* Polity Press: Cambridge.

Bauman, Z. (2002) Foreword: individually, together. In Beck, U. and Beck-Gernsheim, E. (eds) *Individualization.* Sage: London, xiii–xx.

BBC (2000) *Global Protest: Battle of Prague.* BBC TV.

BBC (2001) *Mostar Bridge.* Lucy Blakstad Productions for the BBC.

BBC (2005) *People Power.* 13 March. Available at: http://news.bbc.co.uk/1/hi/programmes/people_power/4338069.stm (accessed 10 November 2014).

BBC (2006a) Bolivia to join Chavez's fight. 3 January. Available at: http://news.bbc.co.uk/1/hi/world/americas/4576972.stm (accessed 10 November 2014).

BBC (2006b) Wenger dismisses Pardew criticism. *BBCSport,* 11 March. Available at: http://news.bbc.co.uk/sport1/hi/football/4793868.stm (accessed 10 November 2014).

BBC (2007) Postcode gangs stalk East End. *BBC News,* 26 January. Available at: http://news.bbc.co.uk/1/hi/england/london/6304345.stm (accessed 10 November 2014).

BBC (2010) Are dying languages worth saving? *BBC News,* 15 September. Available at:

www.bbc.co.uk/news/magazine-11304255 (accessed 10 November 2014).

BBC (2011) New Australian passports allow third gender option. *BBC News*. Available at: www.bbc.co.uk/news/world-asia-pacific-14926598 (accessed 10 November 2014).

BBC (2012a) Cardiff City FC confirm rebranding with new red shirts. *BBC News*, 6 June. Available at: www.bbc.co.uk/news/uk-wales-south-east-wales-18337392 (accessed 10 November 2014).

BBC (2012b) Guerilla stickers appear on the Tube. *BBC News*, 10 October. Available at: www.bbc.co.uk/news/uk-england-london-19895637 (accessed 10 November 2014).

BBC (2012c) Guerrilla sticker craze hits London's Tube. *BBC News*, 12 October. Available at: www.bbc.co.uk/news/uk-england-london-19858746 (accessed 10 November 2014).

BBC (2013) Cardiff City owner Vincent Tan talks of name change. *BBC News*, February. Available at: www.bbc.co.uk/sport/0/football/21604568 28 (accessed 10 November 2014).

BBC World News (2004) Mostar bridge opens with splash. 23 July. Available at: http://news.bbc.co.uk/1/hi/world/europe/3919047.stm (accessed 10 November 2014).

Beck, L. (2007) Beijing to evict 1.5 million for Olympics. *Reuters*. Available at: www.reuters.com/article/2007/06/05/us-olympics-beijing-housing-idUSPEK12263220070605 (accessed 10 November 2014).

Beck, U. (1992) *Risk Society: Towards a New Modernity*. Sage: London.

Beck, U. (1994) The reinvention of politics: towards a theory of reflexive modernisation. In Beck, U., Giddens, A., and Lash, S., *Reflexive Modernization: Politics, Tradition and Aesthetics in the Modern Social Order*. Polity Press: Cambridge.

Beck, U. (1995) *Ecological Politics in an Age of Risk*. Polity Press: Cambridge.

Beck, U. and Beck-Gernsheim, E. (1995) *The Normal Chaos of Love*. Polity Press: Cambridge.

Beck, U. and Beck-Gernsheim, E. (2002) *Individualization*. Sage: London.

Beery, J.(2012) State, capital and spaceships: a terrestrial geography of space tourism. *Geoforum*, 43 (1), 25–34.

Bekoff, M. (2009) Animal emotions, wild justice and why they matter: grieving magpies, a pissy baboon, and empathic elephants. *Emotion, Space & Society*, 2 (2), 82–85.

Bell, C. and Roberts, H. (1984) *Social Researching: Politics, Problems, Practice*. Routledge and Kegan Paul: London.

Bell, D. and Valentine, G. (1995) Introduction: orientations. In Bell, D. and Valentine, G. (eds) *Mapping Desire: Geographies of Sexualities*. Routledge: London, 1–27.

Bell, D. and Valentine, G. (eds) (1995) *Mapping Desire: Geographies of Sexualities*. Routledge: London and New York.

Bell, D., Binnie, J., Cream, J., and Valentine, G. (2001) *Pleasure Zones: Bodies, Cities*. Syracuse: Syracuse University Press.

Benedictus, L. (2005) Every race, colour, nation and religion on earth. *The Guardian*, 21 January. Available at: www.guardian.co.uk/uk/2005/jan/21/britishidentity.race1 (accessed 10 November 2014).

Benioff, D. (2003) *25th Hour*. New English Library: London.

Bentham, J. (1843). *The Works of Jeremy Bentham (Vol. 1)*. Edinburgh, Scotland: William Tait.

Bertolas, R. J. (1998) Cross cultural environmental perception of wilderness. *Professional Geographer*, 50 (1), 98–111.

Bey, H. (1991) *T.A.Z The Temporary Autonomous Zone: Ontological Anarchy and Poetic Terrorism*. Autonomedia: New York.

Bhabha, H. (1990) *The Location of Culture*. Routledge: London.

Binnie, J. (1997) Coming out of geography: towards a queer epistemology? *Environment & Planning D, Society & Space*, 15 (223), 237.

Binnie, J., Edensor, T., Holloway, J., Millington, S., and Young, C. (2007) Mundane mobilities, banal travels. *Social and Cultural Geography*, 8 (2), 165–174.

Bircham, E. (2001) Foreword. In Bircham, E. and Charlton J. (eds) *Anti Capitalism: A Guide to the Movement*. Bookmarks: London, 1–3.

Bissell, D. (2009) Conceptualising differently-mobile passengers: geographies of everyday encumbrance in the railway station. *Social and Cultural Geographies*, 10 (2), 173–195.

Bissell, D. (2010) Passenger mobilities: affective atmospheres and the sociality of public transport. *Environment and Planning D: Society and Space*, 28, 270–289.

Bissell, D. (2014) Encountering stressed bodies: slow creep transformations and tipping points of commuting mobilities. *Geoforum*, 51, 191–201.

Bissell, D., Adey, P., and Laurier, E. (2011) Introduction to the special issue on geographies of the passenger. *Journal of Transport Geography*, 19, 1007–1009.

Blacker, T. (2008) Nimbyism should be applauded, not despised. *The Independent*, 4 January. Available at: www.independent.co.uk/opinion/commentators/terenceblacker/terence-blacker-nimbyism-should-be-applauded-not-despised-768090.html (accessed 10 November 2014).

Blunt, A. and Wills, J. (2000) *Dissident Geographies: An Introduction to Radical Ideas and Practice*. Prentice Hall: New York.

Boas, F. (1911) *Handbook of American Indian Languages*, vol. 1. Bureau of American Ethnology, Bulletin 40, Government Print Office (Smithsonian Institution, Bureau of American Ethnology): Washington.

Boas, F. (1912) The history of the American Race. *Annals of the New York Academy of Sciences*, 21, 177–183.

Bondi, L. (1990) Feminism, postmodernism and geography: space for women? *Antipode*, 22 (2), 156–167.

Bondi, L. (1998) Sexing the city. In R. Fincher and J. M. Jacobs (eds) *Cities of Difference*. Guilford Press: New York, 177–200.

Bondi, L. (2005a) Making connections and thinking through emotions: between geography and psychotherapy. *Transactions of the Institute of British Geographers*, 30, 433–448.

Bondi, L. (2005b) The place of emotions in research: from partitioning emotion and reason to the emotional dynamics of research relationships. In Davidson, J., Bondi, L., and Smith, M. (eds) *Emotional Geographies*. Ashgate: Aldershot, 231–246.

Bondi, L. and Smith, M. (2005) *Emotional Geographies*. Ashgate: Aldershot.

Bondi, L, Davidson, J., and Smith, M. (2005) Introduction: geography's 'emotional turn'. In Davidson, J., Bondi, L., and Smith, M. (eds) *Emotional Geographies*. Ashgate: Aldershot, 1–18.

Bordo, S. (1993) *Unbearable Weight: Feminism, Western Culture, and the Body*. University of California Press: Berkeley.

Bourdieu, P. (1977) *Outline of a Theory of Practice*. Cambridge University Press: Cambridge.

Bourdieu, P. (1991) *Language and Symbolic Order*. Polity Press: Cambridge.

Bowden, M. (eds) *Geographies of the Mind: Essays in Historical Geosophy*. Oxford University Press: New York.

Boyle, D. (2000) *Funny Money. In Search of Alternative Cash*. Flamingo: London.

Bradley, D. (2002) Language attitudes: the key factor in language maintenance. In Bradley, D. and Bradley, M. (eds) *Language Endangerment and Language Maintenance*. Routledge Curzon: London, 1–10.

Braidotti, R. (1994) *Nomadic Subjects. Embodiment and Sexual Difference in Contemporary Feminist Theory*. Columbia University Press: New York.

Branigan, T. (2008) Beijing residents stage protest over Olympic eviction. *The Guardian*, 4 August. Available at: www.guardian.co.uk/

world/2008/aug/04/china.olympics2008 (accessed 10 November 2014).

Branigan, T. and Scott, M. (2008) Elderly women sentenced to year's labour over Olympics protest. *The Guardian*, 20 August. Available at: www.guardian.co.uk/sport/2008/aug/20/olympics2008.china1 (accessed 10 November 2014).

Breitbart, M. (1998) 'Dana's mystical tunnel': young people's design for survival and change in the city. In Skelton, T. and Valentine, G. (eds) *Cool Places: Geographies of Youth Cultures*. Routledge: London, 305–327.

Brenzinger, M., Heine, B., and Sommer, G. (1991) Language death in Africa. In Robins, R. and Uhlenbeck, E. (eds) *Endangered Languages*. Berg: Oxford, 19–44.

Bristol Pound (2014) Bristol Pound: our city, our money. *Bristol Pound*. Available at: http://bristolpound.org (accessed 10 November 2014).

Brown, K. (2014) Leave only footprints? How traces of movement shape the appropriation of space. *Cultural Geographies*. DOI: 1474474014558987

Brown, M. (2013) Fourth plinth at Trafalgar Square gets (no sniggering) a 14ft blue cock. *The Guardian*, 14 January. Available at: www.theguardian.com/artanddesign/2011/jan/14/trafalgar-square-cock-installation (accessed 10 November 2014).

Brown, P. (2002) Peace but no love as Northern Ireland divide grows ever wider. *The Guardian*, 4 January. Available at: www.guardian.co.uk/uk/2002/jan/04/northernireland.paulbrown (accessed 10 November 2014).

Browne, K. (2006) 'A right geezer-bird (manwoman)': the sites and sights of 'female' embodiment. *ACME: An International E-Journal for Critical Geographies*, 5 (2), 121–143.

Bureau of Labor Statistics (2009) *Economic New Release*. Available at: www.bls.gov/news.release/wkyeng.t02.htm (accessed 25 November 2009).

Burgess, J. and Gold, J. (eds) (1982) *Valued Environments*. Allen & Unwin: London.

Burgess, J. and Gold, J. (1985) *Geography, the Media and Popular Culture*. Croom Helm: London.

Burgess, R. (1984) *In the Field: An Introduction to Field Research*. Routledge: London.

Büscher, M. and Urry, J. (2009) Mobile methods and the empirical. *European Journal of Social Theory*, 12 (1), 99–116.

Büscher, M., Urry, J., and Witchger, K. (eds) (2011) *Mobile Methods*. Routledge: London & New York.

Butler, J. (1993) *Bodies That Matter: On the Discursive Limits of 'Sex'*. Routledge: London.

Buttel, F. (2003) Some observations on the anti-globalisation movement. *Australian Journal of Social Issues*, 38 (1), 95–117.

Buttimer, A. (1998) Geography's contested stories: changing states-of-the-art. *Tijdschrift voor Economische en Sociale Geografie*, 89 (1), 90–99.

Buttimer, A. and Seamon, D. (1980) *The Human Experience of Space and Place*. Croom Helm: London.

Calvino, I. (1986) *Invisible Cities*. Martin Secker and Warburg: London.

Cameron, D. (1998) Performing gender identity: young men's talk and the construction of heterosexual of masculinity. In Coates, J. (ed.) *Language and Gender: A Reader*. Blackwell: Oxford, 270–284.

Cameron, D. (2006) David Cameron's speech to CSJ Kids symposium. *The Guardian*, 10 July. Available at: www.guardian.co.uk/politics/2006/jul/10/conservatives.law (accessed 10 November 2014).

Cameron, D. (2010) Globish: how the English language became the world's language by Robert McCrum. *The Guardian*, 5 June. Available at: www.theguardian.com/books/2010/jun/05/globish-robert-mccrum-review (accessed 10 November 2014).

Capital FM (no date) Get your leek on. Available at: www.capitalfm.com/southwales/on-air/breakfast/show-blog/get-your-leek/ (accessed 10 November 2014).

Carroll, R. (2011) Whimper no more: Peru national anthem bound for glory. *The Guardian*, 26 April. Available at: www.guardian.co.uk/world/2011/apr/26/peru-national-anthem-rewrite (accessed 10 November 2014).

Carter, K. and Aulette, J. (2009) Creole in Cape Verde: language, identity and power. *Ethnography*, 10, 213–236.

Casciani, D. (2005) Are the hoodies the goodies? *BBC News Channel*, 1 July. Available at: http://news.bbc.co.uk/1/hi/magazine/4639235.stm (accessed 10 November 2014).

Casey, E. (2000) *Remembering: A Phenomenological Study*, 2nd edn. Indiana University Press: Bloomington.

Casey, E. (2001) Between geography and philosophy: what does it mean to be in the place-world? *Annals of the Association of American Geographers*, 91 (4), 683–693.

Castree, N. (2001) Socialising nature: theory, practice, and politics. In Castree, N. and Braun, B. (eds) *Social Nature: Theory, Practice. Politics*. Blackwell: Oxford, 1–21.

Castree, N. (2004) Commentary. Nature is dead! Long live nature! *Environment & Planning A*, 36, 191–194.

Castree, N. (2005) *Nature*. Routledge: Abingdon.

Castree, N. and Braun, B. (eds) (2001) *Social Nature: Theory, Practice, Politics*. Blackwell: Oxford.

Castree, N. and MacMillan, T. (2001) Dissolving dualisms: actor-networks and the reimagination of nature. In Castree, N. and Braun, B. (eds) *Social Nature*. Blackwell: Oxford, 208–224.

Chatterton, P. (2006) 'Give up activism' and change the world in unknown ways: or, learning to walk with others on uncommon ground. *Antipode*, 38 (2), 259–281.

Cherrier, H. and Murray, J. (2002) Drifting away from excessive consumption: a new social movement based on identity construction. *Advances in Consumer Research*, 29, 245–249.

Christie, M. (2005) *Jump Britain*. Carbon Media for Channel 4.

Clarke, G. (2003) City of tall stories. In Abse, D. et al. (eds) *Cardiff Central: Ten Writers Return to the Welsh Capital*. Gomer: Llandysul, 51–64.

Cleto, F. (ed.) (1999) *Camp: Queer Aesthetics and the Performing Subject. A Reader*. Edinburgh University Press: Edinburgh.

Clifford, N. and Valentine, G. (eds) (2003) *Key Methods in Geography*. Sage: London.

Cloke, P. and Johnson, R. (eds) (2005) *Spaces of Geographical Thought: Deconstructing Human Geography's Binaries*. Sage: London.

Cloke, P., Philo, C., and Sadler, D. (1991) *Approaching Human Geography: An Introduction to Contemporary Theoretical Debates*. Paul Chapman: London.

Cloke, P., Crang, P., and Goodwin, M (eds) (1999) *Introducing Human Geographies*. Arnold: London.

Cohen, S. (1979) *Folk Devils and Moral Panics: The Creation of Mods and Rockers*. Basil Blackwell: London.

Cohen, S. (2010) Chasing a myth? Searching for 'self' through lifestyle travel. *Tourist Studies*, 10 (2), 117–133.

Comaroff, J. and Comaroff, J. (2000) Naturing the nation: aliens, apocalypse and the post-colonial state. *Hagar: International Social Science Review*, 1 (1), 321–361.

Community Cohesion Panel (2004) *The End of Parallel Lives? The Report of the Community Cohesion Panel*. Home Office: London.

Community Cohesion Review Team (2001) *Community Cohesion: A Report of the Independent Review Team*. Home Office: London.

Connell, R. W. (1995) *Masculinities*. University of California Press: Berkeley.

Connelly, J. and Smith, G. (2003) *Politics and the Environment: From Theory to Practice*. Routledge: London and New York.

Conquergood, D. (1998) Beyond the text: toward a performative cultural politics. In

Dailey, S. J. (ed.) *The Future of Performance Studies. Visions and Revisions*. Annadale: National Communication Association, 25–36.

Conrad, J. (1926) Geography and some explorers. In *Last Essays*. Dent: London, 1–31.

Convery, I., Bailey, C., Mort, M., and Baxter, J. (2005) Death in the wrong place? Emotional geographies of the UK 2001 foot and mouth disease epidemic. *Journal of Rural Studies*, 21 (1), 99–109.

Cook, I. (1997) Participant observation. In Flowerdew, R. and Martin, D. (eds) *Methods in Human Geography: A Guide for Students Doing a Research Project*. Longman: Harlow, 127–150.

Cook, I. (2001) You want to be careful you don't end up like Ian: he's all over the place. In Moss, P. (ed.) *Placing Autobiography in Geography*. Syracuse University Press: Syracuse, NY, 99–120.

Cook, I. (2004) Follow the thing: papaya. *Antipode*, 36 (4), 642–664.

Cook, I. and Crang, M. (1995) *Concepts and Techniques in Modern Geography: Doing Ethnographies*. CATMOGS: Durham.

Coonen, A. (2007) Women will rule okay in Chinese tourist town. *The Independent*, 27 April. Available at: www.independent.co.uk/news/world/asia/women-will-ruleok-in-chinese-tourist-town-446361.html (accessed 10 November 2014).

Cosgrove, D. (1989) Geography is everywhere: culture and symbolism in human landscapes. In Gregory, D. and Welford, R. (eds) *Horizons in Human Geography*. Macmillan: Basingstoke, 118–135.

Cosgrove, D. (1994) Worlds of meaning: cultural geography and the imagination. In Foote, K. E., Hugill, P., Mathewson, K., and Smith, J. (eds) *Re-reading Cultural Geography*. University of Texas Press: Austin, 387–395.

Cosgrove, D. and Domosh, M. (1993) Author and authority: writing the new cultural geography. In Duncan, J. and Ley, D. (eds) *Place/Culture/Representation*. Routledge: New York, 25–38.

Cosgrove, D. and Duncan, J. (1993) On 'The reinvention of cultural geography' by Price and Lewis. *Annals of the Association of American Geographers*, 83 (3), 515–519.

Cosgrove, D. and Jackson, P. (1987) New directions in cultural geography. *Area*, 19 (2), 95–101.

Coupland, N. (2012) Bilingualism on display: the framing of Welsh and English in Welsh public spaces. *Language in Society*, 41, 1–27.

Cox, K. (2002) *Political Geography. Territory, State and Society*. Blackwell: Oxford.

Crang, M. (1998) *Cultural Geography*. Routledge: London.

Crang, M. (2003) Qualitative methods: touchy, feely, look-see? *Progress in Human Geography*, 27 (4), 494–504.

Cresswell, T. (1996) *In Place/Out of Place: Geography, Ideology and Transgression*. University of Minnesota Press: Minneapolis.

Cresswell, T. (2000) Falling down: resistance as diagnostic. In Sharp, J. et al. (eds) *Entanglements of Power: Geographies of Domination/Resistance*. Routledge: London and New York, 260–273.

Cresswell, T. (2004) *Place: A Short Introduction*. Blackwell: Oxford.

Cresswell, T. (2006) *On the Move. Mobility in the Modern Western World*. Routledge: London and New York.

Cresswell, T. (2010a) Mobilities I: catching up. *Progress in Human Geography*, 35, 550–558.

Cresswell, T. (2010b) New cultural geography – an unfinished project? *Cultural Geographies*, 17, 169–174.

Cresswell, T. (2010c) Towards a politics of mobility. *Environment & Planning D. Society & Space*, 28, 17–31.

Cresswell, T. (2011) Mobilities II: still. *Progress in Human Geography*. Available at: http://phg/sagepub.com/content/early/2011/09/22/0309132511423349 (accessed 10 November 2014).

Cronon, W. (1995) *Uncommon Ground: Toward Reinventing Nature*. W.W. Norton & Co: New York.

Cronon, W. (1997) The trouble with wilderness: or getting back to the wrong nature. In Miller, C. and Rothman, H. (eds) *Out of the Woods: Essays in Environmental History*. University of Pittsburgh Press: Pittsburgh.

Crutzen, P. (2006) The 'Anthropocene'. In Ehlers, E. and Krafft, T. (eds) *Earth System Science in the Anthropocene*. Springer: Berlin, 13–18.

Curry, P. (2003) Re-thinking nature: towards an ecopluralism. *Environmental Values*, 12, 337–360.

Curtis, J., Shiau, E., Lowery, B., Sloane, D., Hennigan, K., and Curtis, A. (2014) The prospects and problems of integrating sketch maps with geographic information systems to understand environmental perception: a case study of mapping youth fear in Los Angeles gang neighborhoods. *Environment and Planning B: Planning and Design*, 41 (2), 251–271.

Darwin, C. (1859) *On the Origin of Species*. John Murray: London.

Darwin, C. (1996) *Natural Selection*. Phoenix: London.

Davidson, D. (2007) Carling Belong ad escapes ban despite volley of complaints. *Brand Republic*, 2 May. Available at: www.brandrepublic.com/News/654173/Carling-Belong-ad-escapes-ban-despite-volley-complaints/ (accessed 10 November 2014).

Davidson, J. (2003) *Phobic Geographies: The Phenomenology of Spatiality and Identity*. Ashgate: Aldershot.

Davidson, J. and Milligan, C. (2004) Editorial embodying emotion sensing space: introducing emotional geographies. *Social and Cultural Geography*, 5 (4), 523–532.

Davidson, J., Bondi, L., and Smith, M. (2005) *Emotional Geographies*. Ashgate: Aldershot.

Davis, K., Davis, J., and Dowler, L. (2004) In motion, out of place: the public space(s) of Tourette Syndrome. *Social Science and Medicine*, 59, 103–112.

de Beauvoir, S. (1972) *The Second Sex*. Penguin: Harmondsworth.

De Certeau, M. (1984) *The Practice of Everyday Life*. University of California Press: Berkeley.

De Saussure, F. (1967) *Cours de linguistique générale*. Payot: Paris.

Delanda, M. (2006) *A New Philosophy of Society. Assemblage Theory and Social Complexity*. Continuum: London.

Demerritt, D. (2001) Being constructive about Nature. In Castree, N. and Braun, B. (eds) *Social Nature: Theory, Practice, Politics*. Blackwell: Oxford, 22–40.

Democratic Central (2008) Bush gives axis of evil speech. *Democratic Central*, 28 January. Available at: www.democraticcentral.com/show (accessed 10 November 2014).

Denison, J. and Markula, P. (2003) *Moving Writing. Crafting Movement in Sports Research*. Peter Lang: New York.

Denzin, N. (1970) *The Research Act*. Aldine Publishing: Chicago.

Denzin, N. (1984) *On Understanding Emotion*. Jossey Bass: San Francisco.

Denzin, N. (2003) Foreword. In Denison, J. and Markula, P. *Moving Writing. Crafting Movement in Sports Research*. Peter Lang: New York, ix–xiii.

Denzin, N. and Lincoln, Y. (eds) (2000) *The Handbook of Qualitative Research*. Sage: Cambridge.

Department of Justice (2009) Available at: www.ojp.usdoj.gov/bjs (accessed 25 November 2009).

Desforges, L. and Jones, R. (2001) Geographies of languages/languages of geography. *Social & Cultural Geography*, 2 (3), 261–264.

Devall, B. and Sessions, G. (1985) *Deep Ecology: Living as if Nature Mattered*. Gibbs Smith: Salt Lake City.

Dewsbury, J.-D. (2000) Performativity and the event: enacting a philosophy of difference.

Environment & Planning D, Society & Space, 18 (4), 473–496.

Dewsbury, J.-D. (2003) Witnessing space: 'knowledge without contemplation'. *Environment & Planning A*, 35, 1907–1932.

Dillard, A. (1975) *Pilgrim at Tinker Creek*. Picador: London.

Dodd, V. (2013) Minorities stopped disproportionally in decade after Macpherson report. *The Guardian*, 22 April. Available at: www.theguardian.com/law/2013/apr/22/ethnic-minority-britons-stop-search-white (accessed 10 November 2014).

Dodds, K. and Atkinson, D. (eds) (2000) *Geopolitical Traditions: A Century of Geopolitical Thought*. Routledge: London.

Doel, M. (1999) *Poststructuralist Geographies: The Diabolical Art of Spatial Science*. Rowman & Littlefield: Lanham.

Doel, M. (2003) Analysing cultural texts. In Clifford, N. and Valentine, G. (eds) *Key Methods in Geography*. Sage: London, 501–514.

Doherty, B. and de Geus, M. (eds) (1996) *Democracy and Green Political Thought*. Routledge: London and New York.

Douglas, M. (1991) *Purity and Danger: An Analysis of the Concepts of Pollution and Taboo*. Routledge: London.

Dovey, K. (2010) *Becoming Places. Urbanism/Architecture/Identity/Power*. Routledge: Abingdon.

Driver, F. (2000) *Geography Militant: Cultures of Exploration and Empire*. Blackwell: Oxford.

Driver, F. and Gilbert, D. (eds) (1999) *Imperial Cities: Landscape, Display and Identity*. Manchester University Press: Manchester and New York.

Duncan, J. (1980) The superorganic in American cultural geography. *Annals of the Association of American Geographers*, 70 (2), 181–198.

Duncan, J. (1994) After the civil war: reconstructing cultural geography as heterotopia. In Foote, K. E., Hugill, P., Mathewson, K., and Smith, J. (eds) *Re-reading Cultural Geography*. University of Texas Press: Austin, 401–408.

Duncan, J. and Duncan, N. (2004) Culture unbound. *Environment & Planning A*, 36, 391–403.

Dunnett, O. (2012) Patrick Moore, Arthur C. Clarke and 'British outer space' in the mid 20th century. *Cultural Geographies*, 19 (4), 505–522.

Ebony, D. (2005) Quinn marble for Trafalgar Square. *Art in America*, 93 (11), 37.

Ehret, C. and Waller Scanlon, L. (1983) *Overcoming Jet Lag*. Berkley Book: New York.

Elden, S. (2013) Secure the volume: vertical geopolitics and the depth of power. *Political Geography*. Available at: http://dx.doi.org/10.1016/j.polgeo.2012.12.009 (accessed 10 November 2014).

Ellis, A. (2014) *Soundspace: The Notting Hill Carnival*. Available at: http://alphaecho.co.uk/soundspace-1.html (accessed 10 November 2014).

Ellis, C. (2004) *The Ethnographic I: A Methodological Novel about Ethnography*. Alta Mira: Walnut Creek.

Ellis, S. (2004) Hold your tongue. *The Observer*, 8 February. Available at: http://books.guardian.co.uk/reviews/politicsphilosophy-andsociety/0,,1143193,00.html (accessed 10 November 2014).

Elmes, S. (2005) *Talking for Britain: A Journey through the Nation's Dialects*. Penguin: London.

Elwood, S. and Martin, D. (2000) 'Placing' interviews: location and scales of power in qualitative research. *Professional Geographer*, 52 (4), 649–657.

Escobales, R. (2013) Skateboarding's South Bank home to be turned into retail units. *The Guardian*, 12 April. Available at: www.guardian.co.uk/uk/2013/apr/12/skateboarding-south-bank?INTCMP=SRCH (accessed 10 November 2014).

Eyles, J. (1988) Interpreting the geographical world: qualitative approaches in geographical research. In Eyles, J. and Smith, D. M. (eds) *Qualitative Methods in Human Geography*. Polity Press: Cambridge, 1–16.

Falls, S. (2013) 'Redneck customs': race and class at the demolition derby. *Leisure Studies*, 32 (4), 429–445.

Farndale, N. (2006) Say no to gallows humour. *The Telegraph*, 11 December. Available at: www.telegraph.co.uk/opinion/main.jhtml?xml=/opinion/2006/11/12/do1205.xml (accessed 10 November 2014).

Fenster, T. and Yiftachel, O. (eds) (1997) *Frontier Development and Indigenous Peoples*. Pergamon: Oxford.

Festival Wing (2013) *The Skateable Space*. Available at: www.thefestivalwing.co.uk/explore/the-skateable-space/ (accessed 10 November 2014).

Fiick, U. (1998) *An Introduction to Qualitative Research*. Sage: London.

Fisher, L. (2013) Debranding: why Coca-Cola's decision to drop its name worked. *The Guardian*, 6 August. Available at: www.theguardian.com/media-network/media-network-blog/2013/aug/06/coke-debranding-name-dropping (accessed 10 November 2014).

Fistell, G. (2012) *A Portrait of Tourette Syndrome*. Available at: https://www.youtube.com/watch?v=WmFl2YtsVqo#t=35 (accessed 10 November 2014).

Flowerdew, R. and Martin, D. (1997) *Methods in Human Geography*. Longman: Harlow.

Foreman, D. and Haywood, B. (1987) *Ecodefense: A Field Guide to Monkeywrenching*. Ned Ludd Nooks: Tucson.

Foucault, M. (1973) *The Order of Things*. Vintage/Random House: New York.

Foucault, M. (1979) *Discipline and Punish: The Birth of the Prison*. New York: Doubleday.

Foucault, M. (1980) *Power/Knowledge*. Pantheon: New York.

Foucault, M. (1984) *History of Sexuality*, vol. 1. Penguin: Harmondsworth.

Fox News (2009) Transcript: Sen. Lieberman on 'FNS'. *Fox News*, 8 November. Available at: www.foxnews.com/story/2009/11/08/transcript-sen-lieberman-on-fns/ (accessed 10 November 2014).

Fox, K. and Hazelton, C. (2013) Skateboarders fight Southbank closure. *The Guardian*, 11 May. Available at: www.theguardian.com/artanddesign/gallery/2013/may/11/skateboarders-fight-southbank-skate-park-closure-gallery (accessed 10 November 2014).

Fox, W. (1990) *Towards a Transpersonal Ecology: Developing New Foundations for Environmentalism*. Shambala: Boston.

France, A. (2008) Bakri slur on Amir. *The Sun*, 25 March, 1, 4, 5.

Frawley, K. (1999) A 'green' vision: the evolution of Australian environmentalism. In Anderson, K. and Gale, F. (eds) *Cultural Geographies*, 2nd edn. Longman: Melbourne, 265–293.

Frizzel, N. (2014) Dad's tattoos are a map of his life. *The Guardian*, 4 January. Available at: www.theguardian.com/lifeandstyle/2014/jan/04/dads-tattoos-map-of-his-life (accessed 10 November 2014).

Fuller, D. (2008) Public geographies: taking stock. *Progress in Human Geography*, 32, 834–844.

Fuller, D. and Askins, K. (2010) Public geographies II: being organic. *Progress in Human Geography*, 34, 654–667.

Gade, D. (2003) Language, identity, and the scriptorial landscape in Quebec and Catalonia. *The Geographical Review*, 93 (4), 429–449.

Gambino, L. (2014) US patent office strips Washington Redskins of 'offensive' trademarks. *The Guardian*, 18 June. Available at: www.theguardian.com/sport/2014/jun/18/washington-redskins-lose-trademark-team-us-patent-office (accessed 10 November 2014).

Game, A. (1997) Sociology's emotions. *CRSA/RSCA*, 34 (4), 385–399.

Garland, J. and Chakraboti, N. (2006) 'Race', space and place: examining identity and cultures of exclusion in rural England. *Ethnicities*, 6 (2), 159–177.

Garrett, B. (2012) Videographic geographies: using digital video for geographic research. *Progress in Human Geography*, 35 (4), 521–541.

Garrett, B. (2014) Undertaking recreational trespass: urban exploration and infiltration. *Transactions of the Institute of British Geographers*, 39 (1), 1–13.

Garrett, B. (no date) *Place Hacking*. Available at: www.placehacking.co.uk/documentaries/ (accessed 10 November 2014).

Garton Ash, T. (2004) *Free World*. Penguin: London.

Gecko (2009) featured in Selway, B. 'I want Samoa', *Wavelength Magazine*, July, 184, 64–72.

Geddes, P. (1912) *Evolution*. Williams & Norgate: London.

Geertz, C. (1973) *The Interpretation of Cultures: Selected Essays*. Basic Books: New York.

Germann Molz, J. (2010) Performing global geographies: time, space, place and pace in narratives of round-the-world travel. *Tourism Geographies: An International Journal of Tourism Space, Place and Environment*, 12 (3), 329–348.

Germann Molz, J. and Morris Paris, C. (2013) The social affordances of flashpacking: exploring the mobility nexus of travel and communication. *Mobilities*. DOI:10.1080/17450101.2013.848605.

Giddens, A. (1991) *Modernity and Self Identity: Self and Society in the Late Modern Age*. Polity Press: Cambridge.

Giddens, A. (1994) Living in a post-traditional society. In Beck, U., Giddens, A., and Lash, S. (eds) *Reflexive Modernization: Politics, Tradition and Aesthetics in the Modern Social Order*. Polity Press: Cambridge.

Gilbert, D. and Driver, F. (2000) Capital and empire: geographies of imperial London. *GeoJournal*, 51, 23–32.

GLA (Greater London Authority) (2005) Press release, 7 July. Available at: www.london.uk/mayor/mayor_statement.070505.jsp (accessed 10 November 2014).

Godlewska, A. and Smith, N. (eds) (1994) *Geography and Empire*. Blackwell: Oxford.

Gogia, N. (2006) Unpacking corporeal mobilities: the global voyages of labour and leisure. *Environment and Planning A*, 38, 359–375.

Gold, J. and Burgess, J. (eds) (1982) *Valued Environments*. Allen & Unwin: London.

Goldsmith, E. (1996) *The Way: An Ecological Worldview*. Themis Books: Totnes.

Gough, I. (2000) *Global Capital, Human Needs and Social Policies*. Palgrave: Basingstoke.

Graeber, R. C. (1988) Aircrew fatigue and circadian rhythmicity. In Wiener, E. and Nagel, D. (eds) *Human Factors in Aviation*. Academic Press: San Diego, CA, 305–343.

Graham, S. and Marvin, S. (2001) *Splintering Urbanism*. Routledge: London.

Gramsci, A. (1971) *The Prison Notebooks*. Lawrence and Wishart: London.

Gramsci, A. (1972) *Selections from the Prison Notebooks of Antonio Gramsci*. Lawrence and Wishart: London.

Grandjean, G. and Connelly, K. (2013) Bitcoin currency. *The Guardian*, 26 April. Available at: www.theguardian.com/technology/video/2013/apr/26/bitcoin-currency-moves-offline-berlin-video (accessed 10 November 2014).

Grant, R. (2003) *Ghost Riders: Travels with American Nomads*. Abacus: London.

Greenberg, A. (1998) *Cause for Alarm: The Volunteer Fire Department in the Nineteenth Century City*. Princeton University Press: Princeton.

Greenhough, B. (2004) Introduction to part 4: materialities and performance. In Thrift, N. and Whatmore, S. (eds) *Cultural Geography: Critical Concepts in the Social Sciences*, vol. 2. Routledge: London and New York, 255–265.

Greenwald, G. (2009) Can attacks on a military base constitute terrorism? *Salon*. Available at: www.salon.com/2009/11/09/terrorism_7/ (accessed 10 November 2014).

Greenwald, G. (2013) Was the London killing of a British soldier 'terrorism'? *The Guardian*,

23 May. Available at: www.theguardian.com/commentisfree/2013/may/23/woolwich-attack-terrorism-blowback (accessed 10 November 2014).

Gregory, D. (2004) *The Colonial Present: Afghanistan, Palestine, Iraq.* Blackwell: Oxford.

Gregory, D. and Walford, R. (eds) (1989) *Horizons in Human Geography.* Macmillan: London.

Griffith Taylor, T. (1947) *Canada: A Study of Cool Continental Environments and Their Effect on British and French Settlement.* Methuen: London.

Griffith Taylor, T. (1961) *Australia: A Study of Warm Environments and Their Effect on British Settlement.* Methuen: London.

Griffiths, N. (2000): *Grits.* Vintage: London.

Groth, P. and Bressi, T. (eds) (1997) *Understanding Ordinary Landscapes.* Yale University Press: New Haven and London.

Gruffudd, P. (1999) Nationalism. In Cloke, P., Crang, P., and Goodwin, M. (eds) *Introducing Human Geographies.* Arnold: London, 199–205.

Guardian, The (2005) Diversity not segregation. *The Guardian,* 21 January. Available at: www.guardian.co.uk/leaders/story/0,,1395161,00.html (accessed 10 November 2014).

Haartsen, T. and Strijker, D. (2010) Rural youth culture: *Keten* in the Netherlands. *Journal of Rural Studies,* 26, 163–172.

Hakim, C. (1987) *Research Design: Strategies and Choices in the Design of Social Research.* Unwin Hyman: London.

Hall, S. (ed.) (1997) *Representation: Cultural Representations and Signifying Practices.* Sage: London.

Hall, S. and du Gay, P. (1996) *Questions of Cultural Identity.* Sage: London.

Halsey, M. and Young, A. (2006) 'Our desires are ungovernable': writing graffiti in urban space. *Theoretical Criminology,* 10 (3), 275–306, 1362–4806.

Hammersley, M. and Atkinson, P. (1995) *Ethnography: Principles into Practice.* Routledge: London.

Hancox, D. (2013) *The Village Against the World.* Verso: London.

Hannam, K., Sheller, M., and Urry, J. (2006) Editorial: mobilities, immobilities and moorings. *Mobilities,* 1 (1), 1–22.

Haraway, D. (1988) Situated knowledges: the science question in feminism and the privilege of partial perspective. *Feminist Studies,* 14 (3), 575–599.

Haraway, D. (1991) *Simians, Cyborgs and Women: The Reinvention of Nature.* Free Association: London.

Harris, J. (2005) Marooned. *The Guardian,* 5 November. Available at: www.guardian.co.uk/world/2005/nov/05/israelandthepalestinians.johnharris (accessed 10 November 2014).

Hart, J. (2008) Dislocated masculinity: adolescence and the Palestinian nation-in-exile. *Journal of Refugee Studies,* 21 (1), 64–81.

Hartshorne, R. (1960) *Perspective on the Nature of Geography.* John Murray: London.

Harvey, D. (1989) *The Condition of Postmodernity.* Basil Blackwell: Oxford.

Hastrup, K. (1992) Writing ethnography: the state of the art. In Okely, J. and Callaway, H. (eds) *Anthropology and Autobiography.* ASA Monographs 29. Routledge. London and New York, 117–133.

Hawkins, F. (1987) *Human Factors in Flight.* Gower Technical Press: Aldershot.

Hawkins, H. (2014) *For Creative Geographies. Geography, Visual Arts and the Making of Worlds.* Routledge: London.

Hay, I. (2003) Ethical practice in geographical research. In Clifford, N. and Valentine, G. (eds) *Key Methods in Geography.* Sage: London, 37–53.

Hayden, D. (1997) Urban landscape history: the sense of place and the politics of space. In Groth, P. and Bressi, T. (eds) *Understanding Ordinary Landscapes.* Yale University Press: New Haven and London, 111–133.

Heathfield A. (2000) Out of sight – forced entertainment and the limits of vision. In Glendinning, H. and Etchells, T. (eds) *Void Spaces*. Site Gallery: Sheffield, 20–23.

Helm, T. (2013) New asbos 'punish children for being children'. *The Observer*, 13 October. Available at: www.theguardian.com/society/2013/oct/13/asbo-ipna-punish-children-unnecessarily (accessed 10 November 2014).

Helmore, E. (2014) Backlash over US border patrol 'bullying' migrants. *The Guardian*, 21 June. Available at: www.theguardian.com/world/2014/jun/21/arizona-border-patrol-anger-grows (accessed 10 November 2014).

Herbert, S. (2000) For ethnography. *Progress in Human Geography*, 24 (4), 550–568.

Hetherington, K. (1997) *The Badlands of Modernity*. Routledge: London.

Hetherington, K. (1998) *Expressions of Identity: Space, Performance, Politics*. Sage: London.

Hieberti, D. and Ley, D. (2003) Assimilation, cultural pluralism, and social exclusion among ethnocultural groups in Vancouver. *Urban Geography*, 24 (1), 16–44.

Higgins, C. (2013a) Big blue cock erected on fourth plinth in London's Trafalgar Square. *The Guardian*, 25 July. Available at: www.theguardian.com/artanddesign/2013/jul/25/big-blue-cock-trafalgar-square (accessed 10 November 2014).

Higgins, C. (2013b) Planning cock-up on Trafalgar Square's fourth plinth Katharina Fritsch's sculpture of a vast blue cockerel has been ruffling feathers. For all the wrong reasons. *The Guardian*, 1 May. Available at: http://m.guardian.co.uk/culture/charlottehigginsblog/2013/may/01/art-sculpture (accessed 10 November 2014).

Hill, A. (2002) Ghetto of hate where conflict is a way of life. *The Observer*, 13 January. Available at: www.guardian.co.uk/uk/2002/jan/13/northernireland (accessed 10 November 2014).

Hil, R. and Bessant, J. (1999) Spaced-out? Young people's agency, resistance and public space. *Urban Policy and Research*, 17 (1), 41–49.

Hinchliffe, S. (2000) Performance and experimental knowledge: outdoor management training and the end of epistemology. *Environment & Planning D, Society & Space*, 18, 575–595.

Hobson, K. (2002) Competing discourses of sustainable consumption: does the 'rationalisation of lifestyles' make sense? *Environmental Politics*, 11 (2), 95–120.

Holliday, R. (2000) We've been framed: visualising methodology. *Sociological Review*, 48 (4), 503–522.

Holloway, L. and Hubbard, P. (2001) *People and Place: The Extraordinary Geographies of Everyday Life*. Prentice Hall: New York.

Holloway, S. and Valentine, G. (eds) (2000) *Children's Geographies and the New Social Studies of Childhood in Children's Geographies: Playing, Living, Learning*. Routledge: London.

Holloway, S., Rice, S., and Valentine, G. (eds) (2003) *Key Concepts in Geography*. Sage: London.

Holton, M. and Riley, M. (2014) Talking on the move: place-based interviewing with undergraduate students. *Area*, 46 (1), 59–65.

Hooper, J. (2008) Uproar as top police cleared of attack on Genoa G8 protesters. *The Guardian*, 14 November, Available at: www.guardian.co.uk/world/2008/nov/14/italy-human-rights-genoa-protestors (accessed 10 November 2014).

Howard, T. and Flennaugh, T. (2011) Research concerns, cautions and considerations on Black males in a 'post-racial' society. *Race Ethnicity and Education*, 14 (1), 105–120.

Hunter, W. C. (2013) The visual representation of border tourism: Demilitarized Zone (DMZ) and Dokdo in South Korea. *International Journal of Tourism Research*. DOI: 10.1002/jtr.1973.

Huntingdon, E. (1913) *Civilization and Climate*. Yale University Press: New Haven and London.

Huntington, E. (1924) *The Character of Races as Influenced by Physical Environment, Natural Selection and Historical Development*. New York: Scribner.

Imrie, R. (2001) Barriered and bounded places and the spatialities of disability. *Urban Studies*, 38, 231–237.

Ingold, T. (1993) The temporality of the landscape. *World Archaeology*, 25 (2), 152–174.

Ingold, T. (1996) Hunting and gathering as ways of perceiving the environment. In Ellen, R. and Fukui, K. (eds) *Redefining Nature: Ecology, Culture & Domestication*. Berg: Oxford, 117–155.

Ingold, T. (2005) *Up, Across and Along*. Available at: www.spacesyntax.tudelft.nl/media/Long%20papers%20I/tim%20ingold.pdf 45–54. (accessed 10 November 2014).

Inside Out (2007) Gangs UK. Available at: www.bbc.co.uk/insideout/london/series11/week3_postcode_gangs.shtml (accessed 10 November 2014).

Inside TV (2014) Superbowl ratings. Available at: http://insidetv.ew.com/2014/02/03/super-bowl-ratings-3/ (accessed 10 November 2014).

ITN (2013) Body advertising. *The Guardian*, 21 July. Available at: www.theguardian.com/world/video/2013/jul/21/tokyo-womens-thighs-advertising-video?view=mobile (accessed 10 November 2014).

Jackson, P. (1989) *Maps of Meaning: An Introduction to Cultural Geography*. Unwin Hyman: London.

Jackson, P. (2004) Local consumption cultures in a globalizing world. *Transactions of the Institute of British Geographers*, 29 (2), 165–178.

Jackson, P. and Penrose, J. (eds) (1993) *Constructions of Race, Place and Nation*. UCL: London.

Jaffe, R. (2012) Talkin' 'bout the ghetto: popular culture and urban imaginaries of immobility. *International Journal of Urban and Regional Research*, 36 (4), 674–688.

Jenkins, S. (2013) NSA surveillance revelations: Osama bin Laden would love this. *The Guardian*, 7 June. Available at: www.theguardian.com/commentisfree/2013/jun/07/nsa-surveillance-osama-bin-laden (accessed 10 November 2014).

Jenks, C. (1996) *Childhood*. London: Routledge.

Johnston, L. (1996) Flexing femininity: female bodybuilders refiguring 'the body'. *Gender, Place and Culture*, 3 (3), 327–340.

Johnston, L. (2002) Man woman. In Cloke, P. and Johnston, R. (eds) *Spaces of Geographical Thought*. Sage: London, 119–141.

Johnston, R. (1986) *On Human Geography*. Blackwell: Oxford.

Johnston, R. (ed.) (1993) *The Challenge for Geography: A Changing World, a Changing Discipline*. Blackwell: Oxford.

Johnston, R. and Sidaway, J. (2004) *Geography and Geographers: Anglo-American Geography since 1945*, 6th edn. Arnold: London.

Johnston, R., Gregory, D., and Smith, D. (1994) *The Dictionary of Human Geography*, 3rd edn. Blackwell: Oxford.

Jones, K. (2007) Narratives of the in-between: teenagers' identities and spatialities in a north Wales town. Unpublished PhD thesis, School of City & Regional Planning, Cardiff University. Available from the author.

Jones, M., Jones, R., and Woods, M. (2004) *An Introduction to Political Geography. Space, Place and Power*. Routledge: London.

Jones, S. (2005) Spray can prankster tackles Israel's security barrier. *The Guardian*, 5 August.

Jones, S. (2013) Skateboarders form village green preservation society for Southbank. *The Guardian*, 10 May. Available at: www.theguardian.com/uk/2013/may/10/skateboarders-village-green-society-southbank (accessed 10 November 2014).

Jordan, G. and Weedon, C. (1995) *Cultural Politics: Class, Gender, Race, and the Postmodern World*. Blackwell: Oxford.

Jordan, T. (2002) *Activism! Direct Action, Hacktivism and the Future of Society*. Reaktion Books: London.

Joseph, P. (2011) *Concrete Circus*. Channel Four: London.

Jury, L. and Bar-Hillel, M. (2014) Skateboarders 'stoked' as Boris Johnson backs them in fight to stay at the South Bank. *Evening Standard*, 16 January. Available at: www.standard.co.uk/news/london/skateboarders-stoked-as-boris-johnson-backs-them-in-fight-to-stay-at-the-south-bank-9063250.html?origin=internalSearch (accessed 10 November 2014).

Kaltenborn, B. P. (1997). Nature of place attachment: a study among recreation home-owners in southern Norway. *Leisure Sciences*, 19, 175–189.

Kanngieser, A. (2012) A sonic geography of voice: towards an affective politics. *Progress in Human Geography*, 36 (3), 336–353.

Kaye, N. (2000) *Site-specific Art: Performance, Place, and Documentation*. Routledge: London.

Keddie, A., Mills, C., and Mills, M. (2008) Struggles to subvert the gendered field: issues of masculinity, rurality and class. *Pedagogy, Culture & Society*, 16 (2), 193–205.

Keirsey, D. and Gatrell, J. (2001) Ideology on the walls: contested space in planned urban areas in Northern Ireland. Working paper, unpublished, School of Environmental Planning, Queen's University, Belfast.

Keith, M. and Pile, S. (eds) (1993) *Place and the Politics of Identity*. Routledge: London.

Kellerman, A. (1993) *Society & Settlement: Jewish Land of Israel in the Twentieth Century*. State University of New York Press: Albany.

Kenny, Z. (2006) Indonesia: protests demand Freeport mine closure. *Alternative*, 21 March. Available at: www.alternatives.ca/article2420.html (accessed 10 November 2014).

Kent, D. (2013) Cardiff fans protest against owner Tan in wake of victory over rivals Swansea. *Daily Mail*, 3 November. Available at: www.dailymail.co.uk/sport/football/article-2486667/Cardiff-City-fans-protest-owner-Vincent-Tan-wake-victory-rivals-Swansea-City.html (accessed 10 November 2014).

Kenworthy Teather, E. (1999) *Embodied Geographies: Spaces, Bodies and Rites of Passage*. Routledge: London.

Khaleeli, H. (2014) India's third gender claims its place in law. *The Guardian*, 16 April. Available at: www.theguardian.com/society/2014/apr/16/india-third-gender-claims-place-in-law (accessed 10 November 2014).

Kim, J. E. and MacDonald, S. (2005) *Undercover in the Secret State*. Dispatches, Channel 4.

Kingsnorth, P. (2003) *One No, Many Yeses: A Journey to the Heart of the Global Resistance Movement*. Free Press: London.

Kirby, K. (1996) *Indifferent Boundaries: Spatial Concepts of Human Subjectivity*. Guilford Press: London and New York.

Kitchen, R. and Tate, N. (2000) *Conducting Research in Human Geography: Theory, Methodology and Practice*. Prentice Hall: Harlow.

Klein, N. (2001) *No Logo*®. Flamingo: London.

Klein, N. (2002) *Fences and Windows: Dispatches from the Front Lines of the Globalisation Debate*. Flamingo: London.

Knopp, L. (2004) Ontologies of place, placelessness, and movement: queer quests for identity and their impact on contemporary geographic thought. *Gender, Place and Culture*, 11, 121–134.

Kobayashi, A. (2013) Critical 'race' approaches. In Nuala C., Johnson, R., Schein, H., and Winders, J. (eds) *The Wiley-Blackwell Companion to Cultural Geography*. Oxford: Wiley Blackwell, 57–72.

Kroeber, A. (1923) *Anthropology*. University of Berkeley: Berkeley, CA.

Kroeber, A. L. (1952) *The Nature of Culture*. University of Chicago Press: Chicago.

Kropotkin, P. (1915) *Mutual Aid: A Factor of Evolution*. Heinemann: London.

Kusenbach, M. (2003) Street phenomenology: the go along as ethnographic research tool *Ethnography*, 4 (3), 455–485.

Kwan, Mei-Po (2004) Beyond difference: from canonical geography to hybrid geographies. *Annals of the Association of American Geographers*, 94 (4), 756–763.

Kymlicka, W. (2003) Multicultural states and intercultural citizens. *Theory and Research in Education*, 1 (2) 147–169.

Lamy, P. (2003) Laying down the law. *The Guardian*, 8 September. Available at: www.guardian.co.uk/environment/2003/sep/08/wto.fairtrade5 (accessed 10 November 2014).

Lapenta, F. (2011) Locative media and the digital visualisation of space, place and information. *Visual Studies* 26 (1), 1–3.

Larrain, P. (2013) *No*. Participant Media: Los Angeles.

Latham, A. (2003) Research, performance, and doing human geography: some reflections on the diaryphotograph, diary-interview method. *Environment & Planning A*, 35, 1993–2017.

Latour, B. (1993) *We Have Never Been Modern*. Harvard University Press: Cambridge, MA.

Latour, B. (2010) An attempt at a 'Compositionist Manifesto'. *New Literary History*, 41, 471–490.

Laughland, O. (2013) People still climbing Uluru despite closure condition being met. *The Guardian*, 9 July. Available at: www.theguardian.com/world/2013/jul/09/people-still-climbing-uluru-closure (accessed 10 November 2014).

Laurier, E. (2001) Why people say where they are during mobile phone calls. *Environment and Planning D: Society and Space*, 19, 485–504.

Laurier, E. and Philo, C. (2003) Possible geographies: a passing encounter in a café. Talk given at the RGS-IBG Annual Conference 2003. Session: Im-passable Geographies. Available at: http://web.ges.gla.ac.uk/~elaurier/cafesite/texts/elaurier006.pdf (accessed 10 November 2014).

Laurier, E. and Philo, C. (2006) Possible geographies: a passing encounter in a café. *AREA*, 38 (4), 353–363.

Lavers, C. (2004) Three cheers for Boro. *The Guardian*, 7 February. Available at: http://books.guardian.co.uk/review/story/0,,1141921,00.html (accessed 10 November 2014).

Lee, M. (1995) *Earth First! Environmental Apocalypse*. Syracuse University Press: Syracuse.

Leeman, J. and Modan, G. (2009) Commodified language in Chinatown: a contextualised approach to linguistic language. *Journal of Social Linguistics*, 1 (3/3), 332–362.

Lefebvre, H. (1968) *Dialectical Materialism*. Cape: London.

Lefebvre, H. (1971) *Everyday Life in the Modern World*. Allen Lane: London.

Lefebvre, M. (1991) *The Production of Space*. Blackwell: Malden.

Leitner, H. and Sheppard, E. (2003) Unbounding critical geographic research on cities: the 1990s and beyond. *Urban Geography*, 24, 510–528.

Leopold, A. (1949) *A Sand County Almanac*. Oxford University Press: New York.

Lewis, R. (2005) London by ethnicity: analysis. *The Guardian*, 21 January. Available at: www.guardian.co.uk/uk/2005/jan/21/british identity3 (accessed 10 November 2014).

Lindberg, D. and Numbers, R. (1986) *God and Nature: Historical Essays on the Encounter between Christianity and Science*. University of California Press: San Francisco.

Lippard, L. (1997) *The Lure of the Local: Senses of Place in a Multicentred Society*. The New Press: New York.

Livingstone, D. (1992) *The Geographical Tradition*. Blackwell: Oxford.

Lloyd, J. and Mitchinson, J. (eds) (2008) *Advanced Banter: The Qi Book of Quotations*. Faber: London.

Locke, K. (2013) Women choose body art over reconstruction after cancer battle. *The Guardian*, 7 August. Available at: www.theguardian.com/society/2013/aug/07/mastectomy-tattoo-breast-cancer (accessed 10 November 2014).

Lombard, M. (2013) Using auto-photography to understand place: reflections from research in urban informal settlements in Mexico. *Area,* 45 (1), 23–32.

Longhurst, R. (2003) Semi structured interviews and focus groups. In Clifford, N. and Valentine, G. (eds) *Key Methods in Geography.* Sage: London, 117–132.

Loomba, A. (1998) *Colonialism/Postcolonialism.* Routledge: London.

Lopez, B. (1998) *About This Life.* Harvill Press: London.

Lorimer, H. (2005) The busyness of 'more-than-representational'. *Progress in Human Geography,* 29, 83–94.

Lorimer, J. (2010) Moving image methodologies for more-than-human geographies. *Cultural Geographies,* 17 (2), 237–258.

Lorimer, J. (2012) Multinatural geographies for the Anthropocene. *Progress in Human Geography,* 36 (5), 593–612.

Low, S. (1992) Symbolic ties that bind: place attachment in the plaza. In Altman, U. and Low, S. (eds) *Place Attachment: Human Behavior and Environment. Advances in Theory & Research,* vol. 2. Plenum Press: New York and London, 165–185.

Lowenthal, D. (ed.) (1967) *Environmental Perception and Behaviour.* University of Chicago Press: Chicago.

Lowie, R. (1917) *Culture and Ethnology.* DC McMurtrie: New York.

Lucas, J. (2012) Why teens riot. *The Guardian,* 14 August. Available at: www.guardian.co.uk/childrens-books-site/2012/aug/14/john-lucas-london-riots-teens (accessed 10 November 2014).

Lukes, S. (1974) *Power: A Radical View.* Macmillan: London.

Lupton, D. (1998) *The Emotional Self.* Sage: London.

MacAskill, E. (2013) Edward Snowden, NSA files source: 'If they want to get you, in time they will'. *The Guardian,* June 9. Available at: www.theguardian.com/world/2013/jun/09/nsa-whistleblower-edward-snowden-why (accessed 10 November 2014).

MacDonald, F. (2007) Anti-Astropolitik – outer space and the orbit of geography. *Progress in Human Geography,* 31 (5), 592–615.

Macdonald, L. (2013) The Anti-social Behaviour, Crime and Policing Bill. *OPINION.* Matrix Chambers: London. Available at: http://reformclause1.org.uk/files/opinion.pdf (accessed 10 November 2014).

MacIntyre, A. (1990) *Three Rival Versions of Moral Inquiry: Encyclopaedia, Geneology and Tradition.* Duckworth & Co. Ltd: London.

Mackay Yarnal, C., Dowler, L., and Hutchinson, S. (2004) Don't let the bastards see you sweat: masculinity, public and private space, and the volunteer firehouse. *Environment & Planning A,* 36, 685–699.

MacKinnon, D. (2002) Rural governance and local involvement: assessing state-community relations in the Scottish Highlands. *Journal of Rural Studies,* 18 (3), 307–324.

Macnaghten, P. and Urry, J. (1997) *Contested Natures.* Sage: London.

Maffesoli, M. (1996) *The Time of the Tribes.* Sage: London.

Manes, C. (1990) *Green Rage: Radical Environmentalism in the Unmaking of Civilisation.* Little, Brown & Co.: Boston.

Mangold, T. (2007) 'Stealth racism' stalks deep South. *BBC News,* 24 May. Available at: http://news.bbc.co.uk/1/hi/programmes/this_world/6685441.stm (accessed 10 November 2014).

Mann, M. (2000) Democracy and ethnic war. *Hagar: International Social Science Review,* 1 (2), 115–134.

Mansson McGinty, A. (2007) Formation of alternative femininities through Islam: feminist approaches among Muslim converts in Sweden. *Women's Studies International Forum,* 30, 474–485.

Manzo, L. (2003) Beyond house and haven: toward a revisioning of emotional relationships with places. *Journal of Environmental Psychology*, 23, 47–61.

Marcus, E. (1995) Ethnography in/of the world system: the emergence of multi-sited ethnography. *Annual Review of Anthropology*, 24, 95–117.

Marx, K. (2003) *Capital: A Critique of Political Economy*, vol. 1, book 1: *The Process of Capitalist Production*. Lawrence and Wishart: London.

Marx, K. and Engels, F. (1984) *Collected Works*. Lawrence and Wishart: London.

Maskit, J. (1998) Something wild? Deleuze and Guattari and the impossibility of wilderness. In Light, A. and Smith J. M. (eds) *Philosophies of Place*. Rowman & Littlefield: Oxford, 265–283.

Massey, D. (1994) *Space, Place and Gender*. Polity Press: Cambridge.

Massey, D. (1995) Thinking radical democracy spatially. *Environment & Planning D, Society & Space*, 13.

Massey, D. (2000) Entanglements of power: reflections. In Sharp, J., Routledge, P., Philo, C., and Paddison, R. (eds) *Entanglements of Power: Geographies of Domination/Resistance*. Routledge: London.

Massey, D. (2004) Geographies of responsibility. *Geografiska Annaler* B 1, 5–18.

Massey, D. (2005) *For Space*. Sage: London.

Massey, D. (2006) Landscape as a provocation. Reflections on moving mountains. *Journal of Material Culture* 11 (1–2), 33–48.

Massey, D. (2007) *World City*. Polity Press: Cambridge.

Massey, D. and Allen, J. (eds) (1984) *Geography Matters!* Open University: Milton Keynes.

Massey, D. and Jess, P. (eds) (1995) *A Place in the World? Places, Cultures and Globalization*. Oxford University Press: Oxford.

Matthews, H., Taylor, M., Percy-Smith, B., and Limb, M. (2000) The unacceptable flaneur: the shopping mall as a teenage hangout. *Childhood*, 7 (3), 279–294.

Maxey, I. (1999) Beyond boundaries? Activism, academia, reflexivity and research, *Area*, 31, 199–208.

May, J. (2000) Of nomads and vagrants: single homelessness and narratives of home as place. *Environment & Planning D, Society & Space*, 18, 737–759.

May, J. (2003) The view from the streets. Geographies of homelessness in the British newspaper press. In Blunt, A. et al. (eds) *Cultural Geography in Practice*. Arnold: London, 23–36.

May, T. (1997) *Social Research: Issues, Methods and Process*. Open University Press: Buckingham and Philadelphia.

McClure, R. (2005) The war of Nelson's pigeons. *The Guardian*, 10 April. Available at: www.guardian.co.uk/g2/story/0,,1750498,00.html(accessed 10 November 2014).

McConnell, F. (2014) Things you wanted to know about trans people and were rude enough to ask. *The Guardian*, 19 March. Available at: www.theguardian.com/lifeandstyle/2014/mar/19/trans-people-rude-questions-sex-lives (accessed 10 November 2014).

McCormack, D. (2003) An event of geographical ethics in spaces of affect. *Transactions of the Institute of British Geographers*, 28, 488–507.

McCormack, D. (2009) Engineering affective atmospheres: on the moving geographies of the 1897 Andree expedition. *Cultural Geographies*, 15 (4), 413–430.

McCrum, R. (2001) *Globish: How the English Language Became the World's Language*. Penguin: London.

McCurry, J. (2008) Protester 'shot' by Japanese coastguard. *The Guardian*, 8 March. Available at: www.guardian.co.uk/environment/2008/mar/08/whaling.activists (accessed 10 November 2014).

McDowell, L. (1989) Women, gender and the organisation of space. In Gregory, D. and Walford, R. (eds) *Horizons in Human Geography*. Macmillan: Basingstoke, 136–151.

McDowell, L. (1994a) Polyphony and pedegogic authority. *Area*, 26 (3), 241–248.

McDowell, L. (1994b) The transformation of cultural geography. In Gregory, D., Martin, R., and Smith, G. (eds) *Human Geography: Society, Space, and Social Science*. Macmillan: Basingstoke.

McDowell, L. (1998) Elites in the City of London: some methodological considerations. *Environment & Planning A*, 30 (12), 2133–2146.

McDowell, L. (1999) *Gender, Place and Identity: Understanding Feminist Geographies*. University of Minnesota Press: Minneapolis.

McDowell, L. (2000) Learning to serve? Employment aspirations and attitudes of young working-class men in an era of labour market restructuring. *Gender, Place and Culture*, 7 (4), 389–416.

McDowell, L. (2004) Masculinity, identity and labour market change: some reflections on the implications of thinking relationally about difference and the politics of inclusion. *Geografiska Annaler*, 86 B (1), 45–66.

McGhee, D. (2003) Moving to 'our' common ground: a critical examination of community cohesion discourse in twenty-first century Britain. *Sociological Review*, 51 (3), 376–404.

McGrath, S. (2012) Many chains to break: the multi-dimensional concept of slave labour in Brazil. *Antipode*, 45 (4), 1005–1028.

McGrath, S. (2013) Fuelling global production networks with slave labour?: migrant sugar cane workers in the Brazilian ethanol GPN. *Geoforum*, 44, 32–43.

McIntyre, E. (2001) *Five Steps to Tyranny*. BBC TV. Available at: www.brightcove.tv/title.jsp?title958764725 (accessed 10 November 2014).

McKay, G. (1996) *Senseless Acts of Beauty: Cultures of Resistance since the Sixties*. Verso: London and New York.

McKay, G. (1998) *DIY Culture: Party & Protest in Nineties Britain*. Verso: London and New York.

McKibben, B. (1989) *The End of Nature*. Bloomsbury: London.

McLean, G. (2005) In the hood. *The Guardian*, G2, 13 May.

McLuhan, M. (1967) *The Medium is the Message*. Allen Lane/Penguin: London.

McLuhan, T. (1996) *Message of Sacred Places: Cathedrals of the Spirit*. Thorsons: London.

McPhee, J. (1989) *The Control of Nature*. Hutchison Radius: London.

Medeiros, V. (2010) Surfing is a fiction. With apologies to Jacques Lacan. *Pocmag*, 1, 95.

Mellor, D. (2003) Contemporary racism in Australia: the experiences of aborigines. *Personality and Social Psychology Bulletin*, 29, 474–486.

Melville, C. (2002) A carnival history. *Open Democracy*, 3 September. Available at: www.opendemocracy.net/artsfestival/article_548.jsp (accessed 10 November 2014).

Merchant, C. (1980) *The Death of Nature*. Harper Collins: New York.

Merriman, P. (2014) Rethinking mobile methods. *Mobilities*, 9 (2), 167–187.

Merriman, P. and Jones, R. (2009) 'Symbols of Justice': the Welsh Language Society's campaign for bilingual road signs in Wales, 1967–1980. *Journal of Historical Geography*, 35, 350–375.

Michaels, S. (2013) Beastie Boys accuse US toymaker of violating their 1987 track *Girls*. *The Guardian*, 25 November. Available at: www.theguardian.com/music/2013/nov/25/beastie-boys-accuse-toy-maker-copyright-girls (accessed 10 November 2014).

Miewald, C. and McCann, E. (2004) Gender struggle, scale, and the production of place in Appalachian coalfields. *Environment & Planning A*, 36, 1045–1064.

Mills, H., Shamash, J., and Cohen, N. (1990) Poll tax riot: wave of violence sweeps through west end. *The Independent*, 2 April, 2.

Mitchell, D. (1995) There's no such thing as culture: towards a reconceptualisation of the idea of culture in geography. *Transactions of the Institute of British Geographers*, 20, 102–116.

Mitchell, D. (2000) *Cultural Geography: A Critical Introduction*. Blackwell: Malden.

Mitchell, D. (2001) Cultural geography is not what it used to be. *Human Geography*, 53 (1), 36–53.

Mitchell, D. (2002) Cultural landscapes: the dialectical landscape: recent landscape research in human geography. *Progress in Human Geography*, 26 (3), 381–389.

Mitchell, D. (2004) *Cloud Atlas*. Hodder & Stoughton: London.

Moby (2014) I left New York for LA because creativity requires the freedom to fail. *The Guardian*, 3 February. Available at: www.theguardian.com/commentisfree/2014/feb/03/leave-new-york-for-los-angeles (accessed 10 November 2014).

Monbiot, G. (1996) Britain's ethnic cleansing. *The Guardian*, 10 October. Available at: www.monbiot.com/archives/1996/10/10/britains-ethnic-cleansing/ (accessed 10 November 2014).

Monbiot, G. (2001) *Captive State: The Corporate Takeover of Britain*. Pan Books: London.

Monbiot, G. (2003) Time for transformation. *The Guardian*, 8 September. Available at: www.guardian.co.uk/environment/2003/sep/08/wto.fairtrade6 (accessed 10 November 2014).

Monbiot, G. (2013) It's business that really rules us now. *The Guardian*, 11 November. Available at: www.theguardian.com/commentisfree/2013/nov/11/business-rules-lobbying-corporate-interests (accessed 10 November 2014).

Monbiot, G. (2014) At last, a law to stop almost anyone from doing almost anything. *The Guardian*, 6 January. Available at: www.theguardian.com/commentisfree/2014/jan/06/law-to-stop-eveyone-everything (accessed 10 November 2014).

Monks, R. (2006) The corporation feature transcript (11/13/2006). Available at: www.thecorporation.com/media/Transcript_finalpt1%20copy.pdf (accessed 10 November 2014).

Moore, R. (2011) 9/11 Ground Zero: why has its rebirth turned sour? *The Guardian*, 31 July. Available at: www.theguardian.com/world/2011/jul/31/new-york-towers-memorial-architecture (accessed 10 November 2014).

Morrissey, M. and Gaffikin, F. (2006) Planning for peace in contested space. *International Journal of Urban and Regional Research*, 30 (4), 873–893.

Moss, S. (2014) Why I wish I'd kept my Welsh accent. *The Guardian*, 9 July. Available at: www.theguardian.com/commentisfree/2014/jul/09/welsh-accent-accentism-rp (accessed 10 November 2014).

Murdoch, J. (2006) *Post-structuralist Geography: A Guide to Relational Space*. Sage: London.

Naess, A. (1989) *Ecology, Community & Lifestyle. Outline of an Ecosophy*. Cambridge University Press: Cambridge.

Nandrea, L. (1999) Reflections. 'Graffiti taught me everything I know about space': urban fronts and borders. *Antipode*, 31 (1), 110–116.

Nash, C. (2000) Performativity in practice: some recent work in cultural geography. *Progress in Human Geography*, 24 (4), 653–664.

Nash, C. (2002) Cultural geography in crisis. *Antipode*, 34 (2), 321–325.

Nash, C. (2006) Toronto's gay village (1969–1982): plotting the politics of gay identity. *Canadian Geographer*, 50 (1), 1–16.

Nerrière, J.-P. and Hon, D. (2005) *Globlish: The World Over*. International Globish Institute: Seattle.

New York Post (2013) Terror slay soldier ID'd as Lee Rigby – survived Afghan war only to be killed by extremist at home. *New York Post*, 23 May. Available at: http://nypost.com/2013/05/23/terror-slay-soldier-idd-as-lee-rigby-survived-afghan-war-only-to-be-killed-by-extremist-at-home/ (accessed 10 November 2014).

Newman, D. and Paasi, A. (1998) Fences and neighbours in the postmodern world: boundary narratives in political geography. *Progress in Human Geography*, 22 (2), 186–207.

Ng, Naw, Weinehall, L., and Öhman, A. (2007) 'If I don't smoke, I'm not a real man': Indonesian teenage boys' views about smoking. *Health Education Research*, 22 (6), 794–804.

Ninjalicious (2004) *Access All Areas*. Infiltration: Toronto.

Nunn, G. (2014) No English word for it? Make up your own, like Shakespeare. Or steal one. *The Guardian*, 31 January. Available at: www.theguardian.com/media/mind-your-language/2014/jan/31/mind-your-language-no-english-words (accessed 10 November 2014).

O'Hara, M. (2004) Self-imposed apartheid. *The Guardian*, 14 April. Available at: www.guardian.co.uk/society/2004/apr/14/northernireland.societyhousing (accessed 10 November 2014).

O'Loughlin, J. and Luc Anselin, L. (1992) Geography of international conflict and cooperation: theory and methods. In Ward, M. D. (ed.) *The New Geopolitics*. Gordon and Breach: Philadelphia, 39–75.

O'Reilly, K. (2003) CF11: Cardiff in six fugues, 1992–2001. In Abse, D. et al. (eds) *Cardiff Central: Ten Writers Return to the Welsh Capital*. Gomer: Llandysul, 27–34.

O'Riordan, T. (1980) *Environmentalism*. Pion: London.

Oakes, T. and Price, P. (eds) (2008) *The Cultural Geography Reader*. Routledge: London and New York.

Okely, J. and Callaway, H. (eds) (1992) *Anthropology and Autobiography*. ASA Monographs 29. Routledge: London and New York.

Oltermann, P. (2013) Something in common: should English be the official language of the EU? *The Guardian*, 24 April. Available at: www.theguardian.com/world/2013/apr/24/europa-english-official-language-eu (accessed 10 November 2014).

Pain, R. (1997) Social geographies of women's fear of crime. *Transactions of the Institute of British Geographers*, 22 (2), 231–244.

Pain, R. (2001) Gender, race, age and fear in the city. *Urban Studies*, 38 (5), 899–913.

Pain, R., Grundy, S., Gill, S., Towner, E., Sparks, G., and Hughes, K. (2005) 'So long as I take my mobile': mobile phones, urban life and geographies of young people's safety. *International Journal of Urban and Regional Research*, 29 (4), 814–830.

Panelli, R. (2004) *Social Geographies: From Difference to Action*. Sage: London.

Papayanis, M. A. (2000) Sex and the revanchist city: zoning out pornography in New York. *Environment & Planning D, Society & Space*, 18, 341–353.

Parekh, B. (2000) *The Future of Multi-ethnic Britain*. Profile Books: London.

Parker, J. (2010) *Let us now praise . . . Jet Lag*. Available at: www.boston.com/bostonglobe/ideas/articles/2010/03/07/jet_lag/Top of Form (accessed 10 November 2014).

Parsons, J. J. (1987) Now this matter of cultural geography. In Kenzer, M. (ed.) *Carl 0. Sauer: A Tribute*. Oregon State University Press for the Association of Pacific Coast Geographers: Corvallis, 153–163.

Pascal, L. (2003) Laying down the law. *The Guardian*, 8 September. Available at: www.guardian.co.uk/environment/2003/sep/08/wto.fairtrade5 (accessed 10 November 2014).

Paul, J. (2014) Post-racial futures: imagining post-racialist anti-racism(s). *Ethnic and Racial Studies*, 37 (4), 702–718.

Pearson, K. (1982) Conflict, stereotypes and masculinity in Australian and New Zealand surfing. *Australian and New Zealand Journal of Sociology*, 18 (2), 117–135.

Peet, R. (1998) *Modern Geographical Thought*. Blackwell: Oxford.

Pepper, D. (2005) Utopianism and environmentalism. *Environmental Politics*, 14 (1), 3–22.

Perlez, J. and Bonner, R. (2005) Below a mountain of wealth, a river of waste. *The New York Times,* 27 December. Available at: www.nytimes.com/2005/12/27/international/asia/27gold.html?pagewanted=print&_r=0 (accessed 10 November 2014).

Phillips, D. (2006) Parallel lives? Challenging discourses of British Muslim self-segregation. *Environment & Planning D, Society & Space,* 24, 25–40.

Pile, S. (1997) Introduction: opposition, political identities and spaces of resistance. In Pile, S. and Keith, M. (eds) *Geographies of Resistance.* Routledge: London, 1–32.

Pile, S. and Keith, M. (eds) (1997) *Geographies of Resistance.* Routledge: London.

Pilger, J. (2001) Spoils of a massacre. 23 July. Available at: www.johnpilger.com/page.asp?partid=299 (accessed 10 November 2014).

Pilkington, D. (2002) *Rabbit Proof Fence.* Miramax Books: London.

Pink, S. (2001) *Doing Visual Ethnography.* Sage: London.

Pink, S. (2006) *The Future of Visual Anthropology: Engaging the Senses.* Routledge: London.

Pink, S. (2011) Sensory digital photography: re-thinking 'moving' and the image. *Visual Studies,* 26 (1), 4–13.

Pocock, D. (1981) *Humanistic Geography and Literature.* Croom Helm: London.

Popic, L. (2005) Should you climb Uluru? *The Guardian,* 17 December. Available at: www.guardian.co.uk/travel/2005/dec/17/climbingholidays.australia.guardiansaturday-travelsection (accessed 10 November 2014).

Porritt, J. (2005) *Capitalism As If the World Matters.* Earthscan: London.

Pratt, G. (2004) Feminist geographies: spatialising feminist politics. In Cloke, P., Crang, P., and Goodwin, M. (eds) *Envisioning Human Geographies.* Hodder Arnold: London, 283–304.

Preston, C. (2000) Environment and belief: the importance of place in the construction of knowledge. *Ethics and the Environment,* 4 (2), 211–218.

Preston, C. (2003) *Grounding Knowledge: Environmental Philosophy, Epistemology, and Place.* University of Georgia Press: Athens and London.

Preston, C. (2005) Restoring misplaced epistemology. *Ethics, Place and Environment,* 8 (3), 373–384.

Price, M. and Lewis. M. (1993) The reinvention of cultural geography. *Annals of the Association of American Geographers,* 83 (1), 1–17.

Putnam, D. (2000) *Bowling Alone: The Collapse and Revival of American Community.* Simon & Schuster: New York.

Pyne, S. (1998) *How the Canyon Became Grand: A Short History.* Penguin: New York.

Pyyry, N. (2013) 'Sensing with' photography and 'thinking with' photographs in research into teenage girls' hanging out. *Children's Geographies.* DOI: 10.1080/14733285.2013.828453.

Radley, A. and Taylor, D. (2003) Images of recovery: photo-elicitation study on the hospital ward. *Qualitative Health Research,* 13 (1), 77–99.

Ramadam, T. (2003) *Western Muslims and the Future of Islam.* Oxford University Press: Oxford.

Rasmussen, C. E. (2006) We're no metrosexuals: identity, place and sexuality in the struggle over gay marriage. *Social & Cultural Geography,* 7 (5), 807–825.

Ratzel, F. (1896) *Völkerkunde.* Macmillan: London.

Razac, O. (2002) *Barbed Wire: A Political History.* Profile Books: London.

Reagan, R. (1987) Tear down this wall. *This History Place.* Available at: www.historyplace.com/speeches/reagan-tear-down.htm (accessed 10 November 2014).

Reed, A. (2002) City of details: interpreting the personality of London. *Royal Anthropological Institute,* 8, 127–141.

Reed-Danahay, D. E. (ed.) (1997) *Auto/Ethnography. Rewriting the Self and the Social.* Berg: Oxford and New York.

Reeves, R. (1999) Inside the world of the global protesters. *The Observer*, 31 October. Available at: www.guardian.co.uk/world/1999/oct/31/globalisation.businessandmedia (accessed 10 November 2014).

Reid, B. (2005) 'A profound edge': performative negotiations of Belfast. *Cultural Geographies*, 12, 485–506.

Relph, E. (1976) *Place and Placelessness.* Pion: London.

Reuters (2011) Occupy Wall Street. *The Guardian*, 6 October. Available at: www.theguardian.com/world/video/2011/oct/06/occupy-wall-street-protest-video (accessed 10 November 2014).

Reuters (2013) Anti G8 protests. *The Guardian*, 15 June. Available at: www.guardian.co.uk/world/video/2013/jun/15/anti-g8-protests-belfast-video (accessed 10 November 2014).

Reuters (2014) Daredevils jump the Freedom Tower. *The Guardian*, 25 March. Available at: www.theguardian.com/sport/video/2014/mar/25/daredevils-jump-new-york-freedom-tower-video (accessed 10 November 2014).

Rich, A. (1986) *Notes Towards a Politics of Location.* W.W. Norton & Company: New York.

Ritzer, G. (ed.) (2002) *McDonaldisation: The Reader.* Pine Forge Press: Thousand Oaks.

Robins, K. (2000) Encountering globalization. In Held, D. and McGrew, A. (eds) *The Global Transformation Reader.* Cambridge: Blackwell, 195–201.

Robinson, M. (2013) *Rebuilding the World Trade Center.* Channel 4. Available at: www.channel4.com/programmes/rebuilding-the-world-trade-center/4od (accessed 10 November 2014).

Robson, L. (2003) A stain on the map. In Abse, D. (ed.) *Cardiff Central. Ten Writers Return to the Welsh Capital.* Llandysul: Gomer, 95–104.

Rodaway, P. (1994) *Sensuous Geographies: Body, Sense & Place.* Routledge: London.

Rodman, M. (2003) Empowering place: multilocality and multivocality. In Low, S. and Lawrence-Zuniga, D. (eds) *The Anthropology of Space & Place: Locating Culture.* Blackwell: Oxford, 204–223.

Rojek, C. (1993) *Ways of Escape: Modern Transformations in Leisure and Travel.* London: Macmillan.

Rosaldo, R. (1989) *Culture and the Truth.* Boston: Beacon.

Rose, G. (1996) Teaching visualised geographies: towards a methodology for the interpretation of visual materials. *Journal of Geography in Higher Education*, 20 (3), 281–294.

Rose, G. (1997) Situating knowledges: positionality, reflexivities and other tactics. *Progress in Human Geography*, 21, 305–320.

Rose, M. (2002) Landscape and labyrinths. *Geoforum*, 33, 455–467.

Ross, G. (2012) *Hunger Games* [film]. Lionsgate.

Rourke, A. (2012) Australia set to recognise Aborigines as first people of continent. *The Guardian*, 20 January. Available at: www.theguardian.com/world/2012/jan/20/australia-aborigines-race-discrimination-referendum (accessed 10 November 2014).

Routledge, P. (1996) Third space as critical engagement. *Antipode*, 28 (4), 399–419.

Rozell, N. (1995) Fly east for bad jet lag. *Alaska Science Forum*, Article 1261, 16 November.

Ruddick, S. (1996) *Young and Homeless in Hollywood: Mapping Social Identities.* Routledge: London and New York.

Rudner, R. (no date) Brainy quote. Available at: www.brainyquote.com/quotes/quotes/r/ritarudner163137.html (accessed 10 November 2014).

Sack, R. D. (1997) *Homo Geographicus: A Framework for Action, Awareness and Moral Concern.* Johns Hopkins University Press: Baltimore.

Sack, R. D. (2004) Place-making and time. In Mels, T. (ed.) *Reanimating Places:*

Re-materialising. Cultural Geography Series. Ashgate: Aldershot, 243–253.

Said, E. W. (1979) *Orientalism*. Vintage Books: New York.

Salon (2014) The 5 best (and 5 worst) ads of Super Bowl 2014. Available at: www.salon.com/2014/02/03/the_5_best_and_5_worst_ads_of_super_bowl_2014/ (accessed 10 November 2014).

Sargisson, L. (2000) *Utopian Bodies and the Politics of Transgression*. Routledge: London.

Sauer, C. (1925) *The Morphology of Landscape*. University of California Press: Berkeley.

Sauer, C. (1931) *Cultural Geography: Encyclopedia of the Social Sciences*, vol. 6. Macmillan: New York, 621–624.

Sauer, C. (1963) Homestead and community on the middle border. In Ottoson, H. W. (ed.) *Land Use Policy in the United States*. University of Nebraska Press: Lincoln, 65–85.

Sauer, C. (1969) *Agricultural Origins and Dispersals: The Domestication of Animals and Foodstuffs*. MIT Press: Cambridge, MA.

Sayer, D. (1991) *Capitalism and Modernity: An Excursus on Marx and Weber*. Routledge: London.

Scarles, C. (2010) Where words fail, visuals ignite. *Tourism Research Annals of Tourism Research*, 37 (4), 905–926.

Schmidt, J. (2001) Redefining Fa'afafine: western discourses and the construction of transgenderism in Samoa. *Intersections: Gender, History and Culture in the Asian Context*, 6. Available at: http://intersections.anu.edu.au/issue6/schmidt.html (accessed 10 November 2014).

SchNEWS (1999a) Bare breasts & rubber bullets, 240, 10 December. Available at: www.schnews.org.uk/archive/news240.htm (accessed 10 November 2014).

SchNEWS (1999b) Schnewsround the inside story from the direct action frontline. Schnews: Brighton.

Schweder, R. (2000) What about 'female genital mutilation'? *Daedalus*, 129 (4), 209–232.

Scott, A. J. (2001) Capitalism, cities and the production of symbolic forms. *Transactions of the Institute of British Geographers*, 26, 11–23.

Scott, J. (1985) *Weapons of the Weak: Everyday Forms of Peasant Resistance*. Yale University Press: New Haven and London.

Seamon, D. (1984) Emotional experiences of the environment. *American Behavioral Scientist*, 27 (6), 757–770.

Searle, A. (2013) Fourth plinth: Katharina Fritsch's cockerel gives Nelson the bird. *The Guardian*, 25 July. Available at: www.theguardian.com/artanddesign/2013/jul/25/fourth-plinth-gives-nelson-the-bird (accessed 10 November 2014).

Seed, J. (1988) *Thinking Like a Mountain: Towards a Council of All Beings*. New Society Publishers: Philadelphia.

Segrott, J. (2001) Language, geography and identity: the case of the Welsh in London. *Social & Cultural Geography*, 2 (3), 281–296.

Semple, E. (1911) *Influences of Geographic Environment*. Henry Holt: New York.

Semple, E. (1942) *Influences of Geographic Environment on the Basis of Ratzel's System of Anthropo-Geography*. Henry Holt: New York.

Semple, E. and Ratzel, F. ([1911] 1941) *Influences of Geographic Environment*. Constable & Company: London.

Sennett, R. (1994) *Flesh and Stone: The Body and the City in Western Civilization*. Norton, New York.

Sharp, J., Routledge, P., Philo, C., and Paddison, R. (eds) (2000) *Entanglements of Power: Geographies of Domination/Resistance*. Routledge: London and New York.

Shehadeh, R. (2012) Palestinian Occupation. *The Guardian*, 9 October. Available at: www.guardian.co.uk/commentisfree/video/2012/oct/09/raja-shehadeh-palestinian-life-occupation-video (accessed 10 November 2014).

Sheller, M. and Urry, J. (2004) Places to play, places in play. In Sheller, M. and Urry, J. (eds)

Tourism Mobilities: Places to Play, Places in Play. Routledge: London, 1–10.

Sheller, M. and Urry, J. (2006) The new mobilities paradigm. *Environment and Planning A*, 38, 207–226.

Shenker, J. (2013) Mediterranean migrant deaths: a litany of largely avoidable loss. *The Guardian*, 3 October. Available at: www.theguardian.com/world/2013/oct/03/mediterranean-migrant-deaths-avoidable-loss (accessed 10 November 2014).

Sherwood, H., Smith, H., Davies, L., and Grant, H. (2014) Europe faces 'colossal humanitarian catastrophe' of refugees dying at sea. *The Guardian*, 2 June. Available at: www.theguardian.com/world/2014/jun/02/europe-refugee-crisis-un-africa-processing-centres (accessed 10 November 2014).

Shields, R. (1990) The 'system of pleasure': liminality and the carnivalesque at Brighton. *Theory, Culture and Society*, 7 (1), 39–72.

Shields, R. (1991) *Places on the Margin: Alternative Geographies of Modernity*. Routledge: London and New York.

Shields, R. (1992a) Spaces for the subject of consumption. In Shields, R. (ed.) *Lifestyle Shopping: The Subject of Consumption*. Routledge: London.

Shields, R. (1992b) The individual, consumption cultures and the fate of community. In Shields, R. (ed.) *Lifestyle Shopping: The Subject of Consumption*. Routledge: London.

Shields, R. (1996) Foreword: masses or tribes? In Maffesoli, M. (ed.) *The Time of the Tribes*. Sage: London.

Shirlow, P. (2005) Belfast: the 'post-conflict' city. *Space and Polity*, 10 (2), 99–107.

Shurmer-Smith, P. (ed.) (2002) *Doing Cultural Geography*. Sage: London.

Sibley, D. (1988) Survey 13: purification of space. *Environment and Planning D: Society and Space*, 6, 409–421.

Sibley, D. (1991) Children's geographies and some problems of representation. *Area*, 23 (3), 269–270.

Sibley, D. (1995a) Families and domestic routines: constructing the boundaries of childhood. In Pile, S. and Thrift, N. (eds) *Mapping the Subject: Geographies of Cultural Transformation*. Routledge: London.

Sibley, D. (1995b) *Geographies of Exclusion*. Routledge: London.

Sibley, D. (1999) Outsiders in society and space. In Anderson, K. and Gale, F. (eds) *Cultural Geographies*, 2nd edn. Longman: Melbourne, 135–151.

Sim, K. (2004) *The Fight For Ground Zero*. An Equinox Special. Locate TV.

Simmel, G. (1950). *The Sociology of Georg Simmel*. Free Press: London.

Simpson, J. (2003) *The Beckoning Silence*. Vintage: London.

Sinclair, I. (2012) *Ghost Milk*. London: Penguin.

Skelton, J. and Valentine, T. (eds) (1997) *Cool Places: Geographies of Youth Culture*. Routledge: London and New York.

Skelton, T (2000) 'Nothing to do, nowhere to go?': teenage girls and 'public' space in the Rhondda Valleys, South Wales. In Holloway, S. and Valentine, G. (eds) *Children's Geographies: Playing, Living, Learning*. Routledge: London.

Smith, A. (1979) *The Wealth of Nations*. Penguin: Harmondsworth.

Smith, M. (1999) To speak of trees: social constructivism, environmental values, and the future of deep ecology. *Environmental Ethics*, 21 (4), 359–376.

Smith, M. (2005) On 'being' moved by nature: geography, emotion and environmental ethics. In Davidson, J., Bondi, L., and Smith, M. (eds) *Emotional Geographies*. Ashgate: Aldershot, 219–230.

Smith, N. (1998) Nature at the millennium: production and reenchantment. In Braun, B. and Castree, N. (eds) *Remaking Reality: Nature at the Millennium*. Routledge: London, 278–285.

Smith, N. (2000) Guest editorials: global Seattle. *Environment & Planning D, Society & Space*, 18, 1–5.

Smith, R. (2002) Baudrillard's nonrepresentational theory: burn the signs and journey without maps. *Environment & Planning D, Society & Space*, 21, 67–84.

Snyder, G. (2000) *The Gary Snyder Reader: Prose, Poetry and Translations, 1952–1998*. Counterpoint Press: New York.

Soja, E. (1996) *Thirdspace: Journeys to Los Angeles and Other Real-and-Imagined-Places*. Blackwell: Maldon.

Soja, E. (1999) Thirdspace: expanding the scope of the geographical imagination. In Massey, D., Allen, J., and Sarre, P. (eds) *Human Geography Today*. Polity Press: Cambridge, 260–278.

Soja, E. and Hooper, B. (1993) The spaces that difference makes: some notes on the geographical margins of the new cultural politics. In Keith, M. and Pile, S. (eds) *Place and the Politics of Identity*. Routledge: London, 103–205.

Solis, B. (2009) Are blogs losing their authority to the statusphere. *Techcrunch*. Available at: http://techcrunch.com/2009/03/10/are-blogs-losing-their-authority-to-the-statusphere/ (accessed 10 November 2014).

Solnit, R. (2006) *A Field Guide to Getting Lost*. Canongate: Edinburgh.

Solnit, R. (2010) *Infinite City. A San Francisco Atlas*. University of California Press: Berkeley.

Solot, M. (1986) Carl Sauer and cultural evolution. *Annals of the Association of American Geographers*, 76 (4), 508–520.

Somé, S. and McSweeney, K. (1996) Assessing sustainability in Burkina Faso. *ILEIA Newsletter*. ETC Leusden: The Netherlands.

Sommer, R. (1982) Territory. In Kaplan, S. and Kaplan, R. (eds) *Humanscape: Environments for People*. Ulrich's Books: Ann Arbor, 267–270.

Soper, K. (1995) *What Is Nature? Culture, Politics and the Non-human*. Blackwell: Oxford.

Spence, E. (2014) Towards a more-than-sea geography: exploring the relational geographies of superrich mobility between sea, superyacht and shore in the Cote d'Azur. *Area*, 46 (2), 203–209.

Spinney, J. (2010) Performing resistance? Re-reading practices of urban cycling on London's South Bank. *Environment and Planning A*, 42, 2914–2937.

Spinney, J. (2011) A chance to catch a breath: using mobile video ethnography in cycling research. *Mobilities*, 6 (2), 161–182.

Spivak, G. C. (1990) *The Post-colonial Critic: Interviews, Strategies, Dialogues*. Routledge: London and New York.

Spurlock, M. (2011) *The Greatest Movie Ever Sold*. Snoot Entertainment.

Srivathsan, A., Pandian, M. S. S., and Radhakrishnan, M. (2005) City in/as graffiti: Chennai's many cities. *Inter-Asia Cultural Studies*, 6 (3), 422–427.

Steinberg, P. E. (2013) Of other seas: metaphors and materialities in maritime regions. *Atlantic Studies*, 10, 156–169.

Stepputat, F. (1999) Politics of displacement in Guatamala. *Journal of Historical Sociology*, 12 (1), 54–80.

Stewart, H. (2005) The new miserables of France. *Buzzle*, 12 November. Available at: www.buzzle.com/editorials/11–12-2005–81237.asp (accessed 10 November 2014).

Stiglitz, J. (2005) *Big Ideas that Changed the World*. Channel 5.

Swyngedouw, E. (1999) Modernity and hybridity: nature, regeneracionismo, and the production of the Spanish waterscape, 1890–1930. *Annals of the Association of American Geographers*, 89 (3), 443–465.

Szersynski, B., Heim, W., and Waterton, C. (eds) (2003) *Nature Performed. Environment, Culture and Performance*. Blackwell: Oxford.

Taylor, I. and Jamieson, R. (1997) 'Proper little mesters'. Nostalgia and protest masculinity in de-industrialised Sheffield. In Westwood, S. and Williams, J. (eds) *Imaginary Cities: Scripts, Signs, Memories*. Routledge: London and New York, 152–177.

Tennyson, A. (1850) *In Memoriam A.H.H.* Moxon: London.

The Argus (2014) Brighton's Banksy 'Kissing Coppers' mural sold in the US for £345,000. 19 February.

Thein, D. (2005) After or beyond feeling? A consideration of affect and emotion in geography. *Area*, 37 (4), 450–456.

Thomashow, M. (1995) *Ecological Identity: Becoming a Reflective Environmentalist.* MIT Press: Cambridge, MA, and London.

Thrift, N. (1996) *Spatial Formations.* Sage: London.

Thrift, N. (1997) The still point. In Pile, S. and Keith, M. (eds) *Geographies of Resistance.* Routledge: London, 124–151.

Thrift, N. (1999) Steps to an ecology of place. In Massey, D., Allen, J., and Sarre, P. (eds) *Human Geography Today.* Polity Press: Cambridge, 295–322.

Thrift, N. (2000) Commodities. In Johnston, R., Gregory, D., and Smith, D. (eds) *The Dictionary of Human Geography.* Blackwell: Oxford, 78–79.

Thrift, N. (2004) Intensities of feeling: towards a spatial politics of affect. *Geografiska Annaler*, 86 (1), 57–78.

Thrift, N. and Dewsbury, J.-D. (2000) Dead geographies and how to make them live again. *Environment & Planning D, Society & Space*, 18, 411–432.

Tilley, C. (1994) *A Phenomenology of Landscape: Places, Paths and Monuments.* Berg: Oxford.

Tolia-Kelly, D. (2010) The geographies of cultural geography I: identities, bodies and race. *Progress in Human Geography*, 34 (3), 358–367.

Tormey, A. (2007) 'Everyone with eyes can see the problem': moral citizens and the space of Irish nationhood. *International Migration*, 45 (3), 69–100.

Toynbee, P. (2001) Who's afraid of global culture? In Hutton, W. and Giddens, A. (eds) *On the Edge.* Vintage: London.

Trivett, V. (2011) *25 Mega Corporations.* Available at: www.businessinsider.com/25-corporations-bigger-tan-countries-2011-6?op=1 (accessed 10 November 2014).

Trow, M. (1957) Comment on observation and interviewing: a comparison. *Human Organization*, 16, 33–35.

Tuan, Y.-F. (1974) *Topophilia: A Study of Environmental Perception, Attitudes, and Values.* Prentice Hall: Englewood Cliffs.

Tuan, Y.-F. (1976a) Geopiety: a theme in man's attachment to nature and to place. In Lowenthal, D. and Bowden, M. (eds) *Geographies of the Mind: Essays in Historical Geography.* Oxford University Press: New York, 11–40.

Tuan, Y.-F. (1976b) Humanistic geography. *Annals of the Association of American Geographers*, 66 (2), 266–276.

Tuan, Y.-F. (1977) *Space and Place: The Perspective of Experience.* Minneapolis, MN: University of Minnesota Press.

Tuan, Y.-F. (1979) *Landscapes of Fear.* Blackwell: Oxford.

Tuan, Y.-F. (2004) Sense of place: its relationship to self and time. In Mels, T. (ed.) *Reanimating Places: Rematerialising Cultural Geography.* Ashgate: Aldershot, 45–56.

Turner, V. (1969) *The Ritual Process: Structure and Antistructure.* Aldine: New York.

Turner, V. (1974) *Dramas, Fields and Metaphors: Symbolic Action in Human Society.* Cornell University Press: Ithaca.

Turner, V. (1982) Liminal to liminoid in play, flow, ritual: an essay in comparative symbology. In Turner, V. (ed.) *From Ritual to Theatre: The Human Seriousness of Play.* Performing Arts Journals Publications: New York, 20–60.

Twigger-Ross, C. and Uzzell, D. (1996) Place & identity processes. *Journal of Environmental Psychology*, 16, 205–220.

UK Parliament (2013) Anti-social Behaviour, Crime and Policing Bill (HL Bill 66). Available at: www.publications.parliament.uk/pa/bills/lbill/2013-2014/0066/lbill_2013-20140066_en_1.htm (accessed 10 November 2014).

Undercurrents (1996) *The Struggle to Maintain a Traditional Romany Community*. The Alternative News Video 6, November, Swan Farm: Oxford.

Upe, R. (2013) Don't kiss, don't swear: rules of a Dubai stopover. Available at: www.smh.com. au/travel/travel-news/dont-kiss-dont-swear-rules-of-a-dubai-stopover-20130326-2grsc. html (accessed 10 November 2014).

Urry, J. (ed.) (2001) *Bodies of Nature*. Sage: London.

Urry, J. (2014) *Offshoring*. Polity: Cambridge.

Valentine, G. (1996a) Angels and devils: moral landscapes of childhood. *Environment & Planning D, Society & Space*, 14, 581–599.

Valentine, G. (1996b) Children should be seen and not heard: the production and transgression of adults' public space. *Urban Geography*, 17 (3), 205–220.

Valentine, G. (1997) Tell me about . . . : using interviews as a research methodology. In Flowerdew, R. and Martin, D. (eds) *Methods in Human Geography*. Longman: Harlow.

Valentine, G. (1999) Being seen and heard? The ethical complexities of working with children and young people at home and at school. *Ethics, Place and Environment*, 2 (2), 141–155.

Valentine, G. (2000) Exploring children and young people's narratives of identity. *Geoforum*, 31, 257–267.

Valentine, G. (2001) *Social Geographies: Space and Society*. Prentice Hall: Harlow.

Valentine, G. (2003) Boundary crossings: transitions from childhood to adulthood. *Children's Geographies*, 1 (1), 37–52.

Valentine, G., Holloway, S., Knell, C., and Jayne M. (2008) Drinking places: young people and cultures of alcohol consumption in rural environments. *Journal of Rural Studies*, 24, 28–40.

Valentine, G., Sporton, D., and Nielsen, K. B. (2008) Language use on the move: sites of encounter, identities and belonging. *Transactions of the Institute of British Geographers*, 376–387.

van Gennep, A. ([1909] 1960) *The Rites of Passage: A Classic Study of Cultural Celebration*. University of Chicago Press: Chicago.

Van Vliet, W. (1983) Exploring the fourth environment: an examination of the home range of city and suburban teenagers. *Environment and Behavior*, 15 (5), 567–588.

Vannini, P. (2011) The techne of making a ferry: a non-representational approach to passengers' gathering taskscapes. *Journal of Transport Geography*, 19, 1031–1036.

Venus Project (2012) Master TVP/RBE Public Quote Repository. Available at: https://docs. google.com/spreadsheets/d/1Z1_0_lP8pUY bzXp1rlVuIOE8eWmv1zH1hZLsg9Mde_k/ edit#gid=0 (accessed 13 January 2015).

Vidal de la Blache, P. (1941) *La personnalité géographique de la France*. Manchester University Press: Manchester.

Vidal, J. (1999) How the young battalions hatched the battle of Seattle. *The Guardian*, 30 November. Available at: www.guardian. co.uk/world/1999/nov/30/wto.johnvidal (accessed 10 November 2014).

Vitterso, J., Vorkinn, M., and Vistad, O. (2001) Congruence between recreational mode and actual behaviour: a prerequisite for optimal experiences? *Journal of Leisure Research*, 3 (2), 137–159.

Vivoni, F. (2009) Spots of spatial desire: skateparks, skateplazas, and urban politics. *Journal of Sports and Social Issues*, 33 (2), 130–149.

Vycinas, V. (1969) *Earth and Gods: An Introduction to the Philosophy of Martin Heidegger*. Martinus Nijhoff: Dordrecht.

Wainwright, O. (2013) Architects sketch a new space for South Bank skaters. *The Guardian*, 9 September. Available at: www.theguardian. com/uk-news/2013/sep/09/south-bank-skaters-hungerford-bridge (accessed 10 November 2014).

Wainwright, O. (2014) 9/11 Memorial Museum: an emotional underworld beneath Ground Zero. *The Guardian*, 14 May. Available at:

www.theguardian.com/artanddesign/2014/may/14/9-11-memorial-museum-new-york (accessed 10 November 2014).

Walker, P. (2013) Black people twice as likely to be charged with drugs possession. *The Guardian*, 21 August. Available at: www.theguardian.com/world/2013/aug/21/ethnic-minorities-likely-charged-drug-possession (accessed 10 November 2014).

Walker, S. (2014) Vladimir Putin: gay people at Winter Olympics must 'leave children alone'. *The Guardian*, 17 January. Available at: www.theguardian.com/world/2014/jan/17/vladimir-putin-gay-winter-olympics-children (accessed 10 November 2014).

Wallace, J. (2003) A (Karl, not Groucho) Marxist in Springfield. In Irwin, W., Conard, M., and Skoble, A. (eds) *The Simpsons and Philosophy: The D'oh of Homer*. Open Court: Chicago, 235–251.

Wallach, B. (2005) *Understanding the Cultural Landscape*. Guilford Press: New York.

Ward, K. (2006) Geography and public policy: towards public geographies. *Progress in Human Geography*, 30, 495–503.

Warde, A. (1994) Consumers, identity and belonging: reflections on some theses of Zygmunt Baumann. In Keat, R., Whitely, N., and Abercrombie, N. (eds) *The Authority of the Consumer*. Routledge: London, 58–74.

Warren, S. (2005). Photography and voice in critical qualitative management research. *Accounting, Auditing and Accountability Journal*, 18(6), 861–882.

Waterhouse, M. (1993) *Heart of the Matter: Roaming Free*. Roger Bolton Productions for BBC TV: Manchester.

Watson, E. (2004) 'What a dolt one is': language learning and fieldwork in geography. *Area*, 36 (1), 59–68.

Watson, J. W. (1983) The soul of geography. *Transactions of the Institute of British Geographers*, 8, 385–399.

Watson, P. (1994) *Ocean Warrior: My Battle to End the Illegal Slaughter on the High Seas*. Key Porter Books: Toronto.

Watts, L. S. (2009) Reflecting absence or presence? Public space and historical memory at Ground Zero. *Space and Culture*, 12, 412–418.

Watts, M. (1999) Commodities. In Cloke, P., Crang, P., and Goodwin, M. (eds) *Introducing Human Geographies*. Arnold: London.

WBRZ (no date) Baton Rouge gangs. Available at: www.wbrz.com/videos/news2-investigates-baton-rouge-gangs/ (accessed 10 November 2014).

Weber, M. (1978) The tensions between ethical religion and art. In Roth, G. and Widdich, C. (eds) *Max Weber, Economy and Society*. University of California Press: Berkeley, 607–610.

Weber, M. (1985) *Theory of Liberty, Legitimacy and Power: New Directions in the Intellectual and Scientific Legacy of Max Weber*. Routledge & Kegan Paul: London.

Weber, M. (1994) *Sociological Writings*. Continuum: New York.

Whatmore, S. (1997) Dissecting the autonomous self: hybrid cartographies for a relational ethics. *Environment & Planning D, Society & Space*, 15 (1), 37–53.

Whatmore, S. (1999) Nature culture. In Cloke, P., Crang, M., and Goodwin, M. (eds) *Introducing Human Geographies*. Arnold: London, 4–11.

Whatmore, S. (2002) *Hybrid Geographies*. Sage: London.

Whatmore, S. (2006) Materialist returns: practising cultural geography in and for a more-than-human world. *Cultural Geographies*, 13, 600–609.

White, R. (1995) 'Are you an environmentalist or do you work for a living?' Work and nature. In Cronon, W. (ed.) *Uncommon Ground: Toward Reinventing Nature*. Norton: London and New York, 171–185.

Willems-Braun, B. (1997) Buried epistemologies: the politics of nature in (post)colonial

British Columbia. *Annals of the Association of American Geographers*, 87 (1), 3–31.

Williams, A. C. (2005) Rosie's last week. In Finch, P. and Davies, G. (eds) *The Big Book of Cardiff*. Bridgend: Seren, 202–210.

Williams, C. (2002) *Sugar & Slate*. Planet: Aberystwyth.

Williams, Z. (2011) Gay rights: a world of inequality. *The Guardian*, 13 September. Available at: www.theguardian.com/world/2011/sep/13/gay-rights-world-of-inequality (accessed 10 November 2014).

Williamson, D. (2004) 'No Travellers' sign highlights need for stronger action. *The Western Mail*, 22 July, 6.

Willis, P. (1977) *Learning to Labour: How Working Class Kids Get Working Class Jobs*. Hutchinson: London.

Willsher, K. (2005) French celebrities desert Sarkozy in wake of attack on urban poor. *The Guardian*, 23 December. Available at: www.guardian.co.uk/world/2005/dec/23/france.topstories3 (accessed 10 November 2014).

Wilson, A. (1991) *The Culture of Nature: North American Landscapes from Disney to the 'Exxon Valdez'*. Between the Lines: Toronto.

Wirth, L. (1927) The ghetto. *The American Journal of Sociology*, 33 (1), 57–71.

Wood, N. and Smith, S. (2004) Instrumental routes to emotional geographies. *Social and Cultural Geography*, 5 (4), 533–548.

Wren, K. (2001) Cultural racism: something rotten in the state of Denmark? *Social and Cultural Geography*, 2 (2), 141–162.

Wright, M. (2007) *Inside Out: Postcode Gangs*. Available at: www.bbc.co.uk/insideout/london/series11/week3_postcode_gangs.shtml (accessed 10 November 2014).

Wrights and Sites (2006) *A Misguide to Anywhere*. Wrights and Sites: Exeter.

Wylie, J. (2007) *Landscape*. Routledge: Abingdon.

Wylie, J. (2010) Cultural geographies of the future, or looking rosy and feeling blue. *Cultural Geographies*, 17, 211–217.

Wyness, M. (2006) *Childhood and Society: An Introduction to the Sociology of Childhood*. Palgrave Macmillan: Basingstoke.

Young, E. (2010) Park 'encourages' Uluru climbers. *Australian Geographic*, 3 May. Available at: www.australiangeographic.com.au/news/2010/05/park-encourages-uluru-climbers (accessed 10 November 2014).

Young, I. (1990) *Justice and the Politics of Difference*. Princeton University Press: Princeton.

Young, I. (2000) *Inclusion and Democracy*. Oxford University Press: Oxford.

Zelinsky, W. (1973) *The Cultural Geography of the United States*. Prentice Hall: New York.

Index

Note: Page numbers followed by 'f' refer to figures, followed by 'n' refer to notes and followed by 't' refer to tables.